WSO

Water Treatment

Grades 3 & 4

American Water Works Association

ABC
Association of Boards of Certification

Printed in the United States of America.

Disclaimer
Many of the photographs and illustrative drawings that appear in this book have been furnished through the courtesy of various product distributors and manufacturers. Any mention of trade names, commercial products, or services does not constitute endorsement or recommendation for use by the American Water Works Association or the US Environmental Protection Agency. In no event will AWWA be liable for direct, indirect, special, incidental, or consequential damages arising out of the use of information presented in this book. In particular, AWWA will not be responsible for any costs, including, but not limited to, those incurred as a result of lost revenue. In no event shall AWWA's liability exceed the amount paid for the purchase of this book.

Library of Congress Cataloging-in-Publication Data
CIP data has been applied for.

ISBN: 9781625761255

000200010272023376

6666 West Quincy Avenue
Denver, CO 80235-3098
303.794.7711

Contents

Foreword

This book is part of the *Water System Operations* (WSO) series. This water operator education series was designed by the American Water Works Association (AWWA) to address core test content on certification exams by operator certification type (treatment or distribution) and certification grade level.

The current books in the series are:

WSO Water Treatment, Grade 1

WSO Water Treatment, Grade 2

WSO Water Treatment, Grades 3 & 4

WSO Water Distribution, Grades 1 & 2

WSO Water Distribution, Grades 3 & 4

Acknowledgments

The WSO series was developed by AWWA with the help of a volunteer steering committee of subject matter experts.

We would like to extend our thanks to the following individuals for their invaluable help and expertise:

- William Lauer, Project Manager, QualQuest, LLC (Lakewood, CO)
- Zac Bertz, Joint Water Commission (Hillsboro, OR)
- Mary Howell, Backflow Management, Inc. (Portland, OR)
- Bob Hoyt, City of Worcester Department of Public Works and Parks (Worcester, MA)
- Ted Kenney, New England Water Works Association (Holliston, MA)
- Darin LaFalam, City of Worcester Department of Public Works and Parks (Worcester, MA)
- Ray Olson, Distribution System Resources (Littleton, CO)
- Paul Riendeau, New England Water Works Association (Holliston, MA)

Additionally, we would like to thank Barbara Martin, AWWA/Partnership for Safe Water (Denver, CO), for her technical assistance and review.

How to Use This Book

Thank you for purchasing this volume in the Water System Operations series. AWWA's WSO series conforms to the latest Association of Boards of Certification (ABC) Need-to-Know criteria. ABC administers water operator certification testing for most of North America. Some states and provinces utilize different certification testing authorities, but this book covers all of the fundamentals every water operator needs to effectively pass their certification test and do their job effectively.

To help you advance in your career as a water operator, the WSO series is divided by subject areas (water treatment and water distribution) and certification grades (1, 2, 3, and 4, following the most common practice among states and provinces). Reference material, including basic science and mathematics concepts and information on water sources and quality, is included throughout all volumes in the series.

New features have been added to help you get the most from this book:

Key words Important terms are highlighted in blue when they are first introduced, with a corresponding definition box in the margin of the page. You can also look up key words in the glossary at the end of the book.

> The first pretreatment provided in most surface water treatment systems is screening. Coarse screens located on an intake structure are usually called *trash racks* or *debris racks*. Their function is to prevent clogging of the intake by removing sticks, logs, and other large debris in a river, lake, or reservoir. Finer screens may then be used at the point where the water enters the treatment system to remove smaller debris that has passed the trash racks.
>
> The two basic types of screens used by water systems are bar screens and wire-mesh screens. Both types are available in models that are manually cleaned or automatically cleaned by mechanical equipment.
>
> **Bar Screens**
>
> Bar screens are made of straight steel bars, welded at both ends to two horizontal steel members. The screens are usually ranked by the open distance between bars as follows:
>
> - Fine: spacing of 1/16 to ½ in. (1.5–13 mm)
> - Medium: spacing of ½ to 1 in. (13–25 mm)
> - Coarse: spacing of 1¼ to 4 in. (32–100 mm)
>
> **screening**
> A pretreatment method that uses coarse screens to remove large debris from the water to prevent clogging of pipes or channels to the treatment plant.
>
> **bar screens**
> A series of straight steel bars, welded at their ends to horizontal steel beams, forming a grid. Bar screens are placed on intakes or in waterways to remove large debris.

Video clips A number of subjects in this book are linked to helpful video clips available on the AWWA WSO website (www.awwa.org/wsovideoclips). The videos present visual, hands-on information to supplement the descriptions in the text.

WATCH THE VIDEO
Sedimentation and Clarifiers (www.awwa.org/wsovideoclips)

End-of-chapter questions To test your understanding of the core concepts and applications in the book, questions are provided at the end of each chapter. The answers are provided at the end of the book, with the steps worked-out for math problems.

End-of-book questions In the higher-level WSO books, questions are included at the end of the book to test your understanding of topics covered in previous grades. Keep in mind that certification tests will cover material from both lower-level and higher-level books.

Visit our website at www.awwa.org/wso to check out the additional books in the series and to access the free online resources associated with this book.

Chapter 1

Basic Microbiology and Chemistry

Microbiology in Water Treatment

Why might a utility choose to employ the multiple-barrier approach to water treatment? Why not just concentrate on maximizing one approach?

Microorganisms vary greatly in their sizes, shapes, surface charges, surface textures, motilities, and chemical sensitivities. One treatment method is seldom sufficient for surface-water utilities. It is common for one or two species, out of the hundreds that might be present, to evade initial sedimentation or coagulation steps. These few (especially algae and cyanobacteria) can continue to the filtration process where they might cause filtration plugging, or they might also evade filtration and enter the clearwell. Efforts are made to eliminate them by chlorination before they enter the distribution system.

If inadequate treatment barriers precede clearwell chlorination, the removal of *Cryptosporidium* oocysts will be ineffective, since these hardy oocysts are resistant to chlorine, as are other less dangerous microbes.

Surface-water utilities often experience dramatic changes in their source-water rain or snowmelt. The resulting variety in water quality usually necessitates multiple barriers of treatment. Utilities that use groundwater seldom experience these changes and therefore have fewer steps of treatment. If only filtration and chlorination, or only chlorination, are employed, breaches of well casings or the contamination of protected lakes can allow microorganisms to readily pass into the distribution system. In fact, since surface-water utilities have become more vigilant with their treatment for *Cryptosporidium* and other pathogens, groundwater utilities have experienced a greater number of *Cryptosporidium* incidents.

Removing or inactivating microorganisms is an important goal of almost any water treatment process. The addition of lime ($Ca(OH)_2$) removes suspended solids and dissolved chemicals that cause hardness. This process of softening the water also helps kill or inactivate pathogenic microorganisms. Chlorine or other disinfectants are added to filtered water specifically to prevent the spread of pathogens. Some organisms, notably the parasites *Giardia lamblia* and *Cryptosporidium parvum*, are resistant to disinfection, but effective coagulation and filtration can help to remove them, as can membrane filtration.

Softening

Water treatment often involves application of chemical principles to remove calcium bicarbonate, the cause of carbonate hardness. This is accomplished by adding slaked lime to the water:

$$\mathrm{CaO} + \mathrm{H_2O} \longrightarrow \mathbf{Ca(OH)_2} + \text{Heat}$$

calcium oxide (lime) — calcium hydroxide (lime slurry or "slaked lime")

$$\mathbf{Ca(OH)_2} + \mathrm{Ca(HCO_3)_2} \longrightarrow \mathbf{2CaCO_3}\downarrow + \mathrm{2H_2O}$$

calcium bicarbonate (carbonate hardness) — calcium carbonate (lime sludge)

$$\mathbf{Ca(OH)_2} + \mathrm{Mg(HCO_3)_2} \longrightarrow \mathbf{CaCO_3}\downarrow + \mathrm{MgCO_3} + \mathrm{2H_2O}$$

(another component of hardness)

$$\mathbf{Ca(OH)_2} + \mathrm{MgCO_3} \longrightarrow \mathbf{CaCO_3} + \mathbf{Mg(OH)_2}\downarrow$$

Note that the downward-pointing arrows (↓) indicate that the compounds are insoluble and precipitate out of the water.

Addition of slaked lime also aids in the removal of noncarbonated hardness:

$$\mathbf{Ca(OH)_2} + \mathrm{MgSO_4} \longrightarrow \mathbf{Mg(OH)_2}\downarrow + \mathrm{CaSO_4}$$

magnesium sulfate (noncarbonate hardness)

$$\mathbf{Na_2CO_3} + \mathrm{CaSO_4} \longrightarrow \mathbf{CaCO_3}\downarrow + \mathrm{Na_2SO_4}$$

sodium carbonate (soda ash)

These chemical reactions raise the water's pH to between 10 and 11.3. The vast majority of viruses, bacteria, protozoa, and multicellular organisms cannot survive above pH 10.5, so water softening is a powerful disinfectant. Carbon dioxide (CO_2) is added after softening to reduce the water's pH.

WATCH THE VIDEOS

TLS: Lime Softening and Operator Chemistry Made Easy II—Lime Softening (www.awwa.org/wsovideoclips)

calcium bicarbonate
A compound that is controlled through water treatment to prevent the formation of calcium carbonate (hardness).

Disinfection

Water treatment utilities use one of a variety of disinfectants to ensure pathogenic microorganisms are killed in their finished water. Removal of microorganisms actually begins with coagulation and flocculation. As these processes create floc of suspended solids, some microorganisms are trapped with those particles and removed when the solids settle out during clarification.

More microorganisms are removed when clarified water passes through granular filters. In rapid sand filters, the first few inches of the granular media trap remaining flocs, algae, and other materials, which in turn capture some microorganisms. The schmutzdecke performs the same function in a slow sand filter.

This physical removal process is very efficient in purifying the water, especially if treatment includes softening. In fact, it may be the most effective way within a conventional treatment plant to remove some pathogens otherwise resistant to disinfection chemicals.

Following filtration, the final stage of treatment usually includes disinfection using chlorine, chlorine dioxide, chloramines, or ozone. Some water treatment plants employ new disinfection processes using ultraviolet light or membrane filtration. Choices among disinfectant practices depend on several factors, such as cost, safety, convenience, effectiveness, and the condition of the water being treated.

Chlorine (Cl_2) is the most widely used disinfectant in water treatment. It is relatively cheap, easy to use, and effective at killing most microorganisms present in the water. In addition, some of the chemical remains in the finished water as a free-chlorine residual that prevents bacterial regrowth in the distribution system.

When chlorine is added to pure water, it reacts as follows:

$$\underset{\text{(chlorine)}}{Cl_2} + \underset{\text{(water)}}{H_2O} \longrightarrow \underset{\text{(hypochlorous acid)}}{HOCl} + \underset{\text{(hydrochloric acid)}}{HCl}$$

The chlorine combines with the water to produce hypochlorous acid (HOCl), a weak acid that easily penetrates into and kills bacteria. This action makes chlorine an effective disinfectant. HOCl is also one of two chlorine compounds that act as free available chlorine residual. However, some of the HOCl dissociates as follows:

$$\underset{\text{(hypochlorous acid)}}{HOCl} \longleftrightarrow \underset{\text{(hydrogen)}}{H} + \underset{\text{(hypochlorite ion)}}{OCl^-}$$

The hydrogen produced in this reaction neutralizes alkalinity and lowers pH, while the hypochlorite ion (OCl^-) is a second type of free available chlorine residual. Its disinfectant action is not as effective as that of HOCl, but it does help to kill microorganisms.

A number of conditions influence the effectiveness of disinfection using chlorine, including pH, water temperature, and contact time. The water's pH, when chlorine is introduced, strongly influences the ratio of HOCl to OCl^-. Low pH values favor formation of HOCl, the more effective free residual, while high pH favors formation of OCl^-. Similarly, low water temperature slightly favors formation of HOCl.

Contact time is important, along with the concentration of chlorine in the water, in determining the effectiveness of these compounds. Either a longer time or a rising concentration increases disinfectant effect.

The chemical reactions that have been described occur when chlorine is added to pure water. Water being treated with disinfectant also contains other substances, however. For example, chlorine reacts with ammonia present in the water to form chloramine compounds:

$$\underset{\text{(ammonia)}}{NH_3} + \underset{\text{(hypochlorous acid)}}{HOCl} \longrightarrow \underset{\text{(monochloramine)}}{NH_2Cl} + H_2O$$

$$NH_2Cl + HOCl \longrightarrow NHCl_2 + H_2O$$

(monochloramine) (hypochlorous acid) (dichloramine)

$$NHCl_2 + HOCl \longrightarrow NCl_3 + H_2O$$

(dichloramine) (hypochlorous acid) (trichloramine)

The water's pH and the amount of ammonia present determine whether one chloramine compound or more than one are formed.

Monochloramine and dichloramine act as disinfectants, although they are not as effective as free available chlorine, such as HOCl. If contact time is sufficient, chloramines can do an acceptable job of disinfection. However, dichloramine and trichloramine compounds may produce taste-and-odor problems.

Disinfection with chlorine requires additional planning for water with high levels of natural organic material. Chlorine reacts with natural organic material to form undesirable disinfection by-products (DBPs). Some of these compounds, especially trihalomethanes (THMs), are regulated as potential carcinogens by the US Environmental Protection Agency (USEPA) under the Disinfectants and Disinfection By-products Rule. To avoid creating these compounds, a utility must minimize residual organic material in the water before disinfection and carefully control chlorine usage. Effective coagulation, flocculation, sedimentation, and filtration remove many organic compounds, so chlorine will not react with the compounds to generate THMs. Chloramine reactions do not produce THMs.

Chlorine dioxide (ClO_2) is another very effective disinfectant, and its reactions produce fewer DBPs than those of chlorine. Chlorine dioxide is also more costly and more hazardous to use, however.

Ozone has become a popular disinfectant, despite its high cost, with utilities that have problems with protozoa. In particular, ozone treatment is one of the few effective methods for inactivating *Cryptosporidium*. After ozonation, however, the water contains no residual disinfectant to protect against regrowth of pathogens in the distribution system. Also, ozone cannot be stored, so it must be generated on-site as needed.

Ultraviolet (UV) disinfection is another technology used by some utilities. Under favorable conditions, it inactivates almost all microorganisms without producing undesirable THMs or other DBPs. Once practical only for small facilities, the technology is now being considered at large water utilities. The primary drawback to this treatment is the potential for the UV light bulbs to become coated with light-obscuring material, which prevents the UV light from reaching and killing the organisms. Careful maintenance is needed to ensure efficient operation. Also, turbidity in the water can shield organisms from the UV light, so this technology is practical only in very clear waters. Finally, UV treatment does not leave any disinfectant residual.

Membrane filtration is an effective technology for physical removal of microorganisms from water. During membrane filtration, water is fed through membranes with pores of a specific size. Membrane-based water treatment processes (listed in order from larger to smaller pore sizes) include microfiltration, ultrafiltration, nanofiltration, and reverse osmosis. Each removes a progressively finer particle. Ultrafiltration is usually sufficient to remove all microorganisms. No disinfectant residual remains in the water after membrane treatment.

ultraviolet (UV) disinfection
Disinfection using an ultraviolet light.

membrane filtration
Water treatment processes in which relatively pure water passes through a porous membrane while particles, molecules, or ions of unwanted matter are excluded.

Chemistry of Lime–Soda Ash Softening

The two ions most commonly associated with **hardness** in water are calcium (Ca^{+2}) and magnesium (Mg^{+2}). Although aluminum, strontium, iron, manganese, and zinc ions can also cause hardness, they are not usually present in large enough concentrations to produce a hardness problem. Chemical precipitation is one of the more common methods used to soften water. The chemicals normally used are lime (calcium hydroxide, $Ca(OH)_2$) and soda ash (sodium carbonate, Na_2CO_3).

There are two types of hardness: (1) **carbonate hardness**, caused primarily by calcium bicarbonate, and (2) **noncarbonate hardness**, caused by the salts of calcium and magnesium, such as calcium sulfate ($CaSO_4$), calcium chloride ($CaCl_2$), magnesium chloride ($MgCl_2$), and magnesium sulfate ($MgSO_4$). Lime is used to remove the chemicals that cause carbonate hardness. Soda ash is used to remove the chemicals that cause noncarbonate hardness.

Carbonate Hardness

During treatment for carbonate hardness, lime is the only softening chemical needed, as illustrated in the following reactions:

To remove calcium bicarbonate:

$$\underset{\text{calcium bicarbonate}}{Ca(HCO_3)_2} + \underset{\text{lime}}{Ca(OH)_2} \longrightarrow \underset{\text{calcium carbonate}}{2CaCO_3\downarrow} + \underset{\text{water}}{2H_2O} \tag{1-1}$$

To remove magnesium bicarbonate:

$$\underset{\text{magnesium bicarbonate}}{Mg(HCO_3)_2} + \underset{\text{lime}}{Ca(OH)_2} \longrightarrow \underset{\text{calcium carbonate}}{CaCO_3\downarrow} + \underset{\text{magnesium carbonate}}{MgCO_3\downarrow} + \underset{\text{water}}{2H_2O} \tag{1-2}$$

then

$$\underset{\text{magnesium carbonate}}{MgCO_3} + \underset{\text{lime}}{Ca(OH)_2} \longrightarrow \underset{\text{calcium carbonate}}{CaCO_3\downarrow} + \underset{\text{magnesium hydroxide}}{Mg(OH)_2\downarrow} \tag{1-3}$$

To remove the magnesium bicarbonate, it takes two separate reactions and twice the lime needed to remove calcium bicarbonate. In Equation 1-1, lime reacts with calcium bicarbonate to form **calcium carbonate**. Calcium carbonate is relatively insoluble and precipitates.

In Equation 1-2, lime reacts with magnesium bicarbonate to form calcium carbonate, which precipitates, and magnesium carbonate, which does not. In addition to being soluble, magnesium carbonate is a form of carbonate hardness. Therefore, the same amount of lime added in Equation 1-2 is called for in Equation 1-3. Thus, twice the lime required to remove calcium bicarbonate is necessary to remove magnesium bicarbonate and carbonate. In Equation 1-3, the additional lime reacts with magnesium carbonate to form calcium carbonate ($CaCO_3$) and magnesium hydroxide ($Mg(OH)_2$), both relatively insoluble materials that will settle out.

If the water originally had no noncarbonate hardness, then further softening would not be needed. However, because $CaCO_3$ and $Mg(OH)_2$ are very slightly soluble, a small amount of hardness remains, usually at least 35 mg/L.

hardness
A characteristic of water, caused primarily by the salts of calcium and magnesium. Causes deposition of scale in boilers, damage in some industrial processes, and sometimes objectionable taste. May also decrease the effectiveness of soap.

carbonate hardness
Hardness caused primarily by bicarbonate.

noncarbonate hardness
Hardness caused by the salts of calcium and magnesium.

calcium carbonate
The principal hardness- and scale-causing compound in water.

Noncarbonate Hardness

For noncarbonate hardness to be removed, soda ash must be added to remove the noncarbonated calcium compounds, and soda ash together with lime must be added to remove the noncarbonate magnesium compounds.

To remove calcium noncarbonate hardness:

$$\underset{\text{calcium sulfate}}{CaSO_4} + \underset{\text{soda ash}}{Na_2CO_3} \longrightarrow \underset{\text{calcium carbonate}}{CaCO_3\downarrow} + \underset{\text{sodium sulfate}}{Na_2SO_4} \tag{1-4}$$

$$\underset{\text{calcium chloride}}{CaCl_2} + \underset{\text{soda ash}}{Na_2CO_3} \longrightarrow \underset{\text{calcium carbonate}}{CaCO_3\downarrow} + \underset{\text{salt}}{2NaCl} \tag{1-5}$$

In both cases, the calcium noncarbonate hardness is removed by soda ash. The calcium sulfate and the calcium chloride acquire CO_3 from soda ash and precipitate as $CaCO_3$. The compounds that remain after softening, Na_2SO_4 in Equation 1-4 and NaCl in Equation 1-5, are salts and do not cause hardness.

To remove magnesium noncarbonate hardness:

$$\underset{\text{magnesium chloride}}{MgCl_2} + \underset{\text{lime}}{Ca(OH)_2} \longrightarrow \underset{\text{magnesium hydroxide}}{Mg(OH)_2\downarrow} + \underset{\text{calcium chloride, which must also be removed}}{CaCl_2} \tag{1-6}$$

$$\underset{\text{calcium chloride from Eq 1-6}}{CaCl_2} + \underset{\text{soda ash}}{Na_2CO_3} \longrightarrow \underset{\text{calcium carbonate}}{CaCO_3\downarrow} + \underset{\text{salt}}{2NaCl} \tag{1-7}$$

$$\underset{\text{magnesium sulfate}}{MgSO_4} + \underset{\text{lime}}{Ca(OH)_2} \longrightarrow \underset{\text{magnesium hydroxide}}{Mg(OH)_2\downarrow} + \underset{\text{calcium sulfate, which must also be removed}}{CaSO_4} \tag{1-8}$$

$$\underset{\text{calcium sulfate from Eq 1-8}}{CaSO_4} + \underset{\text{soda ash}}{Na_2CO_3} \longrightarrow \underset{\text{calcium carbonate}}{CaCO_3\downarrow} + \underset{\text{sodium sulfate}}{Na_2SO_4} \tag{1-9}$$

The removal of magnesium noncarbonate hardness with lime forms calcium noncarbonated hardness, which must then be removed with soda ash.

The final reaction related to softening involves carbon dioxide (CO_2), a gas that is found in dissolved form in most natural waters. Unless CO_2 is removed prior to softening (for example, by aeration), it will consume some of the lime added.

$$\underset{\text{carbon dioxide}}{CO_2} + \underset{\text{lime}}{Ca(OH)_2} \longrightarrow \underset{\text{calcium carbonate}}{CaCO_3\downarrow} + \underset{\text{water}}{H_2O} \tag{1-10}$$

If CO_2 is present, then enough lime must be added at the beginning of soft ing to allow this reaction to take place and still leave enough lime to complete ı softening reactions. Whenever CO_2 is present, the CO_2 reaction occurs before tı softening reactions take place.

Chemical softening takes place at a high pH. To precipitate calcium carbon ate, a pH of about 9.4 is necessary; the precipitation of magnesium hydroxide requires a pH of 10.6. In both cases, the necessary pH is achieved by adding the proper amount of lime.

Recarbonation

Recarbonation is the reintroduction of carbon dioxide into the water either during or after lime–soda ash softening. When hard water is treated by conventional lime softening, the water becomes supersaturated with calcium carbonate and may have a pH of 10.4 or higher. This very fine, suspended calcium carbonate can deposit on filter media, cementing together the individual media grains (encrustation) and depositing a scale in the transmission and distribution system piping (postprecipitation). To prevent these problems, carbon dioxide is bubbled into the water, lowering the pH and removing calcium carbonate as follows:

$$\underset{\text{calcium carbonate (in suspension)}}{CaCO_3} + \underset{\text{carbon dioxide (gas)}}{CO_2(g)} + \underset{\text{water}}{H_2O} \longrightarrow \underset{\text{calcium bicarbonate}}{Ca(HCO_3)_2} \quad (1\text{-}11)$$

This type of recarbonation is usually performed after the coagulated and flocculated waters are settled but before they are filtered, thereby preventing the suspended $CaCO_3$ from being carried out of the sedimentation basin and cementing the filter media.

When the excess-lime technique is used to remove magnesium, a considerable amount of lime remains in the water. The result is a water that is undesirably caustic and high in pH. Carbon dioxide introduced into the water reacts as follows:

$$\underset{\text{excess lime}}{Ca(OH)_2} + \underset{\text{carbon dioxide (gas)}}{CO_2(g)} \longrightarrow \underset{\text{calcium carbonate (precipitate)}}{CaCO_3\downarrow} + \underset{\text{water}}{H_2O} \quad (1\text{-}12)$$

This form of recarbonation is performed after coagulation and flocculation but before final settling. Carbon dioxide reacts with the excess lime, removing the cause of the caustic, high-pH condition and, incidentally, removing the calcium that added to the hardness. The product, calcium carbonate, is removed by the filtration process.

It is important to select the correct carbon dioxide dosage. If too much CO_2 is added, the following can happen:

$$\underset{\text{excess lime}}{Ca(OH)_2} + \underset{\text{carbon dioxide (gas)}}{2CO_2(g)} \longrightarrow \underset{\text{calcium bicarbonate}}{Ca(HCO_3)_2} \quad (1\text{-}13)$$

recarbonation
The reintroduction of carbon dioxide into the water, either during or after lime–soda ash softening.

Notice that the excess lime combines with the excessive CO_2 to form carbonate hardness. In the excess-lime method, there is too much calcium hydroxide present, and the reaction in Equation 1-13 would significantly increase water hardness. In the methods using lower lime doses, the conversion to calcium bicarbonate may not significantly increase water hardness.

Chemistry of Ion Exchange Softening

The ion exchange process of water softening uses the properties of certain materials (termed *cation exchange materials*) to exchange the hardness-causing cations of calcium and magnesium for nonhardness-causing cations of sodium. The most common cation exchange materials are synthetic polystyrene resins. Each resin particle is a BB-sized, transparent, amber-colored sphere. Each of these insoluble resin spheres contains sodium ions that are released into the water in exchange for hardness ions of calcium and magnesium. When properly operated, ion exchange is completely effective in removing all hardness, whether carbonate or noncarbonate.

The two reactions involved in the cation exchange softening process are as follows:

$$Ca^{+2} + Na_2X \longrightarrow CaX + 2Na^{+} \quad (1\text{-}14)$$

Ca^{+2}	+	Na_2X	→	CaX	+	$2Na^{+}$	
hardness cation		cation exchange resin		spent resin		sodium released to treated water in exchange for calcium	(1-14)

$$Mg^{+2} + Na_2X \longrightarrow MgX + 2Na^{+} \quad (1\text{-}15)$$

Mg^{+2}	+	Na_2X	→	MgX	+	$2Na^{+}$	
hardness cation		cation exchange resin		spent resin		sodium released to treated water in exchange for magnesium	(1-15)

The letter X is used to represent the exchange resin. Before softening, the resin with the nonhardness cations of sodium appears in the equations as Na_2X. Although this is not a chemical compound, it does behave somewhat like one. The sodium cations ("Na_2" in the resin, "$2Na^{+}$" after release) are released into the water just as the sodium in Na_2SO_4 would be released when that compound is dissolved in water. Although the resin "X" acts similarly to an anion such as the sulfate ion (SO_4^{-2}), the resin is actually an insoluble organic material that does not react chemically as SO_4^{-2} would. Instead, it functions more like a "parking lot" for exchangeable cations. The terms CaX and MgX represent the same resin after the exchange has been made.

As shown in the equations, calcium and magnesium hardness ions are removed from the water onto the surface of the resin. In exchange, the resin releases sodium ions. Note that one hardness ion (Mg^{+2} or Ca^{+2}) with a charge of +2 is exchanged for two sodium ions, each having a charge of +1, a total charge of +2. Hence, a +2 charge is exchanged for a +2 charge, an electrically equivalent exchange.

Equations 1-14 and 1-15 can be expanded to show the reactions of the specific hardness-causing compounds (similar to the equations shown for lime–soda ash softening):

ion exchange process
A process used to remove hardness from water that depends on special materials known as resins. The resins trade nonhardness-causing ions (usually sodium) for the hardness-causing ions, calcium and magnesium. The process removes practically all the hardness from water.

$$Ca(HCO_3)_2 + Na_2X \longrightarrow CaX + 2NaHCO_3 \quad (1\text{-}16)$$

$$CaSO_4 + Na_2X \longrightarrow CaX + Na_2SO_4 \quad (1\text{-}17)$$

$$CaCl_2 + Na_2X \longrightarrow CaX + 2NaCl \quad (1\text{-}18)$$

$$Mg(HCO_3)_2 + Na_2X \longrightarrow MgX + 2NaHCO_3 \quad (1\text{-}19)$$

$$MgSO_4 + Na_2X \longrightarrow MgX + Na_2SO_4 \quad (1\text{-}20)$$

$$MgCl_2 + Na_2X \longrightarrow MgX + 2NaCl \quad (1\text{-}21)$$

In each reaction, the anion originally associated with the hardness cation stays in the softened water; after softening, these anions are associated with the sodium cations released by the resin. Hence, the softened water contains sodium bicarbonate ($NaCHO_3$), sodium sulfate (Na_2SO_4), and sodium chloride (NaCl). These compounds do not cause hardness and are present in such small concentrations that they do not cause tastes. Unlike lime–soda ash softening, ion exchange softening operates the same for both carbonate and noncarbonate hardness. Both types are removed by the same exchange reactions.

After most of the sodium ions are removed from the exchange resin in the softening process, the resin must be **regenerated** in order to restore its softening capacity. That is, the exchange process must be reversed, with the hardness cations of calcium and magnesium being forced out of the resin and replaced by cations of sodium. This reverse exchange is achieved by passing a strong brine solution (a concentrated solution of common table salt) through the resin bed. The two ion exchange regeneration reactions are shown below:

$$CaX + 2NaCl \longrightarrow CaCl_2 + Na_2X \quad (1\text{-}22)$$

$$MgX + 2NaCl \longrightarrow MgCl_2 + Na_2X \quad (1\text{-}23)$$

When sodium is taken back into the exchange resin, the resin is again ready to be used for softening. The calcium and magnesium, released during regeneration, are carried to disposal by the spent brine solution.

Properly maintained and operated, cation exchange removes *all* hardness. Water of zero hardness is corrosive, so the final step in ion exchange softening is to mix a portion of the unsoftened water with the softened effluent to provide water that is still relatively soft, but that contains enough hardness to be noncorrosive (stable).

Scaling and Corrosion Control

Scaling and corrosion are closely related problems in water treatment. They may be thought of as being at opposite ends of a hypothetical stability scale, as shown in Figure 1-1.

The objective of scale and corrosion control is to stabilize the water, thus preventing both scale formation and corrosion. The stable range is relatively narrow, requiring careful monitoring during treatment in order to avoid under- or overshooting the stable range.

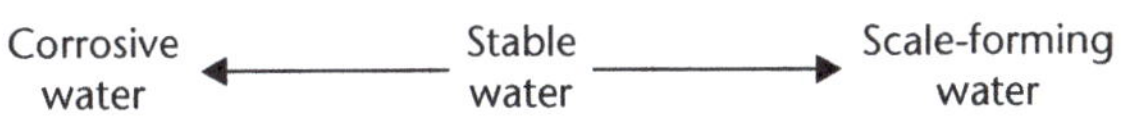

Figure 1-1 Hypothetical stability scale

regeneration
The process of reversing the ion exchange softening reaction of ion exchange materials. Hardness ions are removed from the used materials and replaced with nontroublesome ions, thus restoring the exchange capacity of the resin for further softening.

Scaling and Corrosion

Scale is the familiar chalky-white deposit frequently found at the bottom of a tea kettle. It is caused by carbonate and noncarbonate hardness constituents in water.

The exact combination of pH and alkalinity that will result in scale (or corrosion) depends on the overall chemical characteristics of the water. Under conditions of pH 7.0–9.5, with corresponding alkalinities of 300–1,500 mg/L, calcium carbonate ($CaCO_3$) will precipitate and form a scale on interior surfaces of pipes and tanks. One of two chemical reactions will occur:

$$Ca^{+2} + CO_3^{-2} \longrightarrow CaCO_3\downarrow \qquad (1\text{-}24)$$

or

$$Ca^{+2} + 2HCO_3^{-} \longrightarrow CaCO_3\downarrow + CO_2 + H_2O \qquad (1\text{-}25)$$

In controlled amounts, this scale is beneficial, forming a protective coating inside pipelines and tanks. However, excessive scaling can reduce the capacity of pipelines and the efficiency of heat transfer in boilers.

Corrosion is the oxidation of unprotected metal surfaces. In water treatment, a primary concern is the corrosion of iron and its alloys. Corrosion of iron and steel products is easily identified by the familiar red rust that forms. Iron is an important element in many metallic pipe materials and process equipment. It exists naturally as iron ore, in stable forms such as hematite (Fe_2O_3), magnatite (Fe_3O_4), iron pyrite (FeS_2), and siderite ($FeCO_3$). Smelting converts this ore to elemental iron, which is then used in the manufacture of pipeline materials and treatment equipment. Elemental iron is unstable and has a strong tendency to return by the oxidation or corrosion process to the more stable ore forms noted above. Advanced cases of iron corrosion create the problem of "red water."

There are various theories on how iron corrosion occurs and several factors known to affect corrosion. The following simplified discussion of corrosion chemistry highlights the influence of pH, alkalinity, dissolved oxygen, and carbon dioxide in the corrosion process. Remember that corrosion is a complex process that can be influenced by other factors as well. For example, iron bacteria (*Crenothrix* and *Leptothrix*) and sulfate-reducing bacteria can be major causes of corrosion. Increases in water temperature and velocity of flow can accelerate corrosion. The softening process can convert a noncorrosive water to a corrosive one. However, certain constituents in water, such as silica, are believed to protect exposed metal surfaces from corrosion.

When iron corrodes, it is converted from elemental iron (Fe) to ferrous ion (Fe^{+2}):

$$Fe \longrightarrow Fe^{+2} + 2\text{ electrons} \qquad (1\text{-}26)$$

The electrons that come from the elemental iron build up on metal surfaces and inhibit corrosion. If the water has a very low pH (lower than that of potable water), then hydrogen ions (H^+) in solution will react with electrons to form hydrogen gas ($H_2\uparrow$):

$$2H^{+} + 2\text{ electrons} \longrightarrow 2H_2\uparrow \qquad (1\text{-}27)$$

The hydrogen gas coats the metal surface and could reduce corrosion; however, the coating is removed, partly by the scrubbing action of moving water and partly by combination with oxygen (O_2) normally dissolved in the water:

$$2H_2 + O_2 \longrightarrow 2H_2O \qquad (1\text{-}28)$$

The metal surface is exposed again and corrosion continues. Failure to protect the metal surface or remove corrosion-causing elements will result in destruction of pipes or equipment.

Once the reaction in Equation 1-26 has occurred, subsequent reactions depend on the chemical characteristics of the water. If water is low in pH, is low in alkalinity, and contains dissolved oxygen, then the ferrous ion reacts with water to form ferrous hydroxide, $Fe(OH)_2$:

$$Fe^{+2} + 2H_2O \longrightarrow Fe(OH)_2 + 2H_2\uparrow \qquad (1\text{-}29)$$

The insoluble ferrous hydroxide immediately reacts with CO_2 present in low-alkalinity water to form soluble ferrous bicarbonate, $Fe(HCO_3)_2$:

$$Fe(OH)_2 + 2H_2CO_3 \longrightarrow Fe(HCO_3)_2 + 2H_2O \qquad (1\text{-}30)$$

Since ferrous bicarbonate is soluble, it detaches from the metal surface and is mixed throughout the water. The dissolved oxygen in the water then reacts with the ferrous bicarbonate to form insoluble ferric hydroxide, $Fe(OH)_3$:

$$4Fe(HCO_3)_2 + 10H_2O + O_2 \longrightarrow Fe(OH)_3 + 8H_2CO_3 \qquad (1\text{-}31)$$

Because this reaction occurs throughout water, ferric hydroxide does not form a protective coating on the metal surface. The ferric hydroxide appears as suspended particles that cause red water.

If water begins with a higher pH and alkalinity (where CO_2 is not present), then the corrosion reaction can be controlled. The ferrous ion shown in Equation 1-26 combines with the hydroxyl alkalinity that is present naturally or induced by lime treatment, forming an insoluble film of ferrous hydroxide on the metal surface:

$$Fe^{+2} + 2(OH^-) \longrightarrow Fe(OH)_2 \qquad (1\text{-}32)$$

If dissolved oxygen is present in the water, it will react with the ferrous hydroxide to form an insoluble ferric hydroxide coating:

$$4Fe(OH)_2 + 2H_2O + O_2 \longrightarrow 4Fe(OH)_3 \qquad (1\text{-}33)$$

Both ferrous and ferric hydroxide are somewhat porous, and although their coatings retard corrosion, they cannot fully protect the pipe. However, the same high-pH and high-alkalinity conditions that cause the rust coating to form also favor the formation of a calcium carbonate coating. Together these coatings protect the pipe from further corrosion.

Chemical Methods for Scale and Corrosion Control

Table 1-1 shows common methods used to control scale and corrosion. The following paragraphs contain brief discussions of each method. Lime, soda ash, and caustic soda are typically used to raise pH and alkalinity. Carbon dioxide and

Table 1-1 Scale and corrosion control methods

	For Control of:	
Method	**Scale**	**Corrosion**
pH and alkalinity adjustment with lime	X	X
Chelation	X	
Sequestering	X	
Controlled calcium carbonate scaling		X
Other protective chemical coatings		X
Softening	X	

sulfuric acid are used to lower pH and alkalinity. However, in situations where stabilization is achieved by pH and alkalinity adjustment, the effects of all chemicals used in water treatment must be taken into account. Alum and ferric sulfate (discussed previously) lower pH and alkalinity, as do chlorine (discussed later) and fluorosilicic acid (used in fluoridation).

pH and Alkalinity Adjustment With Lime

For each milligram per liter of lime added, approximately 0.56 mg/L of carbon dioxide is removed. Carbon dioxide is in the form of carbonic acid (H_2CO_3) when dissolved in water. The following equation indicates the chemical reaction that takes place:

$$H_2CO_3 + Ca(OH)_2 \longrightarrow CaCO_3 + 2H_2O \qquad (1\text{-}34)$$

As carbon dioxide (carbonic acid) is removed, pH increases.

For each milligram per liter of lime added, the alkalinity of the treated water will increase by about 1.28 mg/L. The reaction is

$$Ca(HCO_3)_2 + Ca(OH)_2 \longrightarrow 2CaCO_3 + 2H_2O \qquad (1\text{-}35)$$

If the lime dosage is too high, excessive scale will form. If it is too low, the water will be corrosive. The **Langelier saturation index** is one measure used to determine the tendency of water to scale or corrode piping and tanks. The index is based on the assumption that every water has a particular pH value for which the water will neither deposit scale nor cause corrosion. This stable condition is termed **saturation**. The pH value, called saturation pH and abbreviated pH_s, varies, depending on calcium hardness, alkalinity, and temperature. Once the pH_s is calculated, the Langelier saturation index is found as follows:

$$\text{Langelier saturation index} = pH - pH_s$$

If the actual pH of the water is less than the calculated pH_s, then the water has a negative Langelier index and may be corrosive. If the actual pH is greater than the calculated pH_s, then the Langelier index is positive and the water is likely to form scale. In either case, the water is unstable. The greater the difference between pH and pH_s, the stronger the tendency for the water to either form scale or cause corrosion. Thus, water with a Langelier index of +0.4 has a stronger scaling tendency than water with an index of +0.1. Similarly, water with a Langelier index of –0.4 has a stronger corrosion tendency than water with an index of –0.2. If the pH and pH_s are equal, then the Langelier saturation index is zero and water is stable. Neither scale formation nor corrosion should occur.

The value of pH_s can be calculated mathematically, but it is simpler to use a graph, as illustrated in Figure 1-2. To find pH_s, the calcium hardness in milligrams per liter as $CaCO_3$, alkalinity in milligrams per liter as $CaCO_3$, pH, and water temperature must be determined. For example, the following information may be used to demonstrate reading the graph:

$$\text{calcium hardness} = 240 \text{ mg/L as } CaCO_3$$

$$\text{alkalinity} = 200 \text{ mg/L as } CaCO_3$$

$$pH = 6.8$$

$$\text{water temperature} = 70°F$$

Enter the graph at a calcium hardness of 240 mg/L, as shown by the dotted line in Figure 1-2. Proceed upward to the 200-mg/L alkalinity curve. Travel across to locate the temperature of 70°F (estimate the temperature location between the

Langelier saturation index
A measure of the tendency of water to scale or corrode pipes, based on pH, conductivity, hardness, alkalinity, and water temperature.

saturation
A stable condition of water in which the water will neither deposit scale nor cause corrosion.

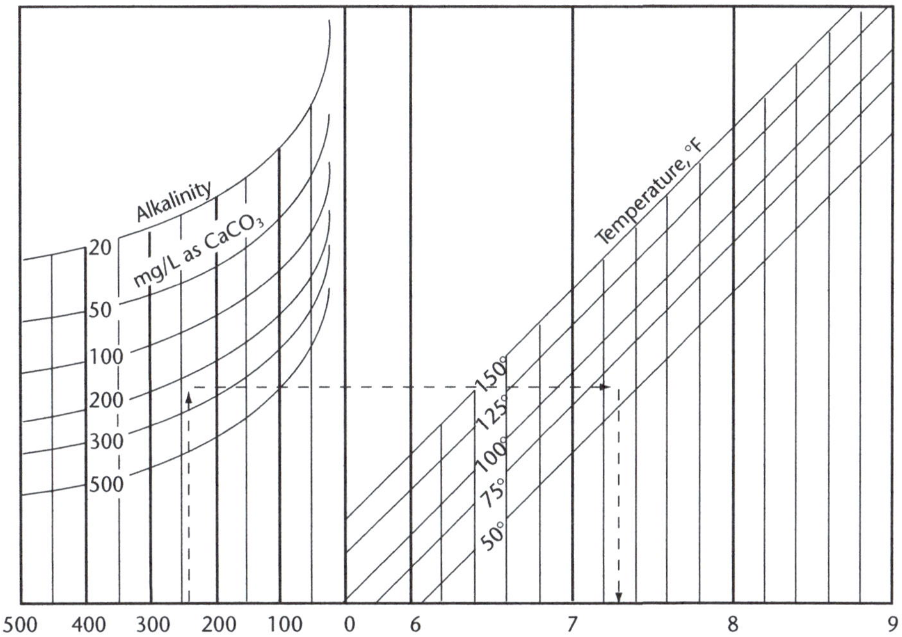

Figure 1-2 Determination of pHs for the Langelier saturation index, given hardness, alkalinity, and temperature (assuming average total dissolved solids of about 500 mg/L)
Courtesy of Ondeo Nalco, Ondeo Nalco Center, Naperville, Illinois, from *Water: The Universal Solvent.*

50°F and 75°F lines). Move downward from the temperature point to the bottom of the graph and identify the pH_s value as 7.3. Finally, calculate the Langelier saturation index as follows:

$$\begin{aligned}\text{Langelier saturation index} &= pH - pH_s \\ &= 6.8 - 7.3 \\ &= -0.5\end{aligned}$$

Since the index is negative, the water is corrosive.

Chelation

Chelation is a chemical treatment process used to control scale formation. The chemical added is known as a chelating agent. It is a water-soluble compound that captures scale-causing ions in solution, preventing precipitation and scale formation. There are several natural organic materials in water that have chelating ability, including humic acid and lignin. When added to water, the chelating agent reacts with calcium ions to keep them in solution and prevent the formation of calcium carbonate scale. In the following equation, the *Y* represents the chelating agent:

$$2Ca^{+2} + Na_4Y \longrightarrow Ca_2Y + 4Na^+ \qquad (1\text{-}36)$$

chelation
A chemical process used to control scale formation, in which a chelating agent "captures" scale-causing ions and holds them in solution, thus preventing them from precipitating out and forming scale.

Sequestration

Sequestration is a chemical addition treatment process that controls scale. The chemical added to the water *sequesters*, or holds in solution, scale-causing ions such as calcium, iron, and manganese, thus preventing them from precipitating and forming scale. Any one of several polyphosphates may be used in this process. The most commonly used is sodium hexametaphosphate, $(NaPO_3)_6$. A common dosage for scale prevention is approximately 0.5 mg/L. Note that sequestration is often related to the deliberate process of adding chemicals to drinking water. Chelation, on the other hand, is often considered a natural process.

sequestration
A chemical process used to control scale formation, in which a sequestering agent holds scale-causing ions in solution, preventing them from precipitating out and forming scale.

Controlled Calcium Carbonate Scaling

One commonly practiced form of corrosion management is controlled calcium carbonate scaling. As stated at the beginning of this section, there are three conditions water can have relative to scale formation and corrosion. Water may be corrosive, stable, or scale-forming. By careful control of the pH and alkalinity adjustment, the condition of water can be altered so that it is slightly scale-forming. Eventually the scale will build up beyond the desired thickness. Adjusting pH and alkalinity to a point that is slightly corrosive will then dissolve the excess scale.

Controlled scaling requires careful and continuous laboratory monitoring because the slightest change in the quality of water may require a change in the amount of lime needed. It is also necessary to monitor the thickness of the calcium carbonate coating developed in the transmission and distribution system pipeline. The coating should be thick enough to prevent corrosion without obstructing the flow of water.

Other Chemical Protective Coatings

There are two other types of chemicals used to create protective coatings in pipelines: (1) polyphosphates and (2) sodium silicate. Polyphosphates include sodium hexametaphosphate, $(NaPO_3)_6$; sodium pyrophosphate, $Na_4P_2O_7$; and a group known as bimetallic glassy phosphates. After being fed into the water, polyphosphates form a phosphate film on interior metal surfaces, protecting them from corrosion. Polyphosphates are also effective as sequestering agents for preventing calcium carbonate scale and for stabilizing dissolved iron and manganese. Dosages of 5–10 mg/L are recommended when treatment is initiated. After 1–2 months' worth of protective film is established, the dosages are reduced and maintained at approximately 1 mg/L.

Sodium silicate $(Na_2Si_4O_9)$, or water-glass, can also be used to control corrosion in water systems. Sodium silicate combines with calcium to form a hard, dense calcium silicate film $(CaSiO_3)$. Dosages vary widely depending on water quality.

Softening

A major problem caused by water hardness is scale formation. Hard waters form calcium carbonate scales in pipelines and boilers. These scales reduce pipeline capacities, lower boiler heat transfer efficiencies, and cause heat exchange tube failures because excessive heating is required to overcome the insulating effects of scale. The results are higher pumping and maintenance costs, longer repair times, shortened equipment life, and higher fuel and power costs.

There are four major water characteristics that control hardness. The tendency to form scale is greater when the hardness, alkalinity, pH, or temperature is increased. Figures 1-3, 1-4, and 1-5 demonstrate how these characteristics are

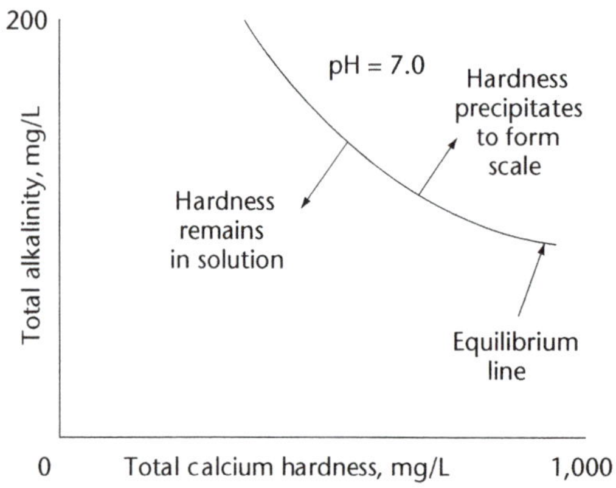

Figure 1-3 Total alkalinity versus total calcium hardness at 77°F (25°C)

Note: Data were computed for one water sample and are not generally applicable.

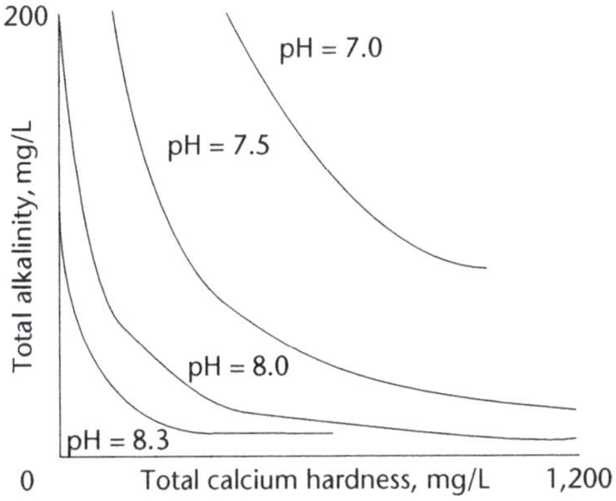

Figure 1-4 Effect of total alkalinity and pH on the amount of total calcium hardness that can be kept in solution

Note: Data were calculated for one water sample and are not generally applicable.

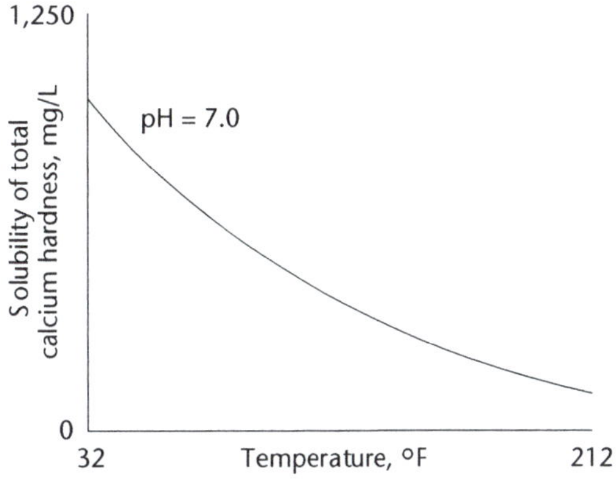

Figure 1-5 Effect of temperature on the amount of calcium that will stay in solution

Note: Data were calculated for one water sample and are not generally applicable.

interrelated. The graphs in the figures were determined for one water sample; they should not be applied in general, but they do illustrate the interaction between the indicated variables.

The 10 chemical reactions involved in water softening were presented earlier in the discussion of lime–soda ash softening. Refer to that material to review hardness removal using lime and soda ash.

Study Questions

1. Which disinfectant would work best against *Cryptosporidium?*
   a. Ozone
   b. Dichloramine
   c. Hypochlorite ion
   d. Hypochlorous acid

2. The pH of a water sample is greater than the pH_s so the Langlier index?
   a. Indicates that the water is corrosive.
   b. Is negative.
   c. Indicates the water is scale forming.
   d. Indicates the water is stable.

3. When water high in calcium bicarbonate is softened by ion exchange ________.
   a. All cations and anions are removed.
   b. Regeneration is usually accomplished by an acid.
   c. Lime is used at a pH of 9.5
   d. Sodium bicarbonate remains in the product water.

4. Lime softening can effectively disinfect water because:
   a. Particles containing microorganisms are settled out in the process.
   b. The pH level is higher than 10.5.
   c. The pH level is below 10.5.
   d. Microorganisms are sensitive to calcium level.

5. A noncarbonate hardness compound is
   a. $CaCO_3$
   b. $MgCO_3$
   c. $Ca(HCO_3)_2$
   d. $MgSO_4$

6. What term refers to the reintroduction of carbon dioxide into water either during or after lime–soda ash softening?
   a. Regeneration
   b. Recarbonation
   c. Replenishment
   d. Recycling

7. ______ is the oxidation of unprotected metal surfaces.
   a. Corrosion
   b. Saturation
   c. Softening
   d. Chelation

8. What is the cause of carbonate hardness?

9. Which two ions are most commonly associated with water hardness?

10. For noncarbonate hardness to be removed, what must be added to remove the noncarbonated calcium compounds?

Chapter 2
Advanced Math for Operators

Pipeline and Hydraulic Calculations

Head

Head is one of the most important measurements in hydraulics. It is used to calculate the hydraulic forces acting in a pipeline and to determine a pump's capacity to overcome or pump against these forces.

Head is a measurement of the energy possessed by the water at any particular location in the water system. In hydraulics, energy (and therefore head) is expressed in units of foot-pounds per pound:

$$\text{head} = \frac{\text{ft-lb}}{\text{lb}}$$

These somewhat cumbersome units of measure cancel out so that head can be expressed in feet:

$$\text{head} = \frac{\text{ft-lb}}{\text{lb}}$$

$$= \text{ft}$$

When head is expressed in feet, as it normally is, the measurement can always be considered to represent the height of water above some reference elevation. The height of water in feet can also be expressed as pressure in pounds per square inch gauge (psig) by dividing the feet-of-water height by 2.31.

head
(1) A measure of the energy possessed by water at a given location in the water system, expressed in feet. (2) A measure of the pressure or force exerted by water, expressed in feet.

pounds per square inch gauge (psig)
Pressure measured by a gauge and expressed in terms of pounds per square inch.

pressure head
A measurement of the amount of energy in water due to water pressure.

Types of Head

There are three types of head, as discussed in detail in the following sections:

- Pressure head
- Elevation head
- Velocity head

Pressure Head Pressure head is a measurement of the amount of energy in water due to *water pressure*. As shown in Figure 2-1, it is the height above the pipeline to which water will rise in a piezometer. The water pressure is easily measured by a

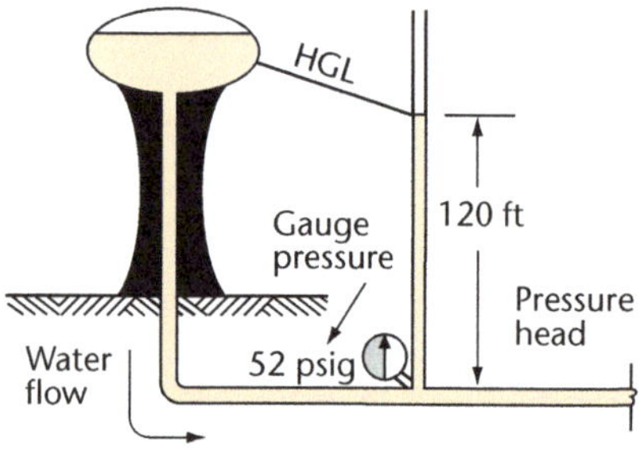

Figure 2-1 Pressure head

pressure gauge, and the resulting pressure reading in pounds per square inch gauge may then be converted to pressure head in feet.

Pressure head describes the vertical distance from the point of pressure measurement to the hydraulic grade line (HGL). So, if you were interested in locating the HGL for a particular pipeline, you would take gauge pressure readings at several critical points along the pipeline, convert those pressure readings to feet of pressure head, and then plot the HGL.

The normal range of pressure heads in water transmission and distribution systems can vary from as little as 50 ft (about 20 psig) to over 1,000 ft (about 450 psig). At certain locations within the treatment plant, pressure heads can be very small, perhaps 2 ft (about 1 psig) or less.

Elevation Head Elevation head is a measurement of the amount of energy that water possesses *because of its elevation*. It is measured as the height in feet from some horizontal reference line or benchmark elevation (such as sea level) to the point of interest in the water system. For example, a reservoir located 500 ft above sea level is said to have an elevation head of 500 ft relative to sea level. The concept is illustrated in Figure 2-2. Elevation head is quite useful in design but has little day-to-day operating significance.

Velocity Head Velocity head is a measurement of the amount of energy in water due to its *velocity*, or *motion*. The greater the velocity, the greater the energy and, therefore, the greater the velocity head. Anything in motion has energy because of that motion.

You might think of the energy of motion in terms of the effort you would need to apply to stop the motion. For example, stopping a car traveling at 55 mph requires that the car's brakes do much more work than if the car were traveling 5 mph, because the faster-moving car has more energy of motion.

hydraulic grade line (HGL)
A line (hydraulic profile) indicating the piezometric level of water at all points along a conduit, open channel, or stream. In an open channel, the HGL is the free water surface.

elevation head
The energy possessed per unit weight of a fluid because of its elevation above some reference point (called the *reference datum*). Elevation head is also called position head or potential head.

velocity head
A measurement of the amount of energy in water due to its velocity, or motion.

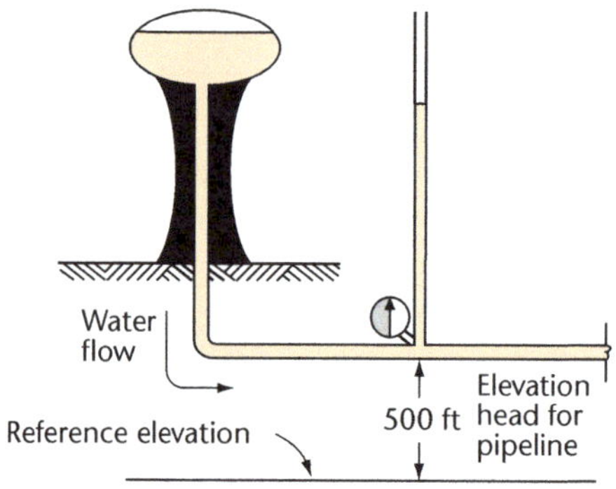

Figure 2-2 Elevation head

Like all other heads, velocity head is expressed in feet. It is determined by multiplying velocity (in feet per second) by itself and then dividing by 64.4 feet per second squared (ft/s^2), as follows:

$$\text{velocity head} = \frac{V^2}{64.4\ \text{ft/s}^2}$$

For example, if the velocity of the water is 30 ft/s, the velocity head would be as follows:

$$\text{velocity head} = \frac{(30)(30)}{64.4}$$

$$= 13.98\ \text{ft}$$

This means that if you pour water from a bucket held 13.98 ft in the air, then the water would be moving 30 ft/s when it hits the ground.

Velocity head is also a measurement of the vertical distance from the HGL to the **energy grade line (EGL)**, as shown in Figure 2-3. EGLs are an advanced concept and will not be discussed further in this basic text.

The velocity, or speed, at which water travels in a water system is usually at least 2 ft/s and generally not more than 10 ft/s. These speeds would produce velocity heads of

$$\text{velocity head at 2 ft/s velocity} = \frac{(2)(2)}{64.4}$$

$$= 0.06\ \text{ft}$$

$$\text{velocity head at 10 ft/s velocity} = \frac{(10)(10)}{64.4}$$

$$= 1.55\ \text{ft}$$

Because velocity heads are so small relative to pressure heads (which usually range from 50 to 1,000 ft), they can usually be ignored in operations without causing a significant error. However, although velocity head can be ignored in most cases, it is important that you recognize that velocity head should be considered whenever it is greater than 1 or 2 percent of the pressure head—that is, whenever it is large enough to be a significant part of the total head. This scenario will usually occur either in pipelines or pump systems when the pressure head is very small (less than 70 ft) and velocities are in the range of 7–10 ft/s.

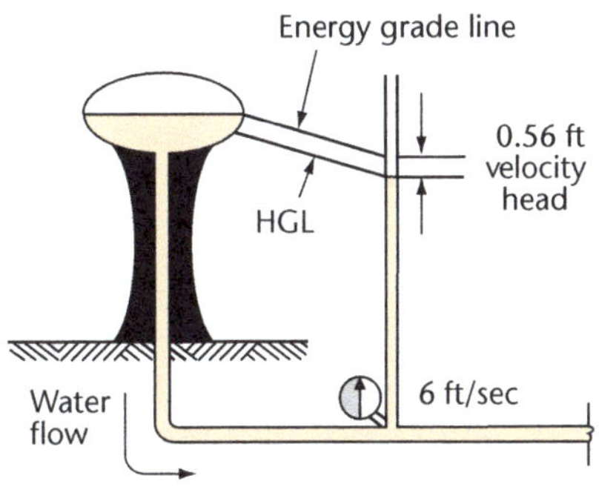

Figure 2-3 Velocity head

energy grade line (EGL)
A line joining the elevations of the energy heads; a line drawn above the hydraulic grade line by a distance equivalent to the velocity head of the flowing water at each section along a stream, channel, or conduit.

Calculating Heads

Of the three types of head discussed, pressure head is the most significant and useful from an operating standpoint. The following example illustrates how pressure heads can be measured and used in operations.

Pressure heads can be measured whether water is standing still or moving. However, dynamic pressure heads will be lower than static heads because of the pressures lost in moving the water.

Example 1

Pressure readings were taken for the system in Figure 2-4A when the valve was open and again when it was closed. Determine pressure heads in feet and show the HGL for both conditions.

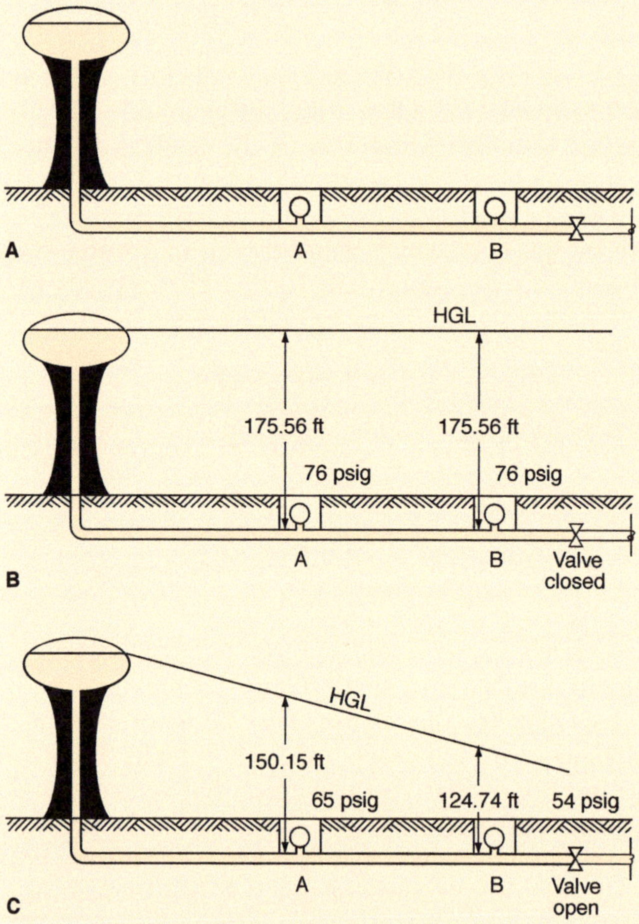

Figure 2-4 Schematic for Example 1

Valve closed: The pressure heads at points A and B are the same, as shown by the gauge readings:

	Valve Open	Valve Closed
Pressure at A	65 psig	76 psig
Pressure at B	54 psig	76 psig

Converting from pounds per square inch gauge to feet of head, you are moving from the smaller box to the larger box (as discussed in the math chapters of the Grade 1 and Grade 2 *WSO Water Treatment* books); therefore you should multiply by 2.31:

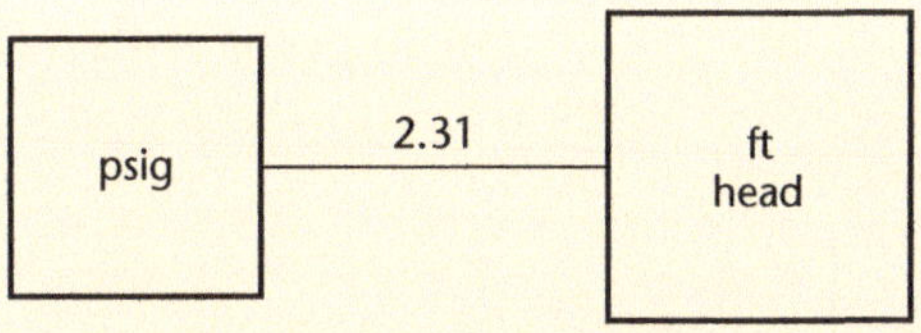

$$\text{pressure head at A and B} = (76 \text{ psig})(2.31 \text{ ft/psig})$$
$$= 175.56 \text{ ft}$$

Valve open: Calculate the pressure heads at points A and B using the same method shown for the closed-valve condition:

$$\text{pressure head at A} = (65 \text{ psig})(2.31 \text{ ft/psig})$$
$$= 150.15 \text{ ft}$$

$$\text{pressure head at B} = (54 \text{ psig})(2.31 \text{ ft/psig})$$
$$= 124.74 \text{ ft}$$

The two pressure heads and the HGL are shown in Figures 2-4B and 2-4C.

The pressure heads of water standing still are higher than those of water moving. The difference represents the amount of energy used by the water in moving from the elevated storage tank to points A and B.

Often in water systems the static pressure head can be found from elevation information that was established by surveyors during construction.

WATCH THE VIDEO
Operator Math II—Head Grade Line (www.awwa.org/wsovideoclips)

Head Loss

Head loss is defined as the amount of energy used up by water moving from one point to another. The two most important categories of head loss are friction head loss and minor head loss.

Friction Head Loss

The inside of a pipe may feel quite smooth, but it is rough from a hydraulic standpoint. The friction caused by water moving over the rough surface causes an energy loss called friction head loss.

The amount of friction head loss occurring in a pipe varies, depending on (1) the roughness of the pipe and (2) the velocity of flow through the pipe.

Pipe roughness varies widely depending on the type of pipe material and the condition of the pipe. Plastic is one of the smoothest pipe materials, and steel is one of the roughest. A smooth interior surface results in lower friction loss; a rough interior surface causes high loss.

The condition of the pipe depends on several factors:

- Type of lining
- Age
- Degree of corrosion or scaling
- Slime growths
- Tuberculation
- Obstructions (such as mud, silt, sand, rocks, sticks)

friction head loss
The head lost by water flowing in a stream or conduit as the result of (1) the disturbance set up by the contact between the moving water and its containing conduit and (2) intermolecular friction.

Smooth linings reduce friction loss. The other factors listed increase friction loss and cause friction loss to vary from point to point within the water system.

The velocity of the moving water also affects friction loss. Without exception, as the velocity increases, friction loss increases, even if the pipe roughness stays the same.

Some of the factors that cause friction loss, such as corrosion, scaling, slime growths, and obstructions, can be controlled by proper operation, maintenance, and treatment of the pipeline. However, the natural roughness of the pipe material or lining, and the natural roughening of that material or lining with age, are factors over which there is little control. Consequently, pipelines must be designed to operate with the additional head needed to overcome friction head loss.

Friction head loss is one of the simplest head losses to calculate. It can be determined from one of many readily available tables. Perhaps the most commonly used are the tables based on the Hazen–Williams formula, in which a smoothness coefficient C represents the smoothness of varying pipe materials of varying ages; the higher the C value, the smoother the pipe. For example, the C value for cast-iron pipe (CIP) may be as high as 140 for the very best new CIP laid perfectly straight. A C value of 130 is the average value for most new CIP. As CIP ages, the C value drops, on the average, to C = 120 at 5 years, C = 110 at 10 years, C = 100 at 17 years, C = 90 at 26 years, C = 80 at 37 years, and C = 75 or less for pipe 40 or more years old. Table 2-1 lists typical smoothness coefficients.

The following example illustrates friction head loss calculations.

Table 2-1 Smoothness coefficients for various pipe materials

	C Value	
Type of Pipe	**New**	**10 Years Old**
Asbestos–cement pipe	140	120–130
Cast-iron pipe	130	100
Reinforced and plain concrete pipe	140	120–130
Ductile iron pipe	130	100
Plastic pipe	150	120–130
Steel pipe	110	100

Example 2

A 12-in.-diameter water transmission line, 10,500 ft long, carries water from the supply reservoir to the treatment plant at a flow rate of 2.0 mgd. The steel transmission line has been in service for about 10 years. Use Table 2-2 to determine the total friction head loss in the line.

To use Table 2-2, you need to know two numbers: the flow rate (2.0 mgd in this problem) and the C value. According to the problem, the pipe is steel, about 10 years old. Table 2-1 shows a smoothness coefficient for 10-year-old steel pipe of C = 100.

With this information, enter Table 2-2 in the first column at the flow rate 2,000,000 gpd. Then move across the table until you reach the column for C = 100. The number shown is 7.6. This is the friction head loss, in feet, for every 1,000-ft length of pipeline.

Table 2-2 Friction loss factors for 12-in. pipe

Discharge		Velocity	Velocity	Loss of Head, ft per 1,000 ft of length						
gpd	ft^3/s	ft/s	Head, ft	C = 140	C = 130	C = 120	C = 110	C = 100	C = 90	C = 80
100,000	0.155	0.20	0.00	0.02	0.02	0.02	0.02	0.03	0.04	0.04
200,000	0.309	0.39	0.00	0.06	0.07	0.08	0.09	0.11	0.13	0.16
300,000	0.464	0.59	0.01	0.12	0.14	0.16	0.19	0.22	0.27	0.34
400,000	0.619	0.79	0.01	0.20	0.24	0.27	0.32	0.38	0.47	0.58
500,000	0.774	0.99	0.02	0.31	0.36	0.41	0.48	0.58	0.71	0.88
600,000	0.928	1.18	0.02	0.44	0.50	0.58	0.68	0.81	0.99	1.23
700,000	1.083	1.38	0.03	0.58	0.66	0.77	0.91	1.08	1.32	1.64
800,000	1.238	1.58	0.04	0.74	0.85	0.99	1.15	1.38	1.68	2.09
900,000	1.392	1.77	0.05	0.92	1.06	1.23	1.45	1.72	2.10	2.61
1,000,000	1.547	1.97	0.06	1.12	1.29	1.50	1.76	2.10	2.57	3.18
1,100,000	1.702	2.17	0.07	1.34	1.54	1.79	2.10	2.50	3.04	3.79
1,200,000	1.857	2.36	0.09	1.58	1.81	2.10	2.47	2.94	3.58	4.45
1,300,000	2.011	2.56	0.10	1.83	2.10	2.43	2.85	3.40	4.14	5.2
1,400,000	2.166	2.76	0.12	2.10	2.40	2.79	3.26	3.90	4.76	5.9
1,500,000	2.321	2.96	0.14	2.39	2.73	3.17	3.71	4.43	5.4	6.7
1,600,000	2.476	3.15	0.15	2.69	3.09	3.58	4.20	5.0	6.1	7.6
1,700,000	2.630	3.35	0.17	3.00	3.45	4.00	4.69	5.6	6.8	8.5
1,800,000	2.785	3.55	0.20	3.33	3.82	4.43	5.2	6.2	7.6	9.4
1,900,000	2.940	3.74	0.22	3.70	4.24	4.92	5.8	6.9	8.4	10.4
2,000,000	3.094	3.94	0.24	4.06	4.65	5.4	6.4	7.6	9.2	11.5
2,200,000	3.404	4.33	0.29	4.85	5.6	6.5	7.6	9.0	10.9	13.7
2,400,000	3.713	4.73	0.35	5.7	6.5	7.6	8.9	10.5	12.8	16.0
2,600,000	4.023	5.12	0.41	6.6	7.6	8.8	10.3	12.3	15.0	18.6
2,800,000	4.332	5.52	0.47	7.6	8.7	10.1	11.9	14.1	17.2	21.5
3,000,000	4.642	5.91	0.54	8.6	9.9	11.5	13.5	16.0	19.4	24.3
3,500,000	5.41	6.89	0.74	11.4	13.2	15.3	17.9	21.3	26.0	32.3
4,000,000	6.19	7.88	0.96	14.5	16.6	19.3	22.6	27.0	33.2	41
4,500,000	6.96	8.87	1.22	18.0	20.6	24.0	28.2	33.6	41.2	51
5,000,000	7.74	9.85	1.50	22.0	25.1	29.2	34.3	41.0	50.0	62
5,500,000	8.51	10.84	1.82	26.5	30.3	35.1	41.4	49.4	60	75
6,000,000	9.28	11.82	2.17	31.1	35.7	41.4	48.8	58	70	88
7,000,000	10.83	13.79	2.96	41.2	47.2	55	65	77	94	116
8,000,000	12.38	15.76	3.86	53	61	71	83	99	121	150
9,000,000	13.92	17.73	4.89	66	75	87	103	122	148	185
10,000,000	15.47	19.70	6.03	81	93	107	126	150	183	228

To determine the total friction head loss asked for in the problem, you must determine how many 1,000-ft lengths there are in 10,500 ft:

$$\frac{10{,}500 \text{ ft}}{1{,}000 \text{ ft}} = 10.5$$

Since the number of 1,000-ft lengths of steel pipe is 10.5, the total friction head loss is

$$\begin{aligned}\text{total friction head loss} &= (\text{loss per } 1{,}000 \text{ ft})(\text{no. of } 1{,}000\text{-ft increments})\\ &= (7.6 \text{ ft})(10.5)\\ &= 79.8 \text{ ft}\end{aligned}$$

Table 2-2 provides other useful information not requested in Example 2 and can be used to solve for several variables relating to friction loss in 12-in. pipe. For example, the third column gives the velocity for the known flow rate (3.94 ft/s for 2.0 mgd), and the fourth column gives the velocity head (0.24 ft). While Table 2-2 is specific to 12-in. pipe, similar tables exist for other sizes of pipe.

Pumping Problems

The rate of flow produced by a pump is expressed as the volume of water pumped during a given period of time.

Pumping Rates

The mathematical equation used in pumping rate problems can usually be determined from the verbal statement of the problem:

VERBAL: What is the pumping rate in "gallons *per* minute"?

MATH: $$\text{pumping rate} = \frac{\text{gallons}}{\text{minutes}}$$

VERBAL: What is the pumping rate in "gallons *per* hour"?

MATH: $$\text{pumping rate} = \frac{\text{gallons}}{\text{hours}}$$

The number of gallons pumped during a period can be determined either by reading a flowmeter or by measuring the number of gallons pumped into or out of a tank.

Example 3

The totalizer of the meter on the discharge side of your pump reads in hundreds of gallons. If the totalizer shows a reading of 108 at 1:00 p.m. and 312 at 1:30 p.m., what is the pumping rate expressed in gallons per minute?

The problem asks for pumping rate in *gallons per minute* (gpm), so the mathematical setup is

$$\text{pumping rate} = \frac{\text{gallons}}{\text{minutes}}$$

To solve the problem, fill in the blanks (number of gallons and number of minutes) in the equation. The total gallons pumped is determined from the totalizer readings:

$$\begin{array}{r} 31{,}200 \text{ gal} \\ -\ 10{,}800 \text{ gal} \\ \hline 20{,}400 \text{ gal} \end{array}$$

The volume was pumped between 1:00 p.m. and 1:30 p.m., for a total of 30 min. From this information, calculate the gallons-per-minute pumping rate:

$$\text{pumping rate} = \frac{20{,}400 \text{ gal}}{30 \text{ min}}$$

$$= 680 \text{ gpm pumping rate}$$

Instead of using totalizer readings to calculate the average pumping rate for a period of a few minutes or hours (as in Example 3), you can read the *instantaneous* pumping rate or instantaneous flow rate—the flow rate or pumping at *one particular moment*—directly from many flowmeters. Other flowmeters require that you perform calculations to determine the instantaneous flow rate, as illustrated in the following example.

Pump Heads

Pump head measurements are used to determine the amount of energy a pump can or must impart to the water. These heads, measured in feet, are a special application of the pressure and velocity heads discussed earlier in the chapter. Specific terms are used to describe heads measured at various points and under different conditions of the pump system.

Suction and Discharge

In pump systems, the words *suction* and *discharge* identify the inlet and outlet sides of the pump. As shown in Figure 2-5, the *suction side* of the pump is the *inlet*, or *low-pressure*, side. The *discharge side* is the *outlet*, or *high-pressure*, side. Any time a pump term includes the words *suction* or *discharge*, you can recognize immediately which side of the pump system is being discussed.

The pump center line is represented by a horizontal line drawn through the center of the pump. This line is important because it is the reference line from which pump head measurements are made.

The terms *suction* and *discharge* help establish the location of the particular pump head measurement. The terms *static* and *dynamic*, however, describe the condition of the system when the measurement is taken. Static heads are measured when the pump is off, and therefore the water is not moving. Dynamic heads are measured with the pump running and water flowing through the system.

instantaneous flow rate
A flow rate of water measured at one particular instant, such as by a metering device, involving the cross-sectional area of the channel or pipe and the velocity of the water at that instant.

pump center line
An imaginary line through the center of a pump.

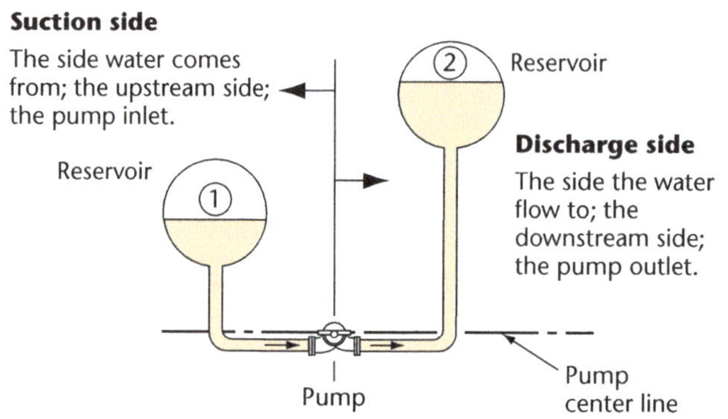

Figure 2-5 Suction and discharge sides of a pump

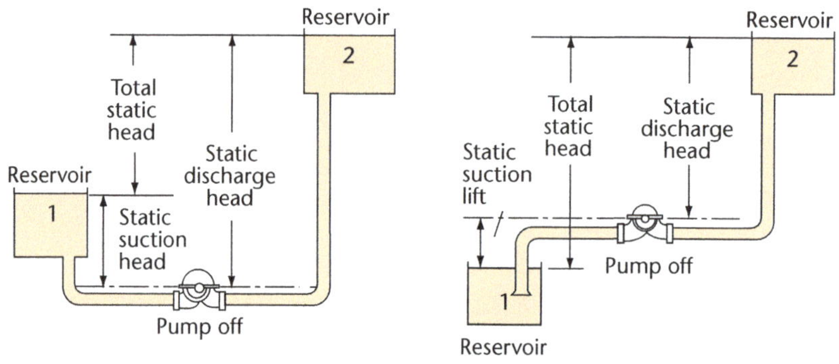

A. Reservoir feeding pump higher than pump

B. Reservoir feeding pump lower than pump

Figure 2-6 Static heads

static suction head
The difference in elevation between the pump center line and the free water surface of the reservoir feeding the pump. In the measurement of static suction head, the piezometric surface of the water at the suction side of the pump is higher than the pump; otherwise, static suction lift is measured.

static suction lift
The difference in elevation between the pump center line of a pump and the free water surface of the liquid being pumped. In a static suction lift measurement, the piezometric surface of the water at the suction side of the pump is lower than the pump; otherwise, static suction head is measured.

static discharge head
The difference in height between the pump center line and the level of the discharge free water surface.

total static head
The total height that the pump must lift the water when moving it from one point to another. The vertical distance from the suction free water surface to the discharge free water surface.

Static Heads

Figure 2-6 illustrates the two basic pumping configurations found in water systems. The labeled vertical distances show the four types of static pump head:

- Static suction head
- Static suction lift
- Static discharge head
- Total static head

In a system where the reservoir feeding the pump is higher than the pump (Figure 2-6A), the difference in elevation (height) between the pump center line and the free water surface of the reservoir feeding the pump is termed **static suction head**. But in a system where the reservoir feeding the pump is lower than the pump (as in Figure 2-6B), the difference in elevation between the center line and the free water surface of the reservoir feeding the pump is termed **static suction lift**. Notice that a single system will have either a static head measurement or a static suction lift measurement, but not both.

Static discharge head is defined as the difference in height between the pump center line and the level of the discharge free water surface.

Total static head is the total height that the pump must lift the water when moving it from *reservoir 1* to *reservoir 2*. It is defined precisely as the vertical distance from the suction free water surface to the discharge free water surface. In Figure 2-6A, total static head is found by subtracting static suction head from static discharge head, whereas in Figure 2-8B, it is the sum of the static suction lift and static discharge head.

Static heads can be calculated from measurements of reservoir and pump elevations or from pressure gauge readings taken when the pump is not running.

Example 4

For Figure 2-7A, locate, label, and calculate the following (in feet):

- Static suction head
- Static discharge head
- Total static head

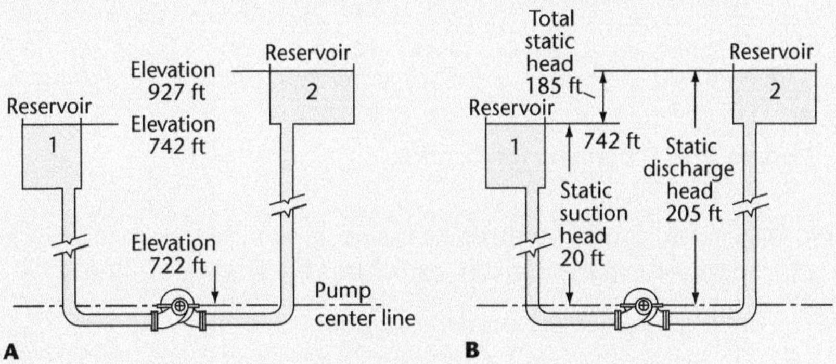

Figure 2-7 Schematic for Example 4

First, locate and label the three types of static head, as shown in Figure 2-7B. Then, using the elevation information given for the problem, calculate each head as follows:

$$\text{static suction head} = 742\text{ ft} - 722\text{ ft}$$
$$= 20\text{ ft}$$

$$\text{static discharge head} = 927\text{ ft} - 722\text{ ft}$$
$$= 205\text{ ft}$$

Total static head can be found in two ways:

$$\text{total static head} = \text{elevation of reservoir 1} - \text{elevation of reservoir 2}$$
$$= 927\text{ ft} - 742\text{ ft}$$
$$= 185\text{ ft}$$

or

$$\text{total static head} = \text{static discharge head} - \text{static suction head}$$
$$= 205\text{ ft} - 20\text{ ft}$$
$$= 185\text{ ft}$$

Example 5

For Figure 2-8A, locate, label, and calculate the following (in meters):

- Static suction lift
- Static discharge head
- Total static head

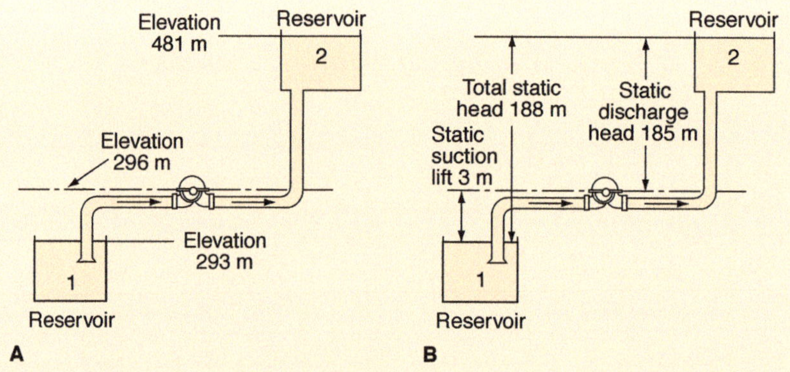

Figure 2-8 Schematic for Example 5

First, locate and label the three heads as shown in Figure 2-8B. Then, using the elevation information given, calculate the heads as follows:

$$\text{static suction lift} = 296 \text{ m} - 293 \text{ m}$$
$$= 3 \text{ m}$$

$$\text{static discharge head} = 481 \text{ m} - 296 \text{ m}$$
$$= 185 \text{ m}$$

$$\text{total static head} = 85 \text{ m} + 3 \text{ m}$$
$$= 188 \text{ m}$$

or

$$\text{total static head} = 481 \text{ m} - 293 \text{ m}$$
$$= 188 \text{ m}$$

If static heads were the only heads involved in pump operations, then any force exerted by the pump greater than the static head would cause water to be pumped. However, this doesn't happen. As soon as the pump is turned on, the pressures and energies within the system change. Measurements made while the pump is running, called dynamic heads, are used to describe the total energy that the pump must develop for pumping to take place.

Dynamic Heads

When water flows through a pipe, the water rubs against the walls of the pipe, and energy is lost to friction. In addition, there is a certain amount of resistance to flow as the water passes through valves, fittings, inlets, and outlets of a piping system. These additional energy losses that the pump must overcome are called *friction head losses* and minor head losses.

minor head losses
The energy losses that result from the resistance to flow as water passes through valves, fittings, inlets, and outlets of a piping system.

Let's look at the effect of these additional energy losses on each side of the pump. Figure 2-9 shows a comparison of the head that exists on the suction side of the pump before and after the pump is turned on. Notice that loss of head is caused by friction losses when the pump is turned on. Part of the static suction head (which could otherwise aid the pump in pumping the water up to reservoir 2) is lost because of friction and minor losses as the water moves from reservoir 1 to the pump.

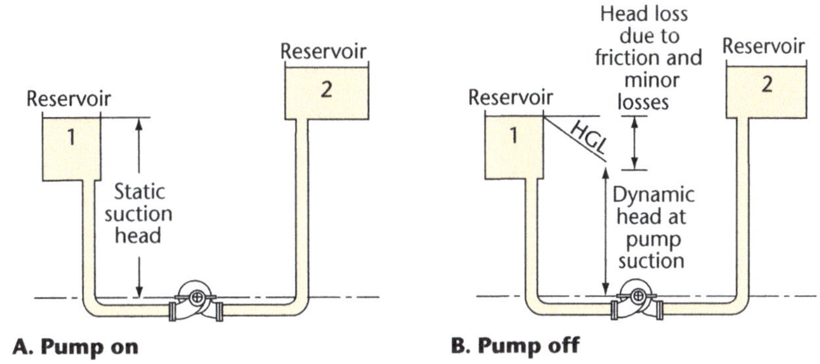

Figure 2-9 Effect of friction and minor head losses on suction head

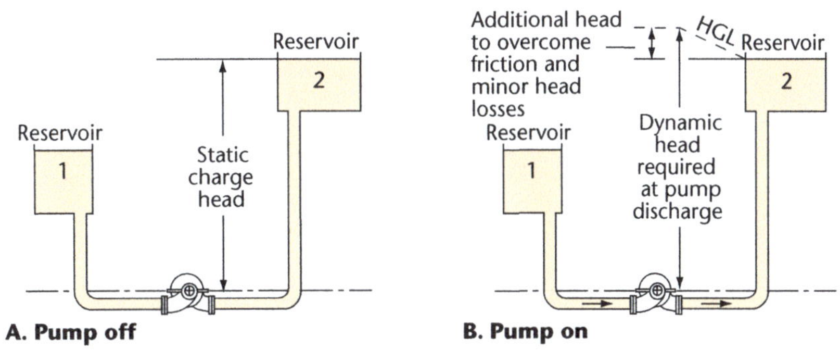

Figure 2-10 Effect of friction and minor head losses on discharge head

Similarly, friction and minor head losses develop on the discharge side of the pump as the water moves from the discharge side of the pump to reservoir 2. These losses create an additional load or head against which the pump must operate (Figure 2-10). If the pump does not add enough energy to overcome the friction and minor head losses that occur on both the suction and discharge sides of the pump, the pump will not be able to move the water.

Now that you have a basic understanding of the reasons for a difference between static and dynamic conditions, consider the four measurements of dynamic head (shown in Figure 2-11):

- Dynamic suction head
- Dynamic suction lift
- Dynamic discharge head
- Total dynamic head

In a system where the HGL on the suction side is higher than the pump (as in Figure 2-11A), the dynamic suction head is measured from the pump center line at the *suction* of the pump to the point on the HGL directly *above* it. In a system where the HGL on the suction side is lower than the pump (as in Figure 2-11B), the dynamic suction lift is measured from the pump center line at the *suction* of the pump to the point on the HGL directly *below* it.

A single system will have a dynamic suction head or a dynamic suction lift, but not both. The location of the HGL, not of the reservoir, determines which condition exists. As shown in Figure 2-12, it is possible to have a dynamic suction lift condition even though the reservoir feeding the pump is above the pump center line; in such cases, the HGL on the suction side is below the pump center line when the pump is running.

The *dynamic discharge* head is measured from the pump center line at the discharge of the pump to the point on the HGL directly *above* it.

dynamic suction head
The distance from the pump center line at the suction of the pump to the point of the hydraulic grade line directly above it.

dynamic suction lift
The distance from the pump center line at the suction of the pump to the point on the hydraulic grade line directly below it.

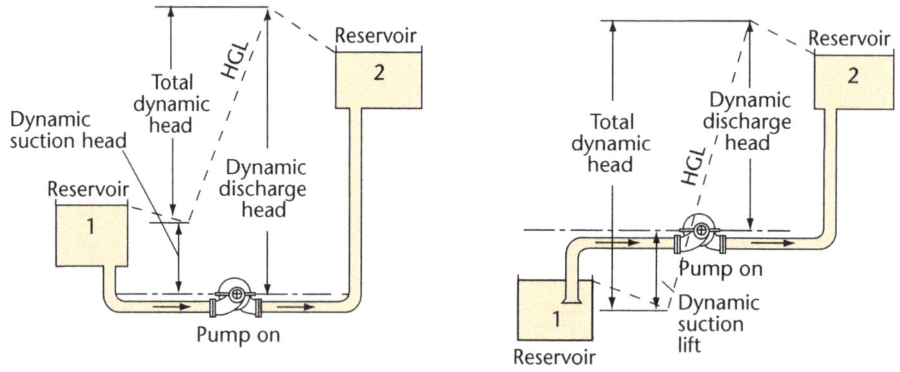

A. Suction side higher than pump

B. Suction side lower than pipe

Figure 2-11 Dynamic heads

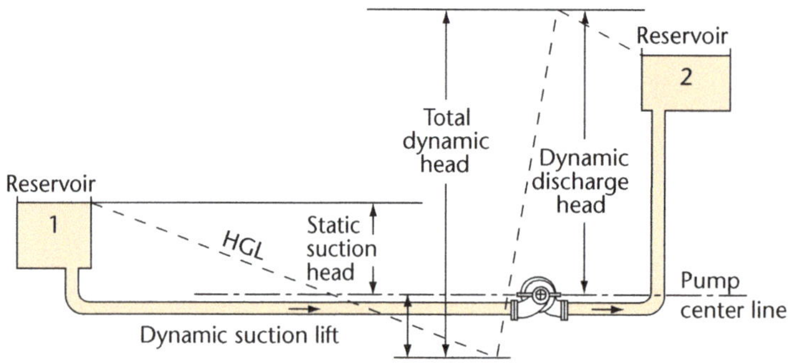

Figure 2-12 Dynamic suction lift for reservoir above pump

The **total dynamic head** is the difference in height between the HGL on the discharge side of the pump and the HGL on the suction side of the pump. This head is a measure of the total energy that the pump must impart to the water to move it from reservoir 1 to reservoir 2.

Dynamic pump heads can be calculated in either of two ways:

- *By adding friction and minor losses to static head.* Static heads can be determined using existing elevation information such as shown in Examples 4 and 5 in this chapter; and friction and minor losses can be found by using tables such as Table 2-2.
- *By direct measurement using pressure gauges.* Pressure gauges cannot measure the velocity head component. As discussed earlier, though, velocity head can often be ignored without significant error.

total dynamic head
The difference in height between the hydraulic grade line (HGL) on the discharge side of the pump and the HGL on the suction side of the pump. This head is a measure of the total energy that a pump must impart to the water to move it from one point to another.

Example 6 illustrates how dynamic pump heads can be measured using pressure gauge information. As noted before, this method of determining heads can also be used to determine static heads by simply turning the pump off before taking the gauge readings.

Example 6

Using the pressure gauge information taken when the pump was running, determine the dynamic heads for the system in Figure 2-13A. Gauge readings are shown in Figure 2-13A in pounds per square inch gauge (psig).

Find the dynamic suction and dynamic discharge heads by converting the pressure gauge readings to feet of head.

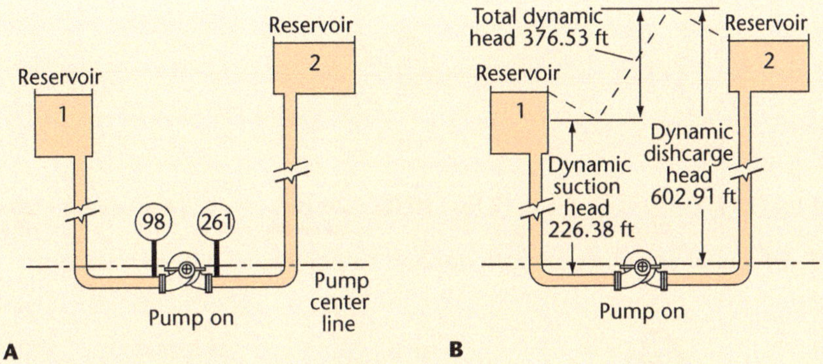

Figure 2-13 Schematic for Example 6

Moving from a smaller box to a larger box, you should multiply by 2.31.

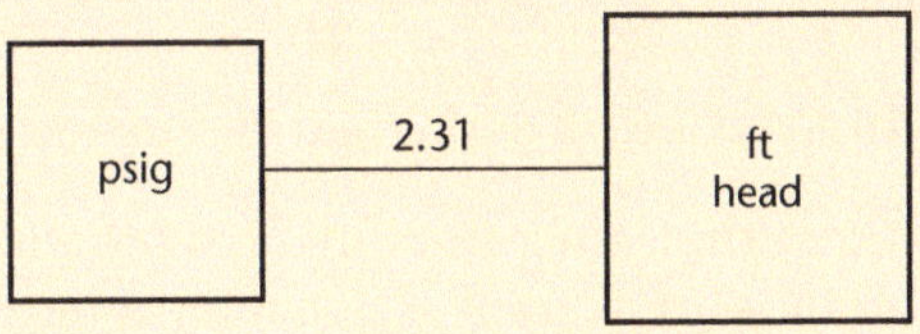

$$\text{feet of head} = (\text{pressure})(2.31 \text{ ft/psig})$$

$$\text{dynamic suction head} = (98 \text{ psig})(2.31 \text{ ft/psig})$$
$$= 226.38 \text{ ft}$$

$$\text{dynamic discharge head} = (261 \text{ psig})(2.31 \text{ ft/psig})$$
$$= 602.91 \text{ ft}$$

The total dynamic head in this problem is the *difference* between the head on the discharge side of the pump and the head on the suction side of the pump (Figure 2-13B).

$$\text{total dynamic head} = 602.91 \text{ ft} - 226.38 \text{ ft}$$
$$= 376.53 \text{ ft}$$

Horsepower and Efficiency

Calculations of pump horsepower and efficiency are made in conjunction with many water transmission, treatment, and distribution operations. The selection of a pump or combination of pumps to provide an adequate pumping capacity will depend on the flow rate required and the effective height through which the flow must be pumped. The effective height is defined as the total feet of head against which the pump must work.

Horsepower

To understand horsepower, you must first understand the technical meaning of the term work. Work is defined as the operation of a force over a specific distance. For example, suppose a 1-lb object is lifted 1 ft; the work done is measured in foot-pounds (ft-lb):

$$(\text{feet})(\text{pounds}) = \text{foot-pounds}$$

horsepower
A standard unit of power equal to 746 W (watts), approximately equal to 33,000 ft-lb/min.

efficiency
The ratio of the total energy output to the total energy input, expressed as percent.

work
The operation of a force over a specific distance.

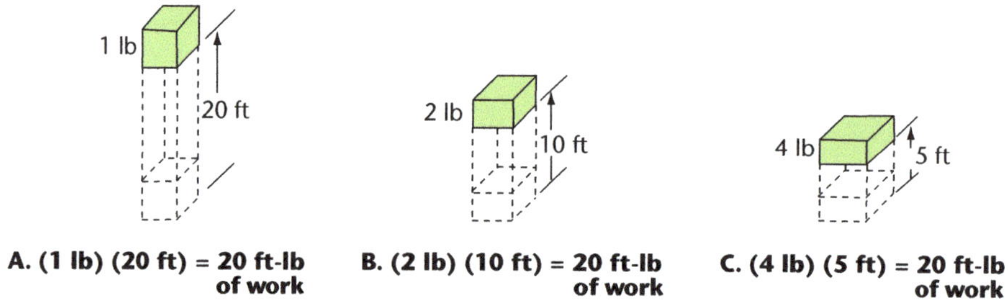

Figure 2-14 Equal amounts of work applied to objects of different weight

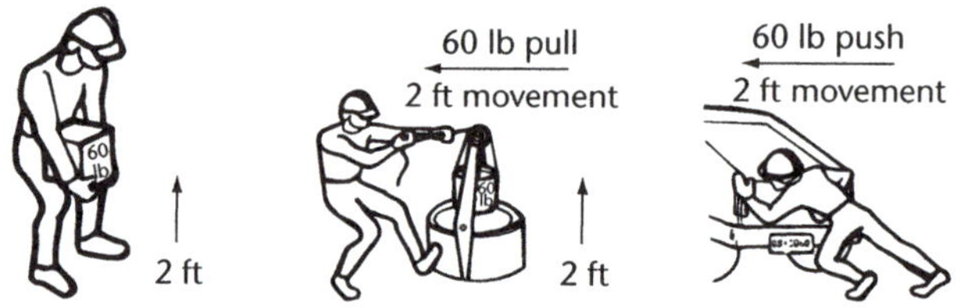

Figure 2-15 Work being performed in different directions

Because it always requires the same amount of work to lift a 1-lb object 1 ft straight up, that amount of work is used as a standard measure: 1 ft-lb. However, work performed on an object of any weight can be measured in foot-pounds (Figure 2-14). Work can be performed in any direction (Figure 2-15). And, as shown in Figure 2-15, work doesn't need to involve lifting; an engine pushing a car along a level highway and a pump pushing water through a level pipeline are both examples of work.

The rate of doing work—that is, the measure of how much work is done in a given time—is called power. To make power calculations, you must know the time required to perform the work. The basic unit for power measurement is foot-pounds per minute (ft-lb/min). Note that this is work performed *per* time. One equation for calculating power in foot-pounds per minute is

$$\left(\begin{array}{c}\text{power in}\\ \text{foot-pounds per minute}\end{array}\right) = \left(\begin{array}{c}\text{head}\\ \text{in feet}\end{array}\right)\left(\begin{array}{c}\text{flow rate in}\\ \text{pounds per minute}\end{array}\right)$$

You will often work with measurements of power expressed in horsepower (hp), which is related to foot-pounds per minute by the conversion equation

$$1 \text{ hp} = 33{,}000 \text{ ft-lb/min}$$

When doing horsepower problems using the basic power equation, you must first convert the measurements of head, flow rate, and power into the proper units:

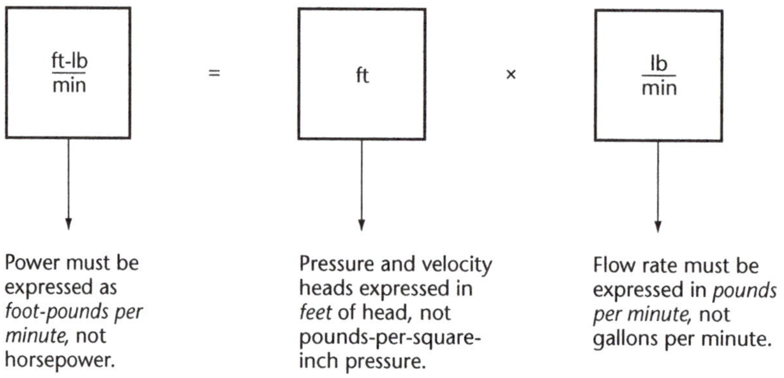

power
The measure of the amount of work done in a given period of time. The rate of doing work. Measured in watts or horsepower.

The power calculated in foot-pounds per minute can be converted to horsepower by dividing by 33,000. (The conversion equation is 1 hp = 33,000 ft-lb/min.) The result is called **water horsepower** (whp), since it is the amount of horsepower required to lift the water. Another equation that can be used to calculate water horsepower is

$$\text{whp} = \frac{\text{(flow rate in gallons per minute)(total head in feet)}}{3{,}960}$$

As an alternative, we will also use this equation in our next example.

Example 7

A pump must pump 2,000 gpm against a total head of 20 ft. What horsepower (water horsepower) will be required to do the work?

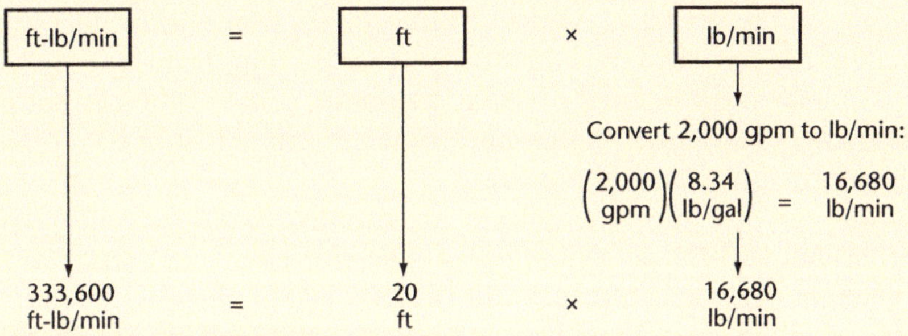

Now convert foot-pounds per minute to horsepower. You know that 1 hp = 33,000 ft-lb/min; in asking, "How much horsepower does this represent?" you are asking, "How many 33,000s are there in 333,600?" Mathematically, this is written as

$$\frac{333{,}600 \text{ ft-lb/min}}{33{,}000 \text{ ft-lb/min/hp}} = 10.11 \text{ hp}$$

As an alternative, you can use the box diagram method to convert foot-pounds per minute to horsepower. You are moving from a larger box to a smaller box. Therefore, division by 33,000 is indicated:

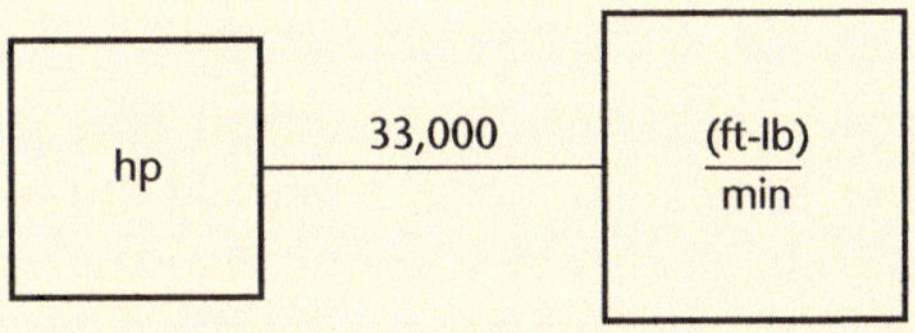

$$\frac{333{,}600 \text{ ft-lb/min}}{33{,}000 \text{ ft-lb/min/hp}} = 10.11 \text{ hp}$$

Efficiency

In the preceding examples, you learned how to calculate water horsepower, the amount of power that must be applied directly to water to move it at a given rate against a given head. Applying the power to the water requires a pump, which in turn is driven by a motor, which is powered by electrical current.

water horsepower
The portion of the power delivered to a pump that is actually used to lift water.

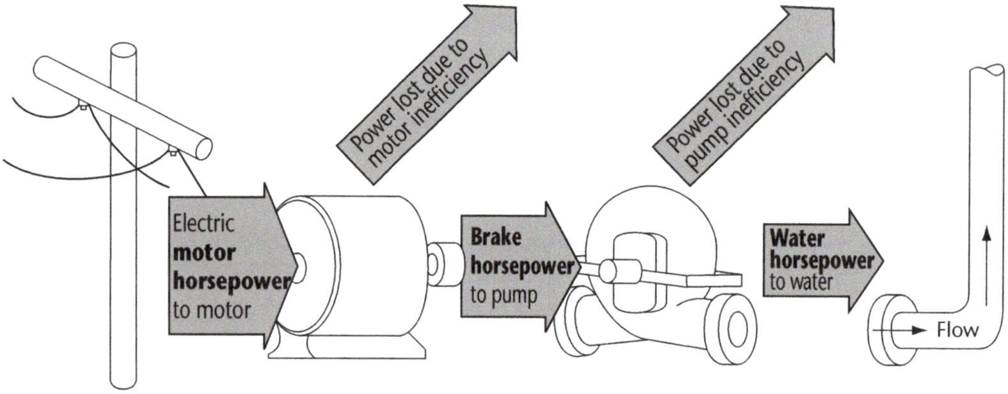

Figure 2-16 Power loss due to motor and pump inefficiency

Neither the pump nor the motor will ever be 100 percent efficient (Figure 2-16). This means that not all of the power supplied by the motor to the pump (called **brake horsepower**, bhp) will be used to lift the water (*water horsepower*); some of the power is used to overcome friction within the pump. Similarly, not all of the power of the electrical current driving the motor (called **motor horsepower**, mhp) will be used to drive the pump; some of the current is used to overcome friction within the motor, and some current is lost in the conversion of electrical energy to mechanical power.

Depending on size and type, pumps are usually 50–85 percent efficient, and motors are usually 80–95 percent efficient. The efficiency of a particular motor or pump is given in the manufacturer's information accompanying the unit.

In some installations, you will know only the combined efficiency of the pump and motor, called the **wire-to-water efficiency**. The wire-to-water efficiency is obtained by multiplying the motor and pump efficiencies together. For example, if a motor is 82 percent efficient and a pump is 67 percent efficient, the overall, or wire-to-water, efficiency is 55 percent (0.82 × 0.67 = 0.55).

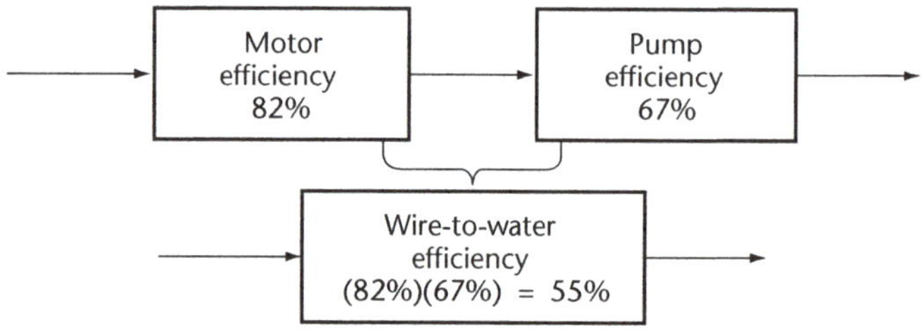

brake horsepower
The power supplied to a pump by a motor.

motor horsepower
The horsepower equivalent to the watts of electric power supplied to a motor.

wire-to-water efficiency
The ratio of the total power input (electric current expressed as *motor horsepower*) to a motor and pump assembly, to the total power output (*water horsepower*); expressed as a percent.

In practical horsepower calculations, you must take into account pump and motor efficiencies so that you can determine the size of pump and motor and the amount of power that will be necessary to move the water at the desired flow rate.

To show how horsepower and efficiency calculations are used, consider a system in which the pump is pumping 800 gpm against a total head of 46.2 ft. The water horsepower required to perform this work is calculated as follows:

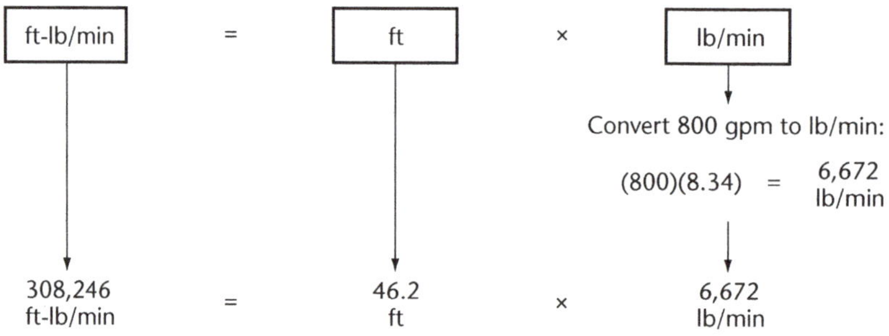

Then convert foot-pounds per minute to horsepower.

$$\frac{308{,}246 \text{ ft-lb/min}}{33{,}000 \text{ ft-lb/min/hp}} = 9.34 \text{ hp}$$

Now, suppose the pump is only 85 percent efficient and the motor 95 percent efficient. To account for these inefficiencies, more than 9.34 whp would have to be supplied to the pump, and even more horsepower would have to be supplied to the motor.

Techniques for calculating the required horsepower are discussed in Example 8. To use the techniques, however, you must know that power input to the motor is usually expressed in terms of electrical power rather than horsepower.

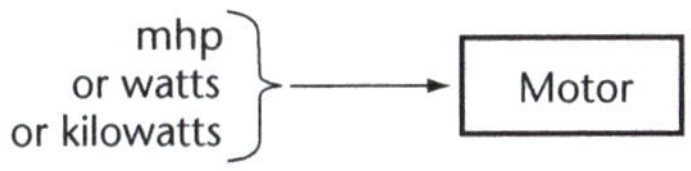

The relationship between horsepower and watts or kilowatts is given below. With these equations, you can convert from one term to the other.

> 1 horsepower = 746 watts power
>
> 1 horsepower = 0.746 kilowatts power

Example 8

If a pump is to deliver 460 gpm of water against a total head of 95 ft, and the pump has an efficiency of 75 percent, what horsepower must be supplied to the pump?

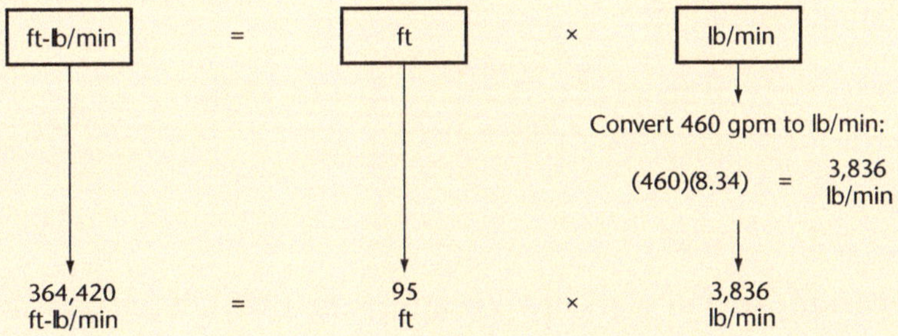

In this problem, brake horsepower is to be calculated. First calculate the water horsepower, based on the work to be accomplished. Then determine the brake horsepower required. Convert foot-pounds per minute to water horsepower:

$$\frac{364{,}420 \text{ ft-lb/min}}{33{,}000 \text{ ft-lb/min/hp}} = 11.04 \text{ whp}$$

This calculation shows that 11.04 whp (water horsepower) is required.

Now determine the amount of horsepower that must be supplied to the pump (brake horsepower):

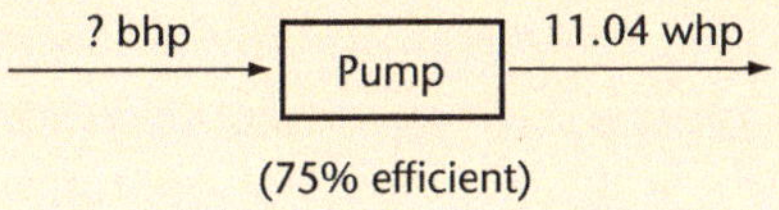

As stated in the problem, the pump is 75 percent efficient. Therefore, the brake horsepower will have to be more than 11.04 hp in order to accomplish the 11.04 hp work. Since the usable brake horsepower (75 percent) must equal 11.04 hp, the equation is

$$(75\%)(\text{bhp}) = 11.04 \text{ whp}$$

Restate the percentage as a decimal number:

$$(0.75)(x \text{ bhp}) = 11.04 \text{ whp}$$

And then solve for the unknown value:

$$(0.75)(x \text{ bhp}) = 11.04$$

$$x = \frac{11.04}{0.75}$$

$$x = 14.72 \text{ bhp}$$

Therefore, the 75 percent efficient pump must be supplied with 14.72 hp to accomplish the 11.04 hp of work:

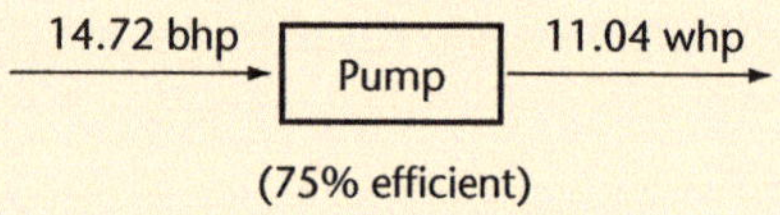

Using the alternative equation, we simply need to insert the efficiency into the denominator, and so the equation becomes:

$$\text{hp} = \frac{\text{gpm} \times \text{head (ft)}}{3{,}960 \times \text{Eff.}}$$

Recalculating Example 8:

$$\text{hp} = \frac{460 \text{ gpm} \times 95 \text{ ft}}{3{,}960 \times 0.75} = 14.71 \text{ hp}$$

Pumping Costs

Power is generally sold in units of kilowatt-hours (kWh). If a motor draws 1 kW of power and runs for 1 hour, then the electric company will charge for 1 kWh. Therefore, to calculate pumping costs, you will need to know the power requirements (power demand) of the motor and the length of time the motor runs. In most situations, you will know, or be able to determine, the total head and flow rate against which the pump is working, so the water horsepower can be calculated. Then, using the wire-to-water efficiency of the motor and pump, you can determine the motor horsepower and the kilowatts of power demand. (Power demand can also be calculated from a meter reading, but that calculation is not covered in this text.) The following examples illustrate how pumping costs can be calculated.

Example 9

The motor horsepower required for a particular pumping job is 25 hp. If your power cost is $0.03/kWh, what is the cost of operating the motor for 1 hour?

First, convert the motor horsepower demand to kilowatt power demand:

$$1 \text{ hp} = 0.746 \text{ kW}$$

Therefore,

$$25 \text{ hp} = (25)(0.746 \text{ kW})$$
$$= 18.65 \text{ kW power demand}$$

Since the kilowatt demand is 18.65 kW, a value of 18.65 kWh is required in 1 hour. Therefore, the cost of operating the pump for an hour is

$$(18.65 \text{ kWh})(\$0.03/\text{kWh}) = \$0.56$$

Example 10

You have calculated that the minimum motor horsepower requirement for a particular pumping problem is 15 mhp. If the cost of power is \$0.025/kWh, what is the power cost in operating the pump for 12 hours?

To determine power costs, as in the previous example, first express the motor horsepower requirement or demand in terms of kilowatts demand:

$$15 \text{ hp} = (15)(0.746 \text{ kW})$$
$$= 11.19 \text{ kW power demand}$$

In 12 hours, the number of kilowatt-hours required to operate the pump is

$$(11.19 \text{ kW})(12 \text{ h}) = 134.28 \text{ kWh}$$

Thus, the power cost associated with operating the pump for 12 hours is

$$(134.28 \text{ kWh})(\$0.025/\text{kWh}) = \$3.36$$

Reading Pump Curves

There is a great deal of information written about pumps, pump performance, and pump curves. The following section covers just those basic topics needed to understand and use pump curves.

Description of Pump Curves

A pump curve is a graph showing the *characteristics* of a particular pump. For this reason, these graphs are commonly called pump characteristic curves. The following characteristics are commonly shown on a pump curve:

- Capacity (flow rate)
- Total head
- Power (brake horsepower)
- Efficiency

If a pump is designed to be driven by a variable-speed motor, then "speed" is also a characteristic shown on the graph. For the purposes of this discussion, it will be assumed that the pump is driven by a constant-speed motor, so the graphs used will have only four curves.

The four pump characteristics (capacity, head, power, and efficiency) are related to each other. This is extremely important, for it is this interrelationship that enables the four pump curves to be plotted on the same graph.

Experience has shown that the *capacity* (flow rate) of a pump changes as the head against which the pump is working changes. Pump capacity also changes as

WATCH THE VIDEO
Operator Math II—Reading Pump Curves (www.awwa.org/wsovideoclips)

pump characteristic curves
A curve or curves showing the interrelation of speed, dynamic head, capacity, brake horsepower, and efficiency of a pump.

the *power* supplied to the pump changes. Finally, pump capacity changes as *efficiency* changes. Consequently, head, power, and efficiency can all be graphed "as a function of" pump capacity. This simply means that *capacity* is shown along the horizontal (bottom) scale of the graph, and *head*, *power*, and *efficiency* (any one or a combination of them) are shown along the vertical (side) scales of the graph. The following material discusses each of these relationships individually.

The H–Q Curve The *H–Q* curve shows the relationship between total head, *H*, against which the pump must operate and pump capacity, *Q*. A typical *H–Q* curve is shown in Figure 2-17. The curve indicates what flow rate the pump will produce at any given total head.

For example, if the total head is 100 ft, the pump will produce a flow of about 1,910 gpm. The *H–Q* curve also indicates that the capacity of the pump *decreases* as the total head *increases*. (Generally speaking, when the force against which the pump must work increases, the flow rate decreases.) The way total head controls the capacity is a characteristic of a particular pump.

The *H–Q* curve also identifies two very important operational facts. Normally on the curve you will find a mark like this: ⅂. The mark defines the **design point**, the head and capacity at which the pump is intended to operate for best efficiency in a particular installation.

The P–Q Curve The *P–Q* curve (Figure 2-18) shows the relationship between power, *P*, and capacity, *Q*. In this figure, pump capacity is measured as gallons per minute, and power is measured as brake horsepower. For example, if you pump at a rate of 1,480 gpm, then the power used to drive the pump is about 58 bhp. This is valuable information. As explained in the preceding section on horsepower calculations, if you know the brake horsepower and the motor efficiency, you can then determine the motor horsepower (the power required to drive the motor).

Knowledge of what power the pump requires is valuable for checking the adequacy of an existing pump and motor system. The knowledge is also important when you find yourself scrambling to rig a temporary system from spare or surplus parts.

The E–Q Curve The *E–Q* curve shows the relationship between pump efficiency, *E*, and capacity, *Q*, as shown in Figure 2-19. (Sometimes the Greek letter eta—η—is used to represent efficiency, and the efficiency-capacity curve is designated the η–*Q* curve.) In sizing a pump system, the engineer attempts to select a pump that will produce the desired flow rate at or near peak pump efficiency.

The more efficient your pump is, the less costly it is to operate. Stated another way, the more efficient your pump is, the more water you can pump for each

design point
The mark on the *H–Q* (head-capacity) curve of a pump characteristics curve that indicates the head and capacity at which the pump is intended to operate for best efficiency in a particular installation.

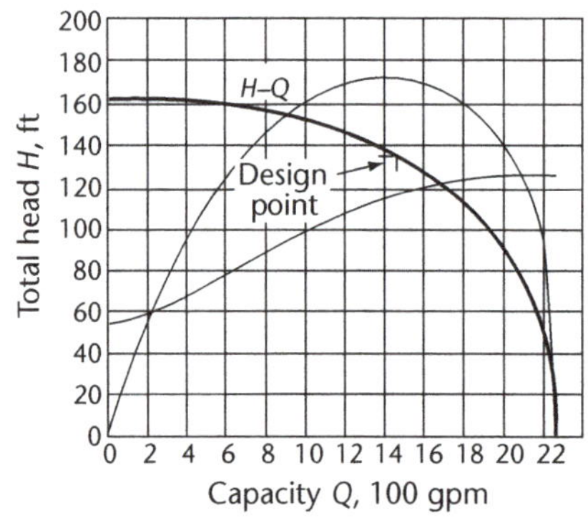

Figure 2-17 *H–Q* curve

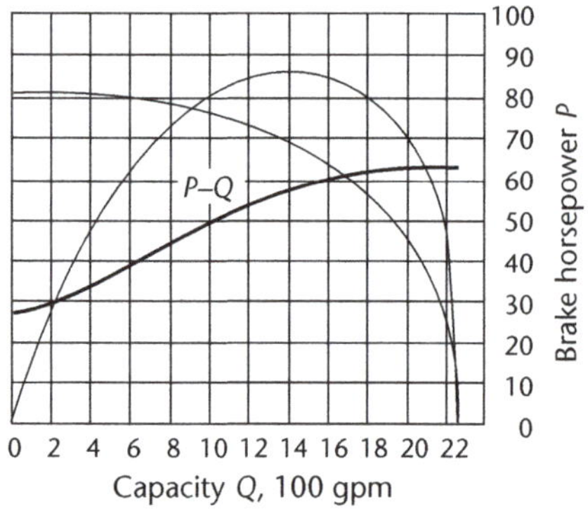

Figure 2-18 *P–Q* curve

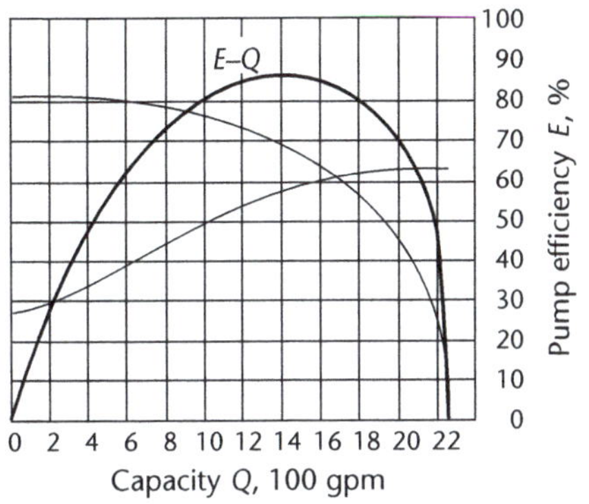

Figure 2-19 *E*–*Q* curve

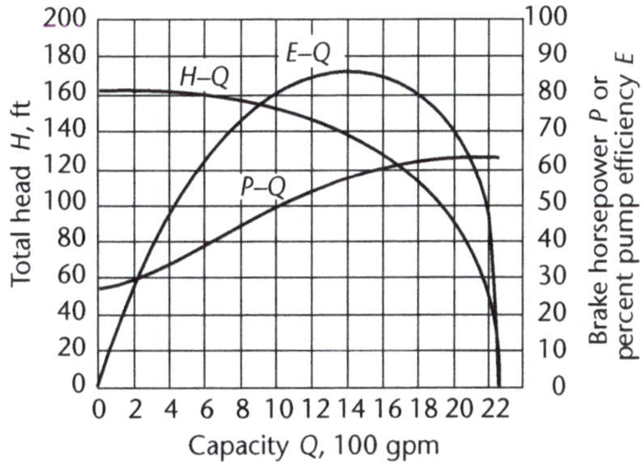

Figure 2-20 Complete pump curve

dollar's worth of power. Knowledge of pump efficiency allows you to compute the cost of pumping water. Therefore, from an operational point of view, pump efficiency is a "must know" item.

Reading the Curve

Figure 2-20 is the complete pump curve. By using the pump curve, you can determine any three of the four pump characteristics (capacity, head, efficiency, and horsepower) when given information about the fourth characteristic. (Normally the total head or the capacity is the known characteristic.) The following examples illustrate how to read pump curves.

Example 11

Using the pump curve given in Figure 2-20, determine the pump head, power, and efficiency when operating at a flow rate of 1,600 gpm.

The key to reading pump curves is to find a certain vertical line and to note where the line intersects all three pump characteristic curves. In this problem, the known characteristic is *capacity* (1,600 gpm). To determine head, power, and efficiency, you should first draw a vertical line upward from the number 16 on the horizontal scale (Figure 2-21).

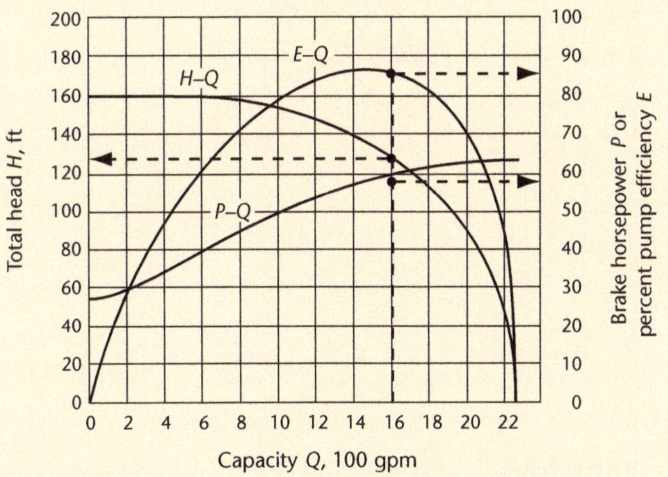

Figure 2-21 Pump curve for Example 11

Note that the vertical line intersects each of the three curves, as indicated by the three dots. Following the line upward, you first reach the intersection with the *P–Q* (power) curve. To determine what the power is at this point, move *horizontally to the right* toward the bhp scale; you should hit the scale at about 60 bhp.

Now return to the vertical line (1,600 gpm) and move to the intersection with the *H–Q* (head) curve. To determine the head at this point, move *horizontally to the left*, toward the total head scale; you should hit the scale at about 128 ft.

Finally, look at the point where the vertical line (1,600 gpm) intersects the *E–Q* (efficiency) curve. To determine the efficiency at this point, move *horizontally to the right* toward the efficiency scale. The reading is about 85 percent.

From the known pump capacity of 1,600 gpm, then, the following characteristics were determined:

$$\text{brake horsepower } P = 60 \text{ bhp}$$

$$\text{pump head } H = 128 \text{ ft}$$

$$\text{pump efficiency } E = 85\%$$

Example 12

A pump operates against a total head of 32 ft. The motor is 78 percent efficient. Power costs $0.03/kWh.

Use the pump curve in Figure 2-22 to determine brake horsepower required under the given conditions. Then calculate the cost of operating the pump for 100 hours.

To locate the brake horsepower that corresponds with a total head of 32 ft, first draw a horizontal line from the value 32 ft on the total head scale until the line intersects the *H–Q* curve. Draw a vertical line through the intersection.

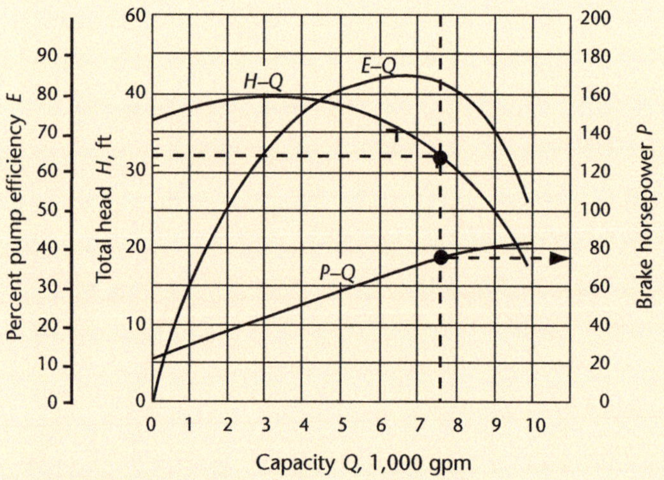

Figure 2-22 Pump curve for Example 12

The vertical line intersects the *P–Q* curve at a point corresponding to about 75 bhp. Using this figure and the motor efficiency given in the problem, calculate the motor horsepower required:

$$(78\%)(x \text{ mhp}) = 75 \text{ bhp}$$

Express the percentage as a decimal number:

$$(0.78)(x \text{ mhp}) = 75 \text{ bhp}$$

And then solve for the unknown value:

$$x = \frac{75}{0.78}$$

$$x = 96.15 \text{ mhp}$$

Once the motor horsepower requirement has been determined, power demand can be expressed in terms of kilowatts (recall that 1 hp = 0.746 kW).

$$96.15 \text{ hp} = (96.15)(0.746 \text{ kW})$$

$$= 71.73 \text{ kW demand}$$

In 100 hours, then, the number of kilowatt-hours used is

$$(7{,}173 \text{ kW})(100\text{h}) = 7{,}173 \text{ kWh}$$

At a cost of \$0.03/kWh, the cost of operating the pump for one day is

$$(7{,}173 \text{ kWh})(\$0.03/\text{kWh}) = \$215.19$$

Chemical Dosage Problems

Lime–Soda Ash Softening Calculations

Two methods for calculating lime and soda ash dosages are the *conventional method* and the *conversion-factor method*. The conventional method, although much longer, is helpful in understanding the chemical and mathematical relationships involved in softening. The conversion-factor method is simpler and quicker and is the more practical method to use in day-to-day operations.

In both methods of calculation, the lime and soda ash dosages required are dependent on the carbonate and noncarbonate constituents in the water. Although hard water will always require lime, it may not always need soda ash. Lime is used to remove carbonate hardness, and both lime and soda ash are used to remove noncarbonate hardness.

To determine whether lime only or both lime and soda ash are needed, total hardness of the water must be compared with total alkalinity. In some areas of the United States, the only alkalinity occurring naturally in water is bicarbonate alkalinity. In these areas, many labs report only the bicarbonate alkalinity concentration or only the total alkalinity concentration, assuming that the operator will be aware that, for the water tested, total alkalinity = bicarbonate alkalinity. However, to be certain of the characteristics of the water tested, an operator should

total hardness
The combined effect of calcium and magnesium.

total alkalinity
The combined effect of hydroxyl alkalinity (OH^-), carbonate alkalinity (CO_3^{-2}), and bicarbonate alkalinity (HCO_3^-).

Table 2-3 Noncarbonate and carbonate hardness based on total hardness and total alkalinity

Laboratory Results	Noncarbonate Hardness	Carbonate Hardness
TH less than TA	0	TH
TH = TA	0	TH
TH greater than TA	TH – TA	TA

request a detailed lab report showing the total alkalinity as well as the bicarbonate, carbonate, and hydroxyl alkalinity.

Table 2-3 shows how to determine the levels of carbonate and noncarbonate hardness based on the laboratory results for total hardness (TH) and total alkalinity (TA).

A *total hardness less than or equal to total alkalinity* indicates two important facts about lime and soda ash dosages:

- The noncarbonate concentration is zero, so only lime treatment will be required.
- The amount of lime normally added to treat magnesium noncarbonate hardness (for example, magnesium chloride, $MgCl_2$) is not required.

The following chemical equation is used to determine lime dosage requirements when TH is less than or equal to TA:

$$\text{lime dosage} = [CO_2] + [\text{total hardness}] + [Mg] + [\text{excess}] \qquad (2\text{-}1)$$

The brackets, [ ], in the equation mean "the concentration of." Therefore, when total hardness is less than or equal to total alkalinity, the lime dosage equals the sum of the concentrations of carbon dioxide, total hardness, and magnesium, plus any excess lime desired.

In most operating situations, more lime will be needed than is indicated by the chemical equations. This additional amount, called "excess lime," is added either to speed the reactions or to force them to react more completely. It is common practice to add from 25 to 50 mg/L as calcium carbonate ($CaCO_3$) of excess lime. An equivalent amount of soda ash (Na_2CO_3) must also be added in order to remove the excess calcium ions added with the lime. Removal of 1 mg/L of excess lime requires 1.06 mg/L as Na_2CO_3 of excess soda ash.

When total hardness is greater than total alkalinity, carbonate and noncarbonate hardness constituents are present in the water. This means that both lime and soda ash will be needed in treating for hardness—lime to remove carbonate hardness constituents and both lime and soda ash to remove noncarbonate hardness constituents in the water. The following chemical equations are used to determine lime and soda ash dosage requirements:

$$\text{lime dosage} = [CO_2] + [HCO_3^-] + [Mg] + [\text{excess}] \qquad (2\text{-}2)$$

$$\text{soda ash dosage} = [\text{total hardness}] - [HCO_3^-] + [\text{excess}] \qquad (2\text{-}3)$$

Thus, the lime dosage equals the sum of the concentrations of carbon dioxide, bicarbonate alkalinity, and magnesium, plus the excess lime desired. The soda ash dosage is equal to the total hardness, minus the bicarbonate alkalinity, plus any excess soda ash desired. Note that when the bicarbonate alkalinity (HCO_3^-) is equal to the total alkalinity (TA), Equation 2-3 can be written

$$\text{soda ash dosage} = [\text{total hardness}] - [\text{TA}] + [\text{excess}] \tag{2-4}$$

And, as shown in Table 2-3, when the total hardness is greater than the total alkalinity, TH – TA = noncarbonate hardness. Under these conditions, Equation 2-3 can be written

$$\text{soda ash dosage} = [\text{noncarbonate hardness}] + [\text{excess}] \tag{2-5}$$

The two ions that most commonly cause hardness in water are calcium and magnesium. Although aluminum, strontium, iron, manganese, barium, and zinc ions can also cause hardness, they are not usually present in large enough concentrations to cause a significant problem. Consequently it is usually assumed, without much sacrifice in accuracy, that

$$[\text{total hardness}] = [\text{Ca}] + [\text{Mg}] \tag{2-6}$$

When only calcium and magnesium are measured in the laboratory, the total hardness can be found from Equation 2-6. When the laboratory measures both total hardness and calcium, Equation 2-6 can be used to determine the magnesium concentration.

In some cases, total hardness, calcium, and magnesium are all measured in the laboratory. When this occurs, total hardness should be used for Equations 2-1 and 2-2, instead of the sum of calcium and magnesium, because total hardness includes the effects of all hardness-causing ions.

Equations 2-1 through 2-6 are the basis for determining lime and soda ash dosages through either the conventional or the conversion-factor method. Before these equations can be used, the concentrations of carbon dioxide, total hardness, magnesium, bicarbonate, and excess lime must all be expressed in "mg/L as $CaCO_3$." Usually, when a laboratory test is made for carbon dioxide, the results are reported in "mg/L as CO_2." Similarly, a magnesium concentration would be reported in "mg/L as Mg," and a bicarbonate alkalinity would be reported in "mg/L as HCO_3^-" or in "mg/L as $CaCO_3$." Since these concentrations are not all expressed in the same units of measure, they cannot be added together. However, each concentration can be converted to "mg/L as $CaCO_3$" (the units commonly used to express hardness). After all laboratory results are converted to these units, they can be used in Equations 2-1 through 2-6.

The conventional and conversion-factor methods are based on the theory that all reactions occur in definite proportions to the compounds involved. More specifically, the reactions occur in definite proportion to the equivalent weights of the reacting ions or compounds. The difference between the two methods is that, in the conversion-factor method, some of the conversion calculations (such as conversion from "mg/L as HCO_3^-" to "mg/L as $CaCO_3$") are shortened by using conversion factors for the part of the calculation that has been completed.

Conventional Method

Given the total hardness, calcium, magnesium, bicarbonate alkalinity, total alkalinity, and free carbon dioxide concentrations from laboratory test results, the conventional method uses equivalent weights to convert all hardness data to milligrams per liter as $CaCO_3$ and then uses the $CaCO_3$ concentrations to calculate the quantity of lime and soda ash needed. This process consists of six steps, which are summarized below and explained in detail in the following examples:

Step 1: Convert all data to concentrations as $CaCO_3$, using the following equivalent weights:

$$Ca = 20$$

$$Mg = 12$$

$$HCO_3^- = 61$$

$$CO_2 = 22$$

$$CaCO_3 = 50$$

For example,

$$5 \text{ mg/L as } CO_2 = (5)\left(\frac{50}{22}\right)$$

$$= 11.36 \text{ mg/L as } CaCO_3$$

Step 2: Determine whether lime only or both lime and soda ash will be required in treating the hardness constituents. Use Table 2-3.

Step 3: Calculate the lime dosage or lime and soda ash dosage (in milligrams per liter as $CaCO_3$). When total hardness is *less than or equal to* total alkalinity, use Equation 2-1:

$$\text{lime dosage} = [CO2\ ] + [\text{total hardness}] + [Mg] + [\text{excess}]$$

When total hardness is *greater than* total alkalinity, use Equations 2-2 and 2-3:

$$\text{lime dosage} = [CO_2] + [HCO_3^-] + [Mg] + [\text{excess}]$$

$$\text{soda ash dosage} = [\text{total hardness}] - [HCO_3^-] + [\text{excess}]$$

If noncarbonate hardness is given, Equation 2-5 can be used for the soda ash dosage:

$$\text{soda ash dosage} = [\text{noncarbonate hardness}] + [\text{excess}]$$

Step 4: Convert lime and soda ash dosages expressed as $CaCO_3$ to dosages expressed as calcium oxide (CaO, also called quicklime), and as sodium bicarbonate (Na_2CO_3), respectively, using the following equivalent weights:

$$CaCO_3 = 50$$

$$CaO = 28$$

$$Na_2CO_3 = 53$$

For example,

$$227 \text{ mg/L lime as } CaCO_3 = (227)\left(\frac{28}{50}\right)$$

$$= 127.12 \text{ mg/L as CaO}$$

Step 5: Adjust dosages for the purity of the lime or soda ash used. For example,

$$(88\%)(\text{actual dosage of } 88\% \text{ pure CaO}) = 127 \text{ mg/L of } 100\% \text{ pure CaO}$$

$$\text{actual dosage of } 88\% \text{ pure CaO} = 144.5 \text{ mg/L}$$

Step 6: Convert these adjusted dosages to pounds per million gallon or pounds per day. For example:

$$(\text{mg/L})(8.34\ \text{lb/gal}) = \text{lb/mil gal}$$

$$(\text{mg/L})(\text{mgd flow})(8.34\ \text{lb/gal}) = \text{lb/d}$$

The following examples illustrate softening calculations using the conventional method.

Example 13

The laboratory returned the following test results:

$$\text{calcium} = 100\ \text{mg/L as Ca}$$

$$\text{total hardness} = 300\ \text{mg/L as } CaCO_3$$

$$\text{bicarbonate alkalinity} = 250\ \text{mg/L as } HCO_3^-$$

$$\text{total alkalinity} = 205\ \text{mg/L as } CaCO_3$$

$$\text{carbon dioxide} = 25\ \text{mg/L as } CO_2$$

Express each concentration as $CaCO_3$.

Using the equivalent weight procedure, you can make the conversions as follows:

$$\text{calcium: } (100\ \text{mg/L})\left(\frac{50}{20}\right) = 250\ \text{mg/L as } CaCO_3$$

$$\text{total hardness: (No conversion needed.)} = 300\ \text{mg/L as } CaCO_3$$

$$\text{bicarbonate alkalinity: } (250\ \text{mg/L})\left(\frac{50}{61}\right) = 204.92\ \text{mg/L as } CaCO_3$$

$$\text{total alkalinity: (No conversion needed.)} = 205\ \text{mg/L as } CaCO_3$$

$$\text{carbon dioxide: } (25\ \text{mg/L})\left(\frac{50}{22}\right) = 56.82\ \text{mg/L as } CaCO_3$$

Example 14

Assume that 675.53 lb/d CaO is required for treating a certain water, as calculated in the previous example. How many pounds per day of 90 percent pure quicklime will actually be required?

Keep in mind that the answer to this problem will be slightly more than 675.53 lb/d. To determine the required dosage of the 90 percent pure chemical that is actually used, insert the information in the equation:

$$(\%\ \text{purity})(\text{actual dosage}) = (\text{calculated dosage})$$

Express the percentage as a decimal:

$$(0.90)(\text{actual dosage}) = 675.53\ \text{lb/d}$$

Then solve for the unknown number:

$$\text{actual dosage} = \frac{675.53}{0.90}$$

$$= 750.58 \text{ lb/d CaO required (90\% pure)}$$

Remember that whenever the purity of the chemical used is less than 100 percent, the actual dosage will be some greater number than the calculated dosage of 100 percent pure chemical.

Recarbonation Calculations

As discussed in Chapter 1, there are three possible reactions involved when carbon dioxide is used to stabilize a lime-softened water. Equation 2-5 shows that calcium carbonate, in suspension, will react with carbon dioxide to form calcium bicarbonate. Equations 2-6 and 2-7 show that excess lime will also react with carbon dioxide, forming either calcium carbonate precipitate or soluble calcium bicarbonate.

$CaCO_3$	+	$CO_2(g)$	+	H_2O	$\longrightarrow$	$Ca(HCO_3)_2$	
calcium carbonate (in suspension)		carbon dioxide (gas)		water		calcium bicarbonate (in suspension)	(2-7)

$Ca(OH)_2$	+	$CO_2(g)$	$\longrightarrow$	$CaCO_3\downarrow$	+	H_2O	
excess lime		carbon dioxide (gas)		calcium carbonate (precipitate)		water	(2-8)

$Ca(OH)_2$	+	$2CO_2(g)$	$\longrightarrow$	$Ca(HCO_3)_2$	
excess lime		carbon dioxide (gas)		calcium bicarbonate (in suspension)	(2-9)

The conditions described by these equations may occur together. Consequently, the carbon dioxide dosage needed for recarbonation is the amount needed to reduce calcium carbonate to a desired level and to eliminate excess lime completely. The amount of calcium carbonate present is measured by the alkalinity test and is assumed to be equal to the carbonate alkalinity. The amount of excess lime present, also measured by the alkalinity test, is assumed equal to the hydroxide alkalinity.

The correct carbon dioxide dosage involves adding 0.44 mg/L of CO_2 for every 1 mg/L of carbonate or hydroxide alkalinity expressed as $CaCO_3$. That is, removing 5 mg/L as $CaCO_3$ of carbonate alkalinity and 5 mg/L as $CaCO_3$ of hydroxide alkalinity will require (5 + 5)(0.44) = 4.4 mg/L of CO_2 as CO_2. Consider Example 15.

Example 15

The following results of water quality analyses were reported for water in the process of treatment. Calculate the correct CO_2 dosage.

Type of Alkalinity	Concentration, mg/L as $CaCO_3$		
	For Raw Water	After Lime–Soda Ash Treatment	Desired After Recarbonation
Bicarbonate	250	0	5
Carbonate	0	35	30
Hydroxide	0	40	0

First, find out how much carbonate and hydroxide alkalinity must be removed. For carbonate alkalinity, subtract the desired concentration from the concentration resulting from the lime–soda ash treatment:

$$\text{carbonate alkalinity to be removed} = 35 \text{ mg/L} - 30 \text{ mg/L}$$
$$= 5 \text{ mg/L as } CaCO_3$$

Do the same for hydroxide alkalinity:

$$\text{hydroxide alkalinity to be removed} = 40 \text{ mg/L} - 0 \text{ mg/L}$$
$$= 40 \text{ mg/L as } CaCO_3$$

Then add the two alkalinities to find the combined amount to be removed:

$$\text{combined alkalinity to be removed} = 5 \text{ mg/L} + 40 \text{ mg/L}$$
$$= 45 \text{ mg/L as } CaCO_3$$

Finally, find the carbon dioxide dosage as follows:

$$CO_2 \text{ dosage} = (\text{alkalinity to be removed})(0.44)$$
$$= (45 \text{ mg/L})(0.44)$$
$$= 19.8 \text{ mg/L}$$

Ion Exchange Softening Calculations

The operation of an ion exchange water softener (more correctly called a cation exchange water softener) involves four types of calculations:

- Hardness removal capacity
- Volume of water softened per cycle
- Length of softening run or cycle
- Regeneration salt requirement

All ion exchange materials (the zeolite types and the resin types) have maximum exchange capacities. The usual capacity ranges are summarized below.

Exchange Material	Exchange Capacity, grains per cubic foot
Zeolites	2,800–11,000
Resins	11,000–35,000

The use of the unit *grain* is very common in this branch of water treatment. It is used in such terms as grains per cubic foot and grains per gallon (gpg). The following conversion factors will be useful in converting these somewhat old-fashioned units to more familiar units:

$$1 \text{ grain} = 64.80 \text{ mg}$$

$$1 \text{ gpg} = 17.12 \text{ mg/L}$$

$$1 \text{ gpg} = 142.86 \text{ lb/mil gal}$$

Example 16

A treatment unit contains 20 ft^3 of cation exchange material that has a rated removal capacity of 20,000 grains per cubic foot. What is the total hardness removal capacity? Give answer in grains.

Since each cubic foot of material can remove 20,000 grains of hardness and there are 20 ft^3 of material, the answer can be found in one multiplication:

$$\text{Total removal capacity} = (20{,}000 \text{ grain/ft}^3)(20 \text{ ft}^3)$$

$$= 400{,}000 \text{ grains removed per cycle}$$

Fluoridation Calculations

There are three chemical compounds commonly used in water fluoridation:

- Sodium fluoride, NaF
- Sodium fluorosilicate, Na_2SiF_6
- Fluorosilicic acid, H_2SiF_6

Since each compound is used for its fluoride ion content, it is important to determine the exact amount of fluoride ion available in a given fluoride compound. This amount depends on two factors:

- The weight of fluoride relative to the total weight of the compound
- The purity of the compound

Example 17

Find the percentage of fluoride ion contained in sodium fluoride, NaF.

Find the molecular weight of sodium fluoride by adding the atomic weights of the individual elements:

$$\text{molecular weight of NaF} = \text{molecular weight of Na} + \text{molecular weight of F}$$

$$= 23 + 19 = 42$$

Then divide the weight of the fluoride ion by the total molecular weight of the compound and multiply by 100, as follows:

$$\text{percent F} = \left(\frac{19}{42}\right)(100)$$

$$= 45\% \text{ fluoride ion (rounded from 45.24)}$$

Example 18

The commercial strength of the sodium fluoride in Example 17 is between 95 and 98 percent. That is, each 100-lb bag of chemical compound delivered to the plant contains 95–98 lb of sodium fluoride and 2–5 lb of impurities. How much fluoride ion is contained in each 100-lb bag?

From Example 17 you know that if it contained pure sodium fluoride, each 100-lb bag would have 45 percent or (0.45)(100) = 45 lb fluoride ion. However, the bags contain slightly impure NaF. Therefore, each bag holds only 95–98 percent of what it would if the NaF were 100 percent pure, or from (45)(0.95) = 42.75 lb of F (43%) to (45)(0.98) = 44.1 lb of F (44%).

Example 19

Find the percentage of fluoride ion contained in sodium fluorosilicate (Na_2SiF_6).

Find the molecular weight of sodium fluorosilicate by adding the atomic weights of the individual elements, as follows:

Na_2:	(2)(23)	=	46
Si:			28
F_6:	(6)(19)	=	114
			188 = molecular weight of Na_2SiF_6

Then divide the total weight of the fluoride ion by the molecular weight of the compound and multiply by 100, as follows:

$$\text{percent F} = \left(\frac{114}{188}\right)(100)$$

$$= 61\%\ \text{fluoride ion (rounded from 60.64)}$$

Study Questions

1. Adequate chlorine (or other disinfectant) contact time is a requirement for surface water treatment plants and is called $C \times T$ value. It determines disinfection sufficiency and is the product of the disinfectant residual times and the effective contact time for the chamber or clearwell in which disinfection takes place. In order to calculate $C \times T$ values, operators need to know the chlorine residual in the chamber, the flow rate of water through the chamber, the baffling factor (T10) of the chamber, and the volume of the chamber. The results are shown in mg-min/L. A disinfection chamber with a chlorine residual of 1.6 mg/L measures 44 ft by 30 ft and has a baffling factor of 0.34. What is the $C \times T$ value at 300 gpm if the chamber holds 8.3 ft of water?

2. How many lb/day of sodium fluorosilicate (Na_2SiF_6) are required given the following parameters?
   - Flow rate is 1,750 gpm.
   - Fluoride desired is 1.20 mg/L.
   - Fluoride in raw water is 0.15 mg/L.
   - Sodium fluorosilicate is 98.1% pure.
   - Fluoride (F) ion percentage is 60.6%.

   a. 34 lb/day, F
   b. 37 lb/day, F
   c. 42 lb/day, F
   d. 48 lb/day, F

3. What is the feed rate for fluorosilicic acid in mL/min given the following data?
   - Flow rate is 11.8 mgd.
   - Fluoride desired is 1.20 mg/L.
   - Fluoride in raw water is 0.20 mg/L.
   - Treatment is with 20.5% solution of H_2SiF_6.
   - Fluoride ion percentage is 79.0%.
   - H_2SiF_6 weighs 9.8 lb/gal.

   a. 33 mL/min, H_2SiF_6
   b. 62 mL/min, H_2SiF_6
   c. 145 mL/min, H_2SiF_6
   d. 163 mL/min, H_2SiF_6

4. A bypass valve would be used with a
   a. gate valve.
   b. butterfly valve.
   c. needle valve.
   d. diaphragm valve.

5. Where ground frost is expected, the riser pipes in a meter pit should be at least how far away from any wall of the meter box to prevent freezing?
   a. 1–2 in.
   b. 2–4 in.
   c. 4–6 in.
   d. 6–8 in.

6. Which type of pipe joint is available in both bolted and boltless flexible pipe joint designs?
   a. Ball and socket joint
   b. Push-on joint
   c. Grooved joint
   d. Shouldered joint

7. Which type of pipe joint permits flexible pipe alignment, is inexpensive to manufacture, and is easy to assemble?
   a. Mechanical joint
   b. Push-on joint
   c. Grooved joint
   d. Ball and socket joint

8. New polyvinyl chloride pipe has a C-value of about
   a. 125.
   b. 125–128.
   c. 135.
   d. 150+.

9. Which type of valve would be particularly useful for throttling the flow of corrosive liquids?
   a. Diaphragm valve
   b. Butterfly valve
   c. Gate valve
   d. Pinch valve

10. Water meter pits are usually used in
    a. areas where flooding will most likely occur.
    b. areas where flooding is very rare.
    c. cold climates.
    d. hot climates.

11. Fuel oil will permeate which type of pipe?
    a. Lead pipe
    b. Polyethylene pipe
    c. Copper pipe
    d. Galvanized pipe

12. An ion exchange softener is treating a flow rate of 125 gpm. What is the operating time in hours if the softener unit treats 297,000 gallons before it requires regeneration?
    a. 37 hr
    b. 39.6 hr
    c. 40 hr
    d. 41.1 hr

13. Determine the psi at the bottom of an alum storage tank if the level of the alum in the tank is 8.95 ft and the density of the alum is 11.32 lb/gal.
    a. 5.13 psi
    b. 5.26 psi
    c. 5.37 psi
    d. 5.41 psi

14. The level in a storage tank drops 4.25 ft in exactly 12 hours. If the tank has a diameter of 50.0 ft and the plant is producing 2.95 mgd, what is the average discharge rate of the treated water discharge pumps in gallons per minute?
    a. 2,090 gpm
    b. 2,100 gpm
    c. 2,120 gpm
    d. 2,140 gpm

15. In describing pump operation, what type of curve shows the relationship between power and capacity?
    a. *E–Q* curve
    b. *H–Q* curve
    c. *P–Q* curve
    d. *E–P* curve

16. List the types of heads used in describing the amount of energy in water.

17. What is head loss?

18. What is static suction lift?

19. What is power?

Chapter 3

USEPA Water Regulations

Regulations that govern US water supply and treatment are developed by the **US Environmental Protection Agency (USEPA)** under the Safe Drinking Water Act (SDWA). Most states administer USEPA regulations after adopting regulations that are no less stringent than federal rules; and in some cases, states have adopted stricter regulations or have developed regulations for additional contaminants not regulated by USEPA.

This chapter discusses current and anticipated USEPA regulations and the challenges that operators face in their efforts to comply with the regulations. Water system operators should consult their local and state regulatory agencies to verify applicable regulations that may be different than the federal regulations listed in this chapter. The chapter concludes with a discussion of selected contaminants that are commonly found in water, their significance, and the methods for their removal.

Types of Water Systems

The SDWA defines a public water system (PWS) as a supply of piped water for human consumption that has at least 15 service connections, or serves 25 or more persons 60 or more days each year. By that definition, private homes, groups of homes with a single water source but having fewer than 25 residents, and summer camps with their own water source that operate less than 60 days per year are not PWSs. They may, however, be subject to state or local regulations. Such systems may also be subject to state and local well construction and water quality requirements.

PWSs are classified into three categories based on the type of customers served:

- *Community PWS:* a system whose customers are full-time residents
- *Nontransient noncommunity PWS:* an entity having its own water supply, serving an average of at least 25 persons who do not live at the location but who use the water for more than 6 months per year

US Environmental Protection Agency
A US government agency responsible for implementing federal laws designed to protect the environment.

- *Transient noncommunity PWS*: an establishment having its own water system, where an average of at least 25 people per day visit and use the water occasionally or for only short periods of time

These classifications are based on the differences in exposure to contaminants experienced by persons using the water. Most chemical contaminants are believed to potentially cause adverse health effects from long-term exposure. Short-term exposure to low-level chemical contamination may not carry the same risk as long-term exposure.

Therefore, the monitoring requirements for both community and noncommunity water systems apply to all contaminants that are considered a health threat. The transient and nontransient noncommunity systems must monitor only for nitrite and nitrate, as well as biological contamination (those sources that pose an immediate threat from brief exposure). The remaining community systems, about 52,000 in the United States, have more stringent and frequent monitoring requirements.

Before examining the specific regulations that govern contaminants, the operator needs to know the difference between the two concepts used in the contaminant monitoring process: the **maximum contaminant level goal (MCLG)** and the **maximum contaminant level (MCL)**:

- The MCLG is set for most substances at a level where there are no known, or anticipated, health effects. For those substances that are suspected carcinogens, the MCLG is set at zero.
- The MCL is set as close as feasible to the MCLG for substances regulated under the SDWA. The MCL is a level that is reasonably and economically achievable. This is the enforceable regulated level. Water systems that exceed an MCL must take steps to install treatment to reduce the contaminant concentration to below the MCL. Where USEPA has found it impractical to set an MCL, a treatment technique has been established instead of an MCL.

maximum contaminant level goal (MCLG)
Nonenforceable health-based goals published along with the promulgation of an MCL. Originally called recommended maximum contaminant levels (RMCLs).

maximum contaminant level (MCL)
The maximum permissible level of a contaminant in water as specified in the regulations of the Safe Drinking Water Act.

With these concepts in mind, the various regulations can be examined. This discussion is not meant to be all-inclusive. Because the regulatory process is an ever-evolving one, the reader is cautioned that some of the stated facts presented in this discussion may have changed since the writing of this chapter. For up-to-date information, it is best to contact the local office of the regulatory authority in the district or state where the utility operates.

Table 3-1 lists the status of USEPA primary drinking water standards at the time this book was prepared. These standards are provided for illustration only and are not intended to be used for regulatory purposes (see the official USEPA regulatory information on the agency website).

Operations personnel are expected to know the regulatory limits for compounds encountered in their water supply. However, the number and variety of regulated substances make it unlikely that operators would know all of the regulatory limits. Operators must rely on current references for the most accurate information. These are available from the regulatory agency responsible for the location of the treatment plant.

Table 3-1 List of contaminants and their MCLs

Contaminant	MCLG,* mg/L†	MCL or TT, mg/L	Potential Health Effects From Ingestion of Water	Sources of Contaminant in Drinking Water
Microorganisms				
Cryptosporidium	Zero	TT‡	Gastrointestinal illness (e.g., diarrhea, vomiting, cramps), Cryptosporidiosis	Human and animal fecal waste
Giardia lamblia	Zero	TT	Gastrointestinal illness (e.g., diarrhea, vomiting, cramps), Giardiasis	Human and animal fecal waste
Heterotrophic plate count (HPC)	N/A	TT	HPC has no health effects; it is an analytic method used to measure the variety of bacteria that are common in water. The lower the concentration of bacteria in drinking water, the better maintained the water system is.	HPC measures a range of bacteria that are naturally present in the environment
Legionella	Zero	TT	Legionnaire's disease, a type of pneumonia	Found naturally in water; multiplies in heating systems
Total coliforms (including fecal coliform and *Escherichia coli* [*E. coli*])	Zero	5.0%**	Not a health threat in itself; it is used to indicate whether other potentially harmful bacteria may be present.††	Coliforms are naturally present in the environment as well as feces; fecal coliforms and *E. coli* only come from human and animal fecal waste.
Turbidity	N/A	TT	Turbidity is a measure of the cloudiness of water. It is used to indicate water quality and filtration effectiveness (e.g., whether disease-causing organisms are present). Higher turbidity levels are often associated with higher levels of disease-causing microorganisms such as viruses, parasites, and some bacteria. These organisms can cause symptoms such as nausea, cramps, diarrhea, and associated headaches.	Soil runoff
Viruses (enteric)	Zero	TT	Gastrointestinal illness (e.g., diarrhea, vomiting, cramps)	Human and animal fecal waste
Disinfection By-products				
Bromate	Zero	0.010	Increased risk of cancer	By-product of drinking water disinfection with ozone
Chlorite	0.8	1.0	Anemia; nervous system effects in infants and young children	By-product of drinking water disinfection chlorine dioxide
Haloacetic acids (HAA5)	N/A‡‡	0.060	Increased risk of cancer	By-product of drinking water disinfection

(continued)

Table 3-1 List of contaminants and their MCLs (continued)

Contaminant	MCLG,* mg/L†	MCL or TT, mg/L	Potential Health Effects From Ingestion of Water	Sources of Contaminant in Drinking Water
Total trihalomethanes (TTHMs)	None***	0.080	Liver, kidney, or central nervous system problems; increased risk of cancer	By-product of drinking water disinfection
Disinfectants				
Chloramines (as Cl_2)	MRDLG=4	MRDL=4.0	Eye/nose irritation; stomach discomfort; anemia	Water additive used to control microbes
Chlorine (as Cl_2)	MRDLG=4	MRDL=4.0	Eye/nose irritation; stomach discomfort	Water additive used to control microbes
Chlorine dioxide (as ClO_2)	MRDLG=0.8	MRDL=0.8	Anemia; nervous system effects in infants and young children	Water additive used to control microbes
Inorganic Chemicals				
Antimony	0.006	0.006	Increase in blood cholesterol; decrease in blood sugar	Discharge from petroleum refineries; fire retardants; ceramics; electronics; new lead-free solder
Arsenic	0	0.010 as of 1/23/06	Skin damage or problems with circulatory systems; possible increased risk of contracting cancer	Erosion of natural deposits; runoff from orchards, runoff from glass and electronics production wastes
Asbestos (fiber >10 micrometers)	7 million fibers per liter (MFL)	7 MFL	Increased risk of developing benign intestinal polyps	Decay of asbestos cement in water mains; erosion of natural deposits
Barium	2	2	Increase in blood pressure	Discharge of drilling wastes; discharge from metal refineries; erosion of natural deposits
Beryllium	0.004	0.004	Intestinal lesions	Discharge from metal refineries and coal-burning factories; discharge from electrical, aerospace, and defense industries
Cadmium	0.005	0.005	Kidney damage	Corrosion of galvanized pipes; erosion of natural deposits; discharge from metal refineries; runoff from waste batteries and paints
Chromium (total)	0.1	0.1	Allergic dermatitis	Discharge from steel and pulp mills; erosion of natural deposits
Copper	1.3	TT†††; action level=1.3	Short-term exposure: gastrointestinal distress; long-term exposure: liver or kidney damage; people with Wilson's disease should consult their personal doctor if the amount of copper in their water exceeds the action level.	Corrosion of household plumbing systems; erosion of natural deposits

Table 3-1 List of contaminants and their MCLs (continued)

Contaminant	MCLG,* mg/L†	MCL or TT, mg/L	Potential Health Effects From Ingestion of Water	Sources of Contaminant in Drinking Water
Cyanide (as free cyanide)	0.2	0.2	Nerve damage or thyroid problems	Discharge from steel/metal factories; discharge from plastics and fertilizer factories
Fluoride	4.0	4.0	Bone disease (pain and tenderness of the bones); children may get mottled teeth	Water additive that promotes strong teeth; erosion of natural deposits; discharge from fertilizer and aluminum factories
Lead	Zero	TT; action level=0.015	Infants and children: Delays in physical or mental development; children could show slight deficits in attention span and learning abilities Adults: Kidney problems; high blood pressure	Corrosion of household plumbing systems; erosion of natural deposits
Mercury (inorganic)	0.002	0.002	Kidney damage	Erosion of natural deposits; discharge from refineries and factories; runoff from landfills and croplands
Nitrate (measured as nitrogen)	10	10	Infants below the age of 6 months who drink water containing nitrate in excess of the MCL could become seriously ill and, if untreated, may die. Symptoms include shortness of breath and blue-baby syndrome.	Runoff from fertilizer use; leaching from septic tanks, sewage; erosion of natural deposits
Nitrite (measured as nitrogen)	1	1	Infants below the age of 6 months who drink water containing nitrite in excess of the MCL could become seriously ill and, if untreated, may die. Symptoms include shortness of breath and blue-baby syndrome.	Runoff from fertilizer use; leaching from septic tanks, sewage; erosion of natural deposits
Selenium	0.05	0.05	Hair or fingernail loss; numbness in fingers or toes; circulatory problems	Discharge from petroleum refineries; erosion of natural deposits; discharge from mines
Thallium	0.0005	0.002	Hair loss; changes in blood; kidney, intestine, or liver problems	Leaching from ore-processing sites; discharge from electronics, glass, and drug factories
Organic Chemicals—Synthetic Organic Chemicals (SOCs)				
2,4,5-TP (Silvex)	0.05	0.05	Liver problems	Residue of banned herbicide
2,4-D	0.07	0.07	Kidney, liver, or adrenal gland problems	Runoff from herbicide used on row crops

(continued)

Table 3-1 List of contaminants and their MCLs (continued)

Contaminant	MCLG,* mg/L†	MCL or TT, mg/L	Potential Health Effects From Ingestion of Water	Sources of Contaminant in Drinking Water
Acrylamide	Zero	TT‡‡‡	Nervous system or blood problems; increased risk of cancer	Added to water during sewage/wastewater treatment
Alachlor	Zero	0.002	Eye, liver, kidney, or spleen problems; anemia; increased risk of cancer	Runoff from herbicide used on row crops
Atrazine	0.003	0.003	Cardiovascular system or reproductive problems	Runoff from herbicide used on row crops
Benzo(a)pyrene (PAHs)	Zero	0.0002	Reproductive difficulties; increased risk of cancer	Leaching from linings of water storage tanks and distribution lines
Carbofuran	0.04	0.04	Problems with blood, nervous system, or reproductive system	Leaching of soil fumigant used on rice and alfalfa
Chlordane	Zero	0.002	Liver or nervous system problems; increased risk of cancer	Residue of banned termiticide
Dalapon	0.2	0.2	Minor kidney changes	Runoff from herbicide used on rights-of-way
Di(2-ethylhexyl) adipate	0.4	0.4	Weight loss; liver problems; possible reproductive difficulties	Discharge from chemical factories
Di(2-ethylhexyl) phthalate (DEHP)	Zero	0.006	Reproductive difficulties; liver problems; increased risk of cancer	Discharge from rubber and chemical
1,2-Dibromo-3-chloropropane (DBCP)	Zero	0.0002	Reproductive difficulties; increased risk of cancer	Runoff/leaching from soil fumigant used on soybeans, cotton, pineapples, and orchards
Dinoseb	0.007	0.007	Reproductive difficulties	Runoff from herbicide used on soybeans and vegetables
Diquat	0.02	0.02	Cataracts	Runoff from herbicide use
Dioxin (2,3,7,8-TCDD)	Zero	0.00000003	Reproductive difficulties; increased risk of cancer	Emissions from waste incineration and other combustion; discharge from chemical factories
Endothall	0.1	0.1	Stomach and intestinal problems	Runoff from herbicide use
Endrin	0.002	0.002	Liver problems	Residue of banned insecticide
Epichlorohydrin	Zero	TT	Increased cancer risk, and over a long period of time, stomach problems	Discharge from industrial chemical factories; an impurity of some water treatment chemicals
Ethylene dibromide	Zero	0.00005	Problems with liver, stomach, reproductive system, or kidneys; increased risk of cancer	Discharge from petroleum refineries

Table 3-1 List of contaminants and their MCLs (continued)

Contaminant	MCLG,* mg/L†	MCL or TT, mg/L	Potential Health Effects From Ingestion of Water	Sources of Contaminant in Drinking Water
Glyphosate	0.7	0.7	Kidney problems; reproductive difficulties	Runoff from herbicide use
Heptachlor	Zero	0.0004	Liver damage; increased risk of cancer	Residue of banned termiticide
Heptachlor epoxide	Zero	0.0002	Liver damage; increased risk of cancer	Breakdown of heptachlor
Hexachloro-benzene	Zero	0.001	Liver or kidney problems; reproductive difficulties; increased risk of cancer	Discharge from metal refineries and agricultural chemical factories
Hexachloro-cyclopentadiene (HEX)	0.05	0.05	Kidney or stomach problems	Discharge from chemical factories
Lindane	0.0002	0.0002	Liver or kidney problems	Runoff/leaching from insecticide used on cattle, lumber, gardens
Methoxychlor	0.04	0.04	Reproductive difficulties	Runoff/leaching from insecticide used on fruits, vegetables, alfalfa, livestock
Oxamyl (Vydate)	0.2	0.2	Slight nervous system effects	Runoff/leaching from insecticide used on apples, potatoes, and tomatoes
Pentachloro-phenol	Zero	0.001	Liver or kidney problems; increased cancer risk	Discharge from wood-preserving factories
Picloram	0.5	0.5	Liver problems	Herbicide runoff
Polychlorinated biphenyls (PCBs)	Zero	0.0005	Skin changes; thymus gland problems; immune deficiencies; reproductive or nervous system difficulties; increased risk of cancer	Runoff from landfills; discharge of waste chemicals
Simazine	0.004	0.004	Problems with blood	Herbicide runoff
Toxaphene	Zero	0.003	Kidney, liver, or thyroid problems; increased risk of cancer	Runoff/leaching from insecticide used on cotton and cattle
Organic Chemicals—Volatile Organic Chemicals (VOCs)				
Benzene	Zero	0.005	Anemia; decrease in blood platelets; increased risk of cancer	Discharge from factories; leaching from gas storage tanks and landfills
Chlorobenzene	0.1	0.1	Liver or kidney problems	Discharge from chemical and agricultural chemical factories
Carbon tetra-chloride	Zero	0.005	Liver problems; increased risk of cancer	Discharge from chemical plants and other industrial activities
o-Dichloro-benzene	0.6	0.6	Liver, kidney, or circulatory system problems	Discharge from industrial chemical factories
p-Dichloro-benzene	0.075	0.075	Anemia; liver, kidney, or spleen damage; changes in blood	Discharge from industrial chemical factories

(continued)

Table 3-1 List of contaminants and their MCLs (continued)

Contaminant	MCLG,* mg/L†	MCL or TT, mg/L	Potential Health Effects From Ingestion of Water	Sources of Contaminant in Drinking Water
1,2-Dichloro-ethane	Zero	0.005	Increased risk of cancer	Discharge from industrial chemical factories
1,1-Dichloro-ethylene	0.007	0.007	Liver problems	Discharge from industrial chemical factories
cis-1,2-Dichloro-ethylene	0.07	0.07	Liver problems	Discharge from industrial chemical factories
trans-1,2-Dichloroethylene	0.1	0.1	Liver problems	Discharge from industrial chemical factories
Dichloromethane	Zero	0.005	Liver problems; increased risk of cancer	Discharge from drug and chemical factories
1,2-Dichloro-propane	Zero	0.005	Increased risk of cancer	Discharge from industrial chemical factories
Ethylbenzene	0.7	0.7	Liver or kidney problems	Discharge from petroleum refineries
Styrene	0.1	0.1	Liver, kidney, or circulatory system problems	Discharge from rubber and plastics factories; leaching from landfills
Tetrachloro-ethylene (PCE)	Zero	0.005	Liver problems; increased risk of cancer	Discharge from factories and dry cleaners
Toluene	1	1	Nervous system, kidney, or liver problems	Discharge from petroleum factories
1,2,4-Trichloro-benzene	0.07	0.07	Changes in adrenal glands	Discharge from textile finishing factories
1,1,1-Trichloro-ethane	0.2	0.2	Liver, nervous system, or circulatory problems	Discharge from metal degreasing sites and other factories
1,1,2-Trichloro-ethane	0.003	0.005	Liver, kidney, or immune system problems	Discharge from industrial chemical factories
Trichloroethylene (TCE)	Zero	0.005	Liver problems; increased risk of cancer	Discharge from metal degreasing sites and other factories
Vinyl chloride	Zero	0.002	Increased risk of cancer	Leaching from PVC pipes; discharge from plastics factories
Xylenes (total)	10	10	Nervous system damage	Discharge from petroleum factories; discharge from chemical factories
Radionuclides				
Alpha particles	None Zero	15 picocuries per liter (pCi/L)	Increased risk of cancer	Erosion of natural deposits of certain minerals that are radioactive and may emit a form of radiation known as alpha radiation
Beta particles and photon emitters	None Zero	4 millirems per year	Increased risk of cancer	Decay of natural and synthetic deposits of certain minerals that are radioactive and may emit forms of radiation known as photons and beta radiation

Table 3-1 List of contaminants and their MCLs (continued)

Contaminant	MCLG,* mg/L†	MCL or TT, mg/L	Potential Health Effects From Ingestion of Water	Sources of Contaminant in Drinking Water
Radium 226 and radium 228 (combined)	None Zero	5 pCi/L	Increased risk of cancer	Erosion of natural deposits
Uranium	Zero	30 µg/L as of 12/8/03	Increased risk of cancer, kidney toxicity	Erosion of natural deposits

*Definitions:

Maximum contaminant level (MCL)—The highest level of a contaminant that is allowed in drinking water. MCLs are set as close to MCLGs as feasible using the best available treatment technology and taking cost into consideration. MCLs are enforceable standards.

Maximum contaminant level goal (MCLG)—The level of a contaminant in drinking water below which there is no known or expected risk to health. MCLGs allow for a margin of safety and are nonenforceable public health goals.

Maximum residual disinfectant level (MRDL)—The highest level of a disinfectant allowed in drinking water. There is convincing evidence that addition of a disinfectant is necessary for control of microbial contaminants.

Maximum residual disinfectant level goal (MRDLG)—The level of a drinking water disinfectant below which there is no known or expected risk to health. MRDLGs do not reflect the benefits of the use of disinfectants to control microbial contaminants.

Treatment Technique (TT)—A required process intended to reduce the level of a contaminant in drinking water.

†Units are in milligrams per liter (mg/L) unless otherwise noted. Milligrams per liter is equivalent to parts per million.

‡USEPA's Surface Water Treatment Rules (SWTRs) require systems using surface water or groundwater under the direct influence of surface water to (1) disinfect their water and (2) filter their water or meet criteria for avoiding filtration so that the following contaminants are controlled at the following levels:

Cryptosporidium (as of Jan. 1, 2002, for systems serving >10,000 and 1/14/05 for systems serving <10,000) 99% removal.

Giardia lamblia: 99.9% removal/inactivation.

Viruses: 99.99% removal/inactivation.

Legionella: No limit, but USEPA believes that if *Giardia* and viruses are removed/inactivated, *Legionella* will also be controlled.

Turbidity: For systems that use conventional or direct filtration, at no time can turbidity (cloudiness of water) go higher than 1 Nephelometric Turbidity Unit (NTU), and samples for turbidity must be less than or equal to 0.3 NTUs in at least 95 percent of the samples in any month. Systems that use filtration other than the conventional or direct filtration must follow state limits, which must include turbidity at no time exceeding 5 NTUs.

HPC: No more than 500 bacterial colonies per milliliter.

Long-Term 1 Enhanced Surface Water Treatment Rule (effective date: Jan. 14, 2005); surface water systems or groundwater under the direct influence of surface water (GWUDI) systems serving fewer than 10,000 people must comply with the applicable Long-Term 1 Enhanced Surface Water Treatment Rule provisions (e.g., turbidity standards, individual filter monitoring, *Cryptosporidium* removal requirements, updated watershed control requirements for unfiltered systems).

Filter Backwash Recycling: The Filter Backwash Recycling Rule requires systems that recycle to return specific recycle flows through all processes of the system's existing conventional or direct filtration system or at an alternate location approved by the state.

**More than 5.0% samples total coliform-positive in a month. (For water systems that collect fewer than 40 routine samples per month, no more than one sample can be total coliform–positive per month.) Every sample that has total coliform must be analyzed for either fecal coliforms or E. coli; if two consecutive total coliform samples are positive and one is also positive for *E. coli* fecal coliforms, the system has an acute MCL violation.

††Fecal coliform and *E. coli* are bacteria whose presence indicates that the water may be contaminated with human or animal wastes. Disease-causing microbes (pathogens) in these wastes can cause diarrhea, cramps, nausea, headaches, or other symptoms. These pathogens may pose a special health risk for infants, young children, and people with severely compromised immune systems.

‡‡Although there is no collective maximum contaminant level goal (MCLG) for this contaminant group, there are individual MCLGs for some of the individual contaminants: trihalomethanes: bromodichloromethane (zero); bromoform (zero); dibromochloromethane (0.06 mg/L). Chloroform is regulated with this group but has no MCLG. Haloacetic acids: dichloroacetic acid (zero); trichloroacetic acid (0.3 mg/L). Monochloroacetic acid, bromoacetic acid, and dibromoacetic acid are regulated with this group but have no MCLGs.

***MCLGs were not established before the 1986 amendments to the SDWA. Therefore, there is no MCLG for this contaminant.

†††Lead and copper are regulated by a treatment technique that requires systems to control the corrosiveness of their water. If more than 10% of tap water samples exceed the action level, water systems must take additional steps. For copper, the action level is 1.3 mg/L; for lead it is 0.015 mg/L.

‡‡‡Each water system must certify in writing to the state (using third-party or manufacturer's certification) that when acrylamide and epichlorohydrin are used in drinking water systems, the combination (or product) of dose and monomer level does not exceed the levels specified, as follows:

Acrylamide = 0.05% dosed at 1 mg/L (or equivalent)

Epichlorohydrin = 0.01% dosed at 20 mg/L (or equivalent)

Disinfection By-product and Microbial Regulations

Drinking water treatment, including use of chemical disinfectants such as chlorine, ozone, and chlorine dioxide, has been an important step in protecting drinking water consumers from exposure to harmful microbial contaminants. However, these chemical disinfectants can also react with organic and inorganic substances in the water to produce by-products that may be harmful to drinking water consumers, particularly some susceptible segments of the population. Therefore, drinking water treatment using chemical disinfectants involves a delicate balancing act—i.e., adding enough disinfectant to control harmful microorganisms but not enough to produce unacceptably high levels of regulated disinfection by-products (DBPs).

USEPA has enacted several regulations impacting microbial control and production of DBPs in groundwater and surface water supplies for small and large public drinking water systems. These rules are referred to collectively as the Microbial/Disinfection By-Products (M/DBP) Rules. Microbial protection for consumers of drinking water from public supplies is provided by provisions of current or pending rules listed below and discussed in more detail later in this chapter:

- Filter Backwash Recycling Rule (FBRR)
- Ground Water Rule (GWR)
- Interim Enhanced Surface Water Treatment Rule (IESWTR)
- Long-Term 1 Enhanced Surface Water Treatment Rule (LT1ESWTR)
- Long-Term 2 Enhanced Surface Water Treatment Rule (LT2ESWTR)
- Stage 1 Disinfectants and Disinfection By-products Rule (Stage 1 DBPR)
- Stage 2 Disinfectants and Disinfection By-products Rule (Stage 2 DBPR)
- Surface Water Treatment Rule (SWTR)
- Total Coliform Rule (TCR)

Provisions of the Disinfectants and Disinfection By-products Rule (DBPR) are intended to protect drinking water consumers against the unintended public health consequences associated with consumption of treated drinking water containing residual disinfectants and DBPs produced from degradation of these residual disinfectants or reaction of disinfectants with organic and inorganic DBP precursors.

More details regarding the DBPR, including the current Stage 1 DBPR and Stage 2 DBPR, are described in this chapter. Also included in the DBPR description is a brief discussion of some currently unregulated DBPs that are being heavily researched and may be the subject of future regulation. In the following discussion, the DBPR will be discussed first, followed by the microbial protection rules (SWTR, GWR, and TCR).

disinfection by-products (DBPs)
A new chemical compound formed by the reaction of disinfectants with organic compounds in water. At high concentrations, many DBPs are considered a danger to human health.

Disinfection By-product Rule (DBPR)

The Stage 1 DBPR and Stage 2 DBPR requirements discussed in the following sections focus first on two specific contaminants (TTHM and HAA5) and then on other aspects of these regulations dealing with control or removal of DBP precursors ("enhanced coagulation"), bromate, chlorite, and residual disinfectants.

Stage 1 DBPR—HAA5 and TTHM Provisions

The Stage 1 DBPR was published in 1998 and established an MCL of 0.080 mg/L for TTHM (the sum of four trihalomethanes, which are chloroform, bromodichloromethane, dibromochloromethane, and bromoform) and 0.060 mg/L for

HAA5 (the sum of five specific haloacetic acids, which are mono-, di-, and trichloroacetic acids plus mono- and dibromoacetic acids). Although the MCLs for TTHM and HAA5 were officially written as 0.080 mg/L and 0.060 mg/L, respectively, the limits are commonly referred to as "80/60," or 80 µg/L and 60 µg/L. While knowing the numerical value of each MCL is important in understanding compliance with the DBPR, it is equally important to understand the methodology, in all its subtleties, used to calculate the compliance value that will be compared to this MCL.

For TTHM and HAA5, the compliance value is determined by monitoring the distribution system. Compliance monitoring locations need to be representative of the distribution system. Systems serving more than 10,000 persons who use surface water sources are required to monitor at least four locations per plant, meaning that distribution systems fed by more than one treatment plant must have at least four monitoring locations designated for each plant entry point.

The compliance monitoring location for systems with only one monitoring point must be representative of maximum residence time in the distribution system. A minimum of one out of every four compliance monitoring locations for systems with more locations must also be representative of maximum residence time. The other locations must be far enough away from the plant entry points to be representative of average residence time in the distribution system.

Unlike acute toxicity risks, for which the exposure could be a single glass of water, cancer risks like those believed to be linked to TTHM and HAA5 involve longer periods of exposure (daily glasses of water spanning decades). For chronic exposures such as these, exposure to an excessively high concentration of a given cancer-causing agent will not necessarily result in the consumer getting cancer from this source. Conversely, a consumer exposed to a lower concentration every day for a lifetime could be more likely to develop cancer. Therefore, regulation of DBPs to reduce cancer risks is *not* based on limiting exposure to a single incident (i.e., not a "single hit"), but rather is aimed at reducing the repeated exposure over time. In other words, DBP exposure needs to be evaluated on an average basis over time.

Under the Stage 1 DBPR, the compliance value for TTHM and HAA5 is determined by calculating a **running annual average (RAA)** during the previous 12 months for each DBP for all monitoring locations at each plant. Most systems are required to monitor quarterly (i.e., 4 times per year), although small groundwater systems (<10,000 persons) may be allowed to sample once a year. Typically, the RAA is based on 4 monitoring locations sampled quarterly, meaning RAA will be the average of 16 monitoring results each for HAA5 and TTHM.

Table 3-2 illustrates one facility's calculations of RAA for HAA5 that were used for Stage 1 compliance (this table also shows calculation of values for

running annual average (RAA)
The average of four quarterly samples at each monitoring location to ensure compliance with the Stage 1 DBPR.

Table 3-2 Example RAA and LRAA calculations for Stage 1 DBPR and Stage 2 DBPR

		Sampling Location, µg/L			
Year	**Quarter**	**A**	**B**	**C**	**D**
1	3rd	52	68	63	66
1	4th	35	42	38	41
2	1st	47	49	42	43
2	2nd	18	42	45	37
LRAA		38	50	47	47
Maximum LRAA		50			
RAA		45			

Stage 2 DBPR, which will be discussed later). It is important to reemphasize that compliance is based solely on the RAA, not on a single quarterly result at any one monitoring location. Consequently, it is *not* correct to refer to a single quarterly monitoring result above 60 µg/L for HAA5 or above 80 µg/L for TTHM as being above the MCL. Therefore, even though several individual monitoring values in Table 3-2 are greater than 60 µg/L, the facility is in compliance with the HAA5 MCL because the RAA is 45 µg/L for HAA5.

Utility personnel should be consistent and rigorous in their use of terminology when dealing with the general public or with state and local health officials, and should ensure that all people participating in these discussions are consistent in applying the MCL only to RAA values and do not make the common mistake of referring to a single quarterly monitoring value as being "above the MCL."

Stage 2 DBPR—HAA5 and TTHM Provisions

The Stage 2 DBPR, published in 2006, is now in effect. This rule tightened requirements for DBPs, but compliance is not achieved by modifying the numerical value of the MCLs or by requiring monitoring of new constituents. Instead, the rule makes compliance more challenging than under the Stage 1 DBPR by (1) changing the way the compliance value is calculated and (2) changing the compliance monitoring locations to sites representative of the greatest potential for THM and HAA formation. These changes were made to ensure uniform compliance with the DBP standards across all areas of the distribution system; that is, compliance is required at each sampling location.

The compliance value in the Stage 2 DBPR is called the *locational running annual average* (LRAA), and it is calculated by separately averaging the four quarterly samples at each monitoring location. Compliance is based on the maximum LRAA value (see Table 3-2). Furthermore, the Stage 2 DBPR included several interim steps that led to the replacement of many existing Stage 1 DBPR monitoring locations with new locations representative of the greatest potential for consumer exposure to high levels of TTHM and HAA5.

The Stage 2 DBPR required that facilities maintain compliance with the Stage 1 DBPR using the existing monitoring locations during the first three years after the final version of the Stage 2 DBPR was published. In the time period between the third and sixth year after the Stage 2 DBPR was published, compliance continues to be based on maintaining 80/60 (TTHM and HAA5) or lower for RAA; it also includes a requirement for maximum LRAA at existing Stage 1 monitoring locations. The long-term goal of the Stage 2 DBPR was to identify locations within the distribution system with the greatest potential for either TTHM or HAA5 formations and then base compliance on the LRAA at or below 80/60 for each of these locations. Many of these locations were identified during the initial distribution system evaluation (IDSE).

The IDSE included monitoring, modeling, and/or other evaluations of drinking water distribution systems to identify locations representative of the greatest potential for consumer exposure to high levels of TTHM and HAA5. The goal of the IDSE was to evaluate a number of potential monitoring locations to justify selection of monitoring locations for long-term compliance (i.e., Stage 2B) with the Stage 2 DBPR.

One item to note regarding the Stage 2 DBPR as it applies to TTHM and HAA5 is that the goal was to find the locations in the distribution system where average annual levels of these DBPs are highest. TTHM formation increases as contact time with free or combined chlorine increases, although formation in the presence of combined chlorine is limited. Therefore, establishing points in the distribution system with the highest potential for TTHM formation is related to knowing the points with maximum water age. Utilities that have not performed a tracer study in the distribution system to determine water age should consider doing so.

By contrast, peak locations for HAA5 are more complicated because microorganisms in biofilm attached to distribution system pipe surfaces can biodegrade HAA5. Consequently, increasing formation of HAA5 over time is offset by biodegradation, eventually reaching a point where HAA5 levels decrease over time, even to the point where they drop to zero. Figure 3-1 shows a gradual reduction in HAA5 formation over time in a distribution system, followed by an eventual decrease of HAA5 as water age increases (water age measured in tracer test). In chloramination systems, HAA5 formation is limited. In fact, ammonium chloride is added as a quenching agent in HAA5 compliance samples in order to halt HAA5 formation prior to analysis (*Standard Methods for the Examination of Water and Wastewater*, latest edition). Therefore, little additional HAA5 formation occurs after chloramination to offset HAA5 biodegradation occurring in the distribution system.

Enhanced Coagulation Requirement of the Stage 1 DBPR and Stage 2 DBPR

The enhanced coagulation requirement has been developed to promote optimization of coagulation processes in conventional surface water treatment systems as required to improve removal of organic DBP precursors. The focus of the SWTR is separate from that of the enhanced coagulation requirement, with the former directed toward optimizing particle removal and the latter toward optimizing removal of natural organic matter (DBP precursors). Both promote efforts by water utilities to properly control and optimize coagulation processes and reduce DBP formation.

Under the enhanced coagulation requirements, treatment plants must remove specific percentages of total organic carbon (TOC) based on their source water TOC and alkalinity levels. Facilities must meet the enhanced coagulation

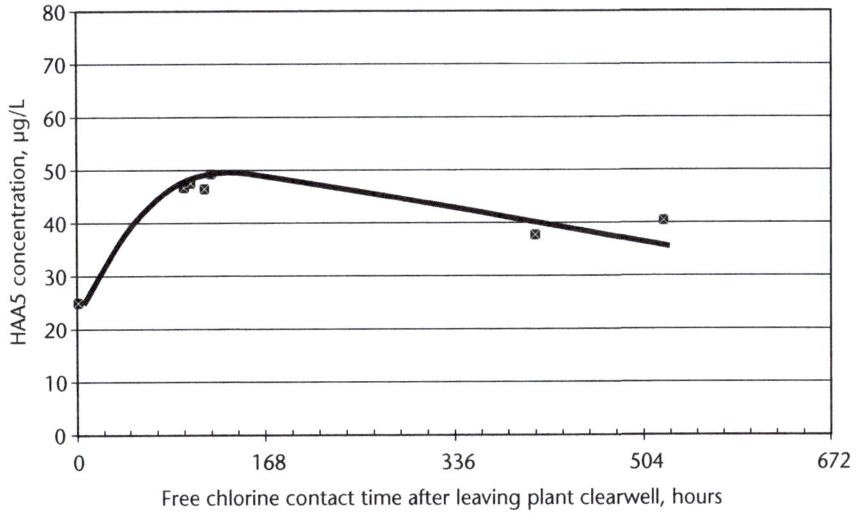

Figure 3-1 Formation and decay of HAA5 in a distribution system (time estimated by fluoride tracer test—T_{100})

requirements unless they meet any of the following exemptions (USEPA Stage 1 DBPR Guidance):

1. The PWS's source water TOC level is <2.0 mg/L, calculated quarterly as an RAA.
2. The PWS's treated water TOC level is <2.0 mg/L, calculated quarterly as an RAA.
3. The PWS's source water TOC level is <4.0 mg/L, calculated quarterly as an RAA; the source-water alkalinity is >60 mg/L (as calcium carbonate [$CaCO_3$]), calculated quarterly as a running annual average; and either the TTHM and HAA5 running annual averages are no greater than 0.040 mg/L and 0.030 mg/L, respectively, or the PWS has made a clear and irrevocable financial commitment to use technologies that will limit the levels of TTHMs and HAA5 to no more than 0.040 mg/L and 0.030 mg/L, respectively.
4. The PWS's TTHM and HAA5 RAAs are no greater than 0.040 mg/L and 0.030 mg/L, respectively, and the PWS uses only chlorine for primary disinfection and maintenance of a residual in the distribution system.
5. The PWS's source water specific ultraviolet absorption at 254 nm (SUVA), prior to any treatment and measured monthly, is ≤2.0 L/mg-m, calculated quarterly as an RAA.
6. The PWS's finished-water SUVA, measured monthly, is ≤2.0 L/mg-m, calculated quarterly as an RAA.

Additionally, alternative compliance criteria for softening systems include the following:

7. Softening that results in lowering the treated water alkalinity to <60 mg/L (as $CaCO_3$), measured monthly and calculated quarterly as an RAA.
8. Softening that results in removing at least 10 mg/L of magnesium hardness (as $CaCO_3$), measured monthly and calculated quarterly as an RAA.

Utilities that cannot meet these avoidance criteria should know their enhanced coagulation endpoint, identified as the coagulant dosage and/or pH value that, when achieved, no longer produces significant TOC reduction. Specifically, when the source water TOC is not reduced by at least 0.3 mg/L with an incremental dosage increase of 10 mg/L alum (or equivalent ferric salt) and the pH value of the source reaches a value listed in Table 3-3, the enhanced coagulation endpoint has been reached.

If a utility is not exempt, a number of steps have to be evaluated relating to TOC removal, alkalinity of source water, range of source water TOC, required TOC removal for given source water characteristics, and several other factors.

Table 3-3 Target pH values for coagulation when TOC removal rates are not sufficient

Alkalinity	pH Value
0–60	5.5
>60–120	6.3
>120–240	7.0
>240	7.5

Bromate

The bromate MCL from the Stage 1 DBPR remained at 0.010 mg/L for the Stage 2 DBPR. Bromate can be present in systems using ozone that have bromide present at the ozone application point. Bromate is also potentially formed during manufacture and storage of sodium hypochlorite. Consequently, systems using ozone for oxidation or disinfection are required to monitor once a month at a distribution system entry point for bromate, but systems without ozone are not required to perform this monitoring. Systems that use ozone and that also add sodium hypochlorite will need to closely monitor the quality of these sodium hypochlorite products for bromate content.

Chlorite

Similar to bromate, chlorite monitoring is required only for systems using chlorine dioxide as an oxidant or disinfectant. Chlorite is a degradation product of chlorine dioxide. Chlorate is also a degradation product of chlorine dioxide but is not currently regulated. Chlorite and chlorate are potential degradation products of sodium hypochlorite, but systems using sodium hypochlorite are not required to monitor for chlorite unless they also use chlorine dioxide.

Monitoring for chlorite is more complicated than for bromate because chlorine dioxide will degrade and chlorite formation will increase over time. Therefore, chlorite monitoring requirements include daily monitoring at the distribution system entry point and monthly samples at three locations in the distribution system (first customer, average residence time, maximum residence time). Unlike the health risks for bromate, TTHM, and HAA5, the risk for chlorite requires compliance based on the average of the three chlorite-monitoring locations each month. The Stage 1 DBPR MCL for chlorite is 1.0 mg/L.

Residual Disinfectants

The maximum residual disinfectant level (MRDL) for combined or total chlorine is 4.0 mg/L as Cl_2. These values are based on the same data used to monitor minimum free and combined chlorine levels in the distribution system as required by the SWTR, using the same monitoring locations used for the TCR. Chlorine dioxide residual also has an MRDL of 0.8 mg/L as ClO_2, based on daily samples at the treatment plant.

WATCH THE VIDEO
WSO Disinfection Byproducts—DBPs Regulations
(www.awwa.org/wsovideoclips)

Surface Water Treatment Rule (SWTR)

IESWTR and LT1ESWTR

The goal of the IESWTR is to limit human exposure to harmful organisms, including *Cryptosporidium*, by promoting achievement of particle and turbidity removal targets for surface water treatment systems. Among the IESWTR requirements that apply to surface water treatment plants are the following:

- Combined filter effluent turbidity must be ≤0.3 ntu for 95 percent of samples collected each month, including none with >1 ntu. Compliance is based on combined filter effluent samples collected at 4-hour intervals during the entire month).

- The utility must monitor each individual filter for turbidity at 15-minute intervals and must report results, including a filter profile (graphical representation of filter performance), if either of the following two conditions are met: (1) turbidity in any filter for two consecutive 15-minute intervals exceeds 1 ntu or (2) turbidity during the first 4 hours of a given filter run exceeds 0.5 ntu for two consecutive 15-minute samples. Results must be reported within 10 days of the end of the month.
- Any newly constructed finished water reservoirs must include covers to keep out dust, debris, birds, etc.
- The utility must complete a sanitary survey every 3 years. Existing surveys conducted after December 1995 can be used if they meet minimum requirements. Variances can be granted to decrease frequency to 5 years. The IESWTR explicitly requires that sanitary surveys include efforts to evaluate and control *Cryptosporidium*, in addition to other target organisms.
- Systems where the average of quarterly TTHM or HAA5 values exceeds 64 and 48 µg/L, respectively, need to complete disinfection profiling and benchmarking. *Profiling* involves determination of **$C \times T$ values** (concentration of disinfectant × contact time) for each segment of the treatment plant (see later discussion). *Benchmarking* involves determining the lowest monthly average during 12-month monitoring of *Giardia* and virus inactivation. This procedure is required for any systems that are considering a major change to their disinfection practice. Consultation with the primacy agency is also required before any disinfection change.
- Turbidity monitoring records must be maintained for a minimum of 3 years.

Facilities in compliance with these requirements, chiefly the turbidity monitoring provisions, are designated by the IESWTR to have provided 2-log virus removal, 2.5-log *Giardia* removal, and 2-log *Cryptosporidium* removal. Literature and other information cited in the IESWTR final rule indicate that these credits are conservative, and most facilities meeting these requirements are probably achieving far greater levels of virus, *Giardia*, and *Cryptosporidium* removal than the minimum credits previously cited. The level of *Cryptosporidium* protection cited is sufficient to meet all requirements of the IESWTR, but the rule requires a total of 3.0 credits for *Giardia* and 4.0 credits for viruses. The additional credits (0.5-log for *Giardia* and 2-log for viruses) are required to be achieved by disinfection with free chlorine, chloramines, ozone, or chlorine dioxide by meeting $C \times T$ requirements described later in this chapter.

$C \times T$ value
The product of the residual disinfectant concentration C, in milligrams per liter, and the corresponding disinfectant contact time T, in minutes. Minimum $C \times T$ values are specified by the Surface Water Treatment Rule as a means of ensuring adequate kill or inactivation of pathogenic microorganisms in water.

Provisions of the IESWTR apply to large systems (>10,000 persons) using surface water sources. However, similar provisions are applied to smaller surface water systems (<10,000 persons), as outlined in the LT1ESWTR. The objectives of the LT1ESWTR and IESWTR are identical, though some of the compliance deadlines and other regulatory provisions are slightly different based on greater financial and personnel resources for larger systems.

Sanitary Surveys

Sanitary surveys are a requirement of the Interim Enhanced Surface Water Treatment Rule (IESWTR). A sanitary survey is "an onsite review of the water source, facilities, equipment, operation, and maintenance of the public water system for the purpose of evaluating the adequacy of such source, facilities, equipment, operation, and maintenance for producing and distributing safe drinking water." Surveys are usually performed by the state primacy agency and are required of all surface water systems and groundwater systems under the direct influence of surface water.

These surveys are typically divided into eight main sections, although some state primacy groups may have more:

1. Water sources
2. Water treatment process
3. Water supply pumps and pumping facilities
4. Storage facilities
5. Distribution systems
6. Monitoring, reporting, and data verification
7. Water system management and operations
8. Operator compliance with state requirements

Sanitary surveys are required on a periodic basis usually every 3 years. Surveys may be comprehensive or focused according to the regulatory agency requirements.

C × T *Requirements*

Every water system that uses surface water as a source must meet treatment technique requirements for the removal and/or inactivation of *Giardia*, viruses, *Legionella*, and other bacteria. Because these pathogens are not easily identified in the laboratory on a routine basis, USEPA has set quality goals in lieu of MCLs in this instance. Meeting SWTR treatment technique goals demonstrates all or part of the required microbial protection, as previously noted, but additional protection is required through the use of approved disinfection treatment chemicals. The effectiveness of disinfection depends on the type of disinfectant chemical used, the residual concentration, the amount of time the disinfectant is in contact with the water, the water temperature, and, when chlorine is used, the pH of the water.

According to USEPA, a combination of the residual concentration, C, of a disinfectant (in milligrams per liter) multiplied by the contact time, *T* (in minutes), can be used as a measure of the disinfectant's effectiveness in killing or inactivating microorganisms. For water plant operators, this means that high residuals held for a short amount of time or low residuals held for a long period of time will produce similar results. Water plants are required to provide this computation daily, and the measure must always be higher than the required minimum value.

Long-Term 2 Enhanced SWTR (LT2ESWTR)

The LT2ESWTR supplements the SWTR requirements contained in the IESWTR for large surface water systems (>10,000 persons) and the LT1ESWTR for small systems (<10,000 persons). Details of the rules can be reviewed in the *Federal Register* or at the USEPA website (http://water.epa.gov/drink/index.cfm). One of the key elements of the LT2ESWTR was the use of *Cryptosporidium* monitoring results to classify surface water sources into one of four USEPA-defined risk levels called "bins." Facilities in the lowest bin (bin 1) are required to maintain compliance with the current IESWTR. Facilities in higher bins (bins 2 to 4) are required to either (1) provide additional *Cryptosporidium* protection from new facilities or programs not currently in use at a facility or (2) demonstrate greater *Cryptosporidium* protection capabilities of existing facilities and programs using a group of USEPA-approved treatment technologies, watershed programs, and demonstration studies, referred to collectively as the *Microbial Toolbox*.

Implementation of the LT2ESWTR was phased over many years according to system size. Four separate size categories were established (schedules 1–4, with 4 being the smallest at <10,000 population) for implementing the rule. The rule for schedule-4 systems allows filtered supplies to perform initial monitoring for fecal coliform to determine if *Cryptosporidium* monitoring is required.

Filter Backwash Recycle Rule (FBRR)

The FBRR currently applies to systems of all sizes and is intended to help utilities minimize potential health risks associated with recycle, particularly associated with respect to *Giardia* and *Cryptosporidium*. Other contaminants of concern in the recycle stream include suspended solids (turbidity), dissolved metals (especially iron and manganese), and dissolved organic carbon. Plants that control recycle will also help minimize operational problems.

Prior to the FBRR, no USEPA regulation governed recycle. Regulations within the United States regarding recycle had been established by the states, if at all. State regulatory approaches varied from a requirement of equalization of two backwashes in Illinois to 80 percent solids removal prior to recycle and maintaining recycle flows at less than 10 percent of raw water flow in California. Virginia discourages recycling.

Key components of the FBRR include (1) recycle must reenter the treatment process *prior* to primary coagulant addition, (2) direct filtration plants must report their recycle practices to the state and may need to treat their recycle streams, and (3) a self-assessment must be done at those plants that use direct recycle (i.e., no separate equalization and/or treatment of recycle stream) and that operate fewer than 20 filters. The goal of the self-assessment is to determine if the design capacity of the plant is exceeded due to recycle practices.

Ground Water Rule (GWR)

USEPA promulgated the final GWR in October 2006 to reduce the risk of exposure to fecal contamination that may be present in PWSs that use groundwater sources. The rule establishes a risk-targeted strategy to identify groundwater systems that are at high risk for fecal contamination. The GWR also specifies when corrective action (which may include disinfection) is required to protect consumers who receive water from groundwater systems from bacteria and viruses.

A sanitary survey is required, by the state primacy agency, at regular intervals depending on the condition of the water system as determined in the initial survey. Systems found to be at high risk for fecal contamination are required to provide 4-log inactivation of viruses. Increased monitoring for fecal contamination indicators may be required by the regulatory authority.

Total Coliform Rule (TCR) and Revised Total Coliform Rule (RTCR)

The objective of the TCR is to promote routine surveillance of distribution system water quality to search for fecal matter and/or disease-causing bacteria. All points in a distribution system cannot be monitored, and complete absence of fecal matter and disease-causing bacteria cannot be guaranteed. The TCR is an attempt to persuade water utilities to implement monitoring programs sufficient to verify that public health is being protected as much as possible, as well as allowing utilities to identify any potential contamination problems in their distribution system. The rule requires monthly sampling at each distribution sampling point.

The TCR, and the RTCR that was finalized in 2013, impact all PWSs. The RTCR requires PWSs that are vulnerable to microbial contamination to identify and fix problems. The RTCR also established criteria for systems to qualify for and stay on reduced monitoring, thereby providing incentives for improved water system operation.

The RTCR rule established an MCLG and an MCL for *Escherichia coli* (*E. coli*) and eliminated the MCLG and MCL for total coliform, replacing it with a treatment technique for coliform that requires assessment and corrective action. The rule establishes an MCLG and an MCL of zero for *E. coli*, a more specific indicator of fecal contamination and potentially harmful pathogens than total coliform. USEPA removed the MCLG and MCL of zero for total coliform. Many of the organisms detected by total coliform methods are not of fecal origin and do not have any direct public health implication.

Under the treatment technique for coliform, total coliform serves as an indicator of a potential pathway of contamination into the distribution system. A PWS that exceeds a specified frequency of total coliform occurrence must conduct an assessment to determine if any sanitary defects exist and, if found, correct them. In addition, a PWS that incurs an *E. coli* MCL violation must conduct an assessment and correct any sanitary defects found.

The RTCR also changed monitoring frequencies. It links monitoring frequency to water quality and system performance and provides criteria that well-operated small systems must meet to qualify and stay on reduced monitoring. It also requires increased monitoring for high-risk small systems with unacceptable compliance history and establishes some new monitoring requirements for seasonal systems such as state and national parks.

The revised rule eliminated monthly public notification requirements based only on the presence of total coliforms. Total coliforms in the distribution system may indicate a potential pathway for contamination but in and of themselves do not indicate a health threat. Instead, the rule requires public notification when an *E. coli* MCL violation occurs, indicating a potential health threat, or when a PWS fails to conduct the required assessment and corrective action.

The rule requires that PWSs collect total coliform samples at sites representative of water quality throughout the distribution system according to a written plan approved by the state or primacy agency. Samples are collected at regular intervals monthly. Positive total coliform samples must be tested for *E. coli*. If any positive total coliform sample is also positive for *E. coli* the state must be notified by the end of the day on which the result was received. Repeat samples are required within 24 hours of any total coliform–positive routine sample. Three repeat samples are required, one at the site of the positive sample and one within five service taps both upstream and downstream of the positive site. Any positive total coliform samples must be tested for *E. coli*. Any positive *E. coli* (EC+) samples must be reported by the end of the day. Any positive total coliform (TC+) samples require another set of repeat samples.

A Level 1 or Level 2 sanitary assessment and corrective action is triggered to occur within 30 days if there is indication of coliform contamination. A Level 1 assessment by the PWS is triggered if more than 5 percent of the routine/repeat monthly samples (if at least 40 are required) are total coliform positive or a repeat sample is not taken for a total coliform positive result. A Level 2 assessment conducted by the state or its representative is triggered if the PWS has an *E. coli* violation or repeated Level 1 assessment triggers.

Major violations of the RTCR are MCL violations and treatment technique violations. A PWS will receive an *E. coli* MCL violation when there is any combination of an EC+ sample result with a routine/repeat TC+ or EC+ sample result, as follows:

E. coli **MCL Violation Occurs With the Following Sample Result Combination**

Routine	Repeat
EC+	TC+
EC+	Any missing sample
EC+	EC+
TC+	EC+
TC+	TC+ (but no *E. coli* analysis)

A PWS will receive a treatment technique violation given any of the following conditions:

- Failure to conduct a Level 1 or Level 2 assessment within 30 days of a trigger.
- Failure to correct all sanitary defects from a Level 1 or Level 2 assessment within 30 days of a trigger or in accordance with the state-approved time frame.
- Failure of a seasonal system to complete state-approved start-up procedures prior to serving water to the public.

Lead and Copper Rule (LCR)

The objective of the LCR is to control corrosiveness of the finished water in drinking water distribution systems to limit the amount of lead (Pb) and copper (Cu) that may be leached from certain metal pipes and fittings in the distribution system. Of particular concern are pipes and fittings connecting the household tap to the distribution system service line at individual homes or businesses, especially because water can remain stagnant in these service lines for long periods of time, increasing the potential to leach Pb, Cu, and other metals. Although the utility is not responsible for maintaining and/or replacing these household connections, they are responsible for controlling pH and corrosiveness of the water delivered to the consumers.

Details of the LCR include the following:

- The LCR became effective December 7, 1992.
- The action level for Pb is 0.015 mg/L and for Cu is 1.3 mg/L.
- A utility is in compliance at each sampling event (frequency discussed below) when <10 percent of the distribution system samples are above the action levels for Pb and Cu (i.e., 90th percentile value for sampling event must be below action level).
- Utilities found not to be in compliance must modify water treatment until they are in compliance. The term *action level* is used rather than *MCL* because noncompliance (i.e., exceeding an action level) triggers a need for modifications in treatment.

After identifying sampling locations and determining initial tap water Pb and Cu levels at each of these locations, utilities must also monitor other water quality parameters (WQPs) at these same locations as needed to monitor and evaluate corrosion control characteristics of treated water. The only exemptions from analysis of these WQPs are systems serving less than 50,000 people for which Pb and Cu levels in initial samples are below action levels.

Pb, Cu, and WQPs are initially collected at 6-month intervals; this frequency can be reduced if action levels are not exceeded and optimal water treatment is maintained. Systems that are in noncompliance and are performing additional corrosion-control activities must continue to monitor at 6-month intervals, plus they must collect WQPs from distribution system entry points every 2 weeks.

Each utility must complete a survey and evaluate materials that comprise their distribution system, in addition to using other available information, to target homes that are at high risk for Pb/Cu contamination.

Revisions to the LCR were enacted in 2007. These clarifications to the existing rule were made in seven areas:

- Minimum number of samples required
- Definitions for compliance and monitoring periods
- Reduced monitoring criteria
- Consumer notice of lead tap water monitoring results
- Advanced notification and approval of long-term treatment changes
- Public education requirements
- Reevaluation of lead service lines

Additional revisions are under consideration for the Lead and Copper Rule. Consult your local regulatory agency for those revisions that are applicable to your system.

Phase I, II, and V Contaminants

The Phase I, II, and V regulations were finalized in 1989, 1992, and 1995, respectively, and include various inorganic and organic contaminants. Sampling and reporting frequency vary with constituent, though sampling is typically required once every 3 years after the initial sampling period. Variances or waivers are possible for a number of constituents based on analytical results and/or a vulnerability assessment.

Public Notification Rule

USEPA has implemented a regulation called the *Public Notification Rule.* This rule is separate from the Consumer Confidence Report (CCR) Rule. The Public Notification Rule includes requirements for reporting certain water quality monitoring violations and other water quality incidents, as well as requirements for the timing, distribution, and language of the public notices. For example, the Public Notification Rule includes requirements that some incidents be reported within 24 hours, others within 30 days, and others included as part of the annual CCR. Some of these reporting requirements are more stringent than those currently required by USEPA. The regulation also includes requirements regarding how notices are to be distributed/broadcast (i.e., TV, radio, newspaper, hand delivery, regular mail, etc.), the format of the notices, the wording of certain items in the notice, and the need to include information in languages other than English.

Public notification according to the rule might include the following:

- Templates, or model notices, to be available for adaptation for certain potential incidents
- Consolidated and updated lists of phone numbers and contacts for government (local, county, state), regulatory agencies, hospitals, radio and TV, newspapers, etc., that should be contacted per requirements of the Public Notification Rule
- Checklists and flow diagrams outlining activities that would need to be completed for certain potential events outlined in the regulation
- Identification of key personnel and what their roles and responsibilities would be to respond as required by the regulation
- A plan to periodically review and update all lists, templates, and other aspects of a response plan every year or when/if the Public Notification Rule is modified by future federal or state regulations

Unregulated Contaminant Monitoring Rule (UCMR)

The 1996 amendments to the SDWA require USEPA to establish criteria for a monitoring program for currently unregulated contaminants to generate data that USEPA can use to evaluate and prioritize contaminants that could potentially be regulated in the future. USEPA has developed three cycles of the UCMR:

1. UCMR1 in 1999
2. UCMR2 in 2007
3. UCMR3 in 2012

Failing to (1) perform required sampling and analysis, (2) use the appropriate analytical procedures, or (3) report these results is a violation of the UCMR. However, the numerical results of these analytical efforts cannot result in a violation because none of the constituents in the UCMR are currently regulated (i.e., no MCLs, action levels, or other standards apply).

Although the UCMR contaminants have no standards associated with them, the data from this monitoring will need to be reported in the annual CCR. Therefore, the CCR will need to address implications of any constituents found above detection limits. Reporting UCMR results in the CCR would also fulfill the notification requirements for "unregulated contaminants" included in the recently promulgated Public Notification Rule.

Note that the UCMR is an ongoing part of the regulatory development process that will be repeated every 5 years. Utilities will be performing similar mandatory sampling for a new list of constituents every 5 years.

UCMR3 was signed by USEPA Administrator Lisa P. Jackson on April 16, 2012. As finalized, UCMR3 requires monitoring for 30 contaminants using USEPA and/or consensus organization analytical methods during 2013–2015. Together, USEPA, states, laboratories, and PWSs will participate in UCMR3.

Operator Certification

Amendments to the 1996 SDWA required USEPA to develop national guidance for operator certification. The final rule was published on February 5, 1999, and became effective on February 5, 2001. State operator certification programs were required to address nine baseline standards, including operator qualifications, certification renewal, and program review. Indirect impacts of the rule on most water utilities include availability of Drinking Water State Revolving Fund (DWSRF) money and perhaps some slight modifications in paperwork/record-keeping requirements.

Arsenic MCL

The MCL for arsenic was reduced from 50 µg/L to 10 µg/L in the *Federal Register* published on January 22, 2001. This was the second time USEPA has established an MCL that was higher than the technically feasible level (3 µg/L), with the first being the uranium rule in 2000. The original SDWA required the MCL to be set as close to the health goal (zero for arsenic and all other suspected carcinogens) as technically feasible. Amendments to the SDWA allowed USEPA the discretion to set the MCL above the technically feasible level.

The final rule, including the revised MCL, became effective 3 years after the rule was published.

Radionuclides Rule

The Radionuclide Rule was published in December 2000. In the final rule, USEPA maintained the gross alpha MCL at 15 pCi/L MCL, 4 mrem/yr for beta emitters, 4 mrem/yr for photon emitters, and 5 pCi/L for combined radium 226 and 228 isotopes, and an MCL for uranium of 30 µg/L.

Analytical Methods

Each of the individual USEPA regulations contains its own information regarding analytical methods approved for compliance monitoring. These and other approved analytical methods are compiled in a final rule titled "Analytical Methods for Chemical and Microbiological Contaminants and Revisions to Laboratory Certification Requirements," published December 1, 1999. These analytical methods were approved for compliance monitoring effective January 3, 2000. The USEPA-approved methods include analytical procedures developed by USEPA, plus procedures developed by others that USEPA endorses, including specific procedures developed by the American Society for Testing and Materials (ASTM) and some specific procedures included in *Standard Methods for the Examination of Water and Wastewater*, published jointly by the American Public Health Association (APHA), AWWA, and the Water Environment Federation (WEF).

Currently, only approved analytical methods can be used for compliance monitoring. In the future, USEPA hopes to implement a performance-based measurement system that will allow utilities to use alternative screening methods instead of requiring only USEPA-approved reference methods. The 1996 SDWA amendments require USEPA to review new analytical methods that may be used for the screening and analysis of regulated contaminants. After this review, USEPA may approve methods that may be more accurate or cost-effective than established methods for compliance monitoring. These screening methods are expected to provide flexibility in compliance monitoring and may be better and/or faster than existing analytical methods.

The approval of new drinking water analytical methods can be announced through an expedited process in the *Federal Register*. This allows laboratories and water systems more timely access to new alternative testing methods than the traditional rule-making process. If alternate test procedures perform the same as or better than the approved methods, they can be considered for approval using the expedited process.

Study Questions

1. If a water system collects at least 40 samples per month for the analyses of total coliforms, what percentage of total coliform positive samples are acceptable for the system to remain in compliance with the maximum contaminant level for total coliforms?
   a. No more than 2%
   b. No more than 3%
   c. No more than 4%
   d. No more than 5%

2. Water systems are required to achieve at least _________ removal and/or inactivation of viruses between a point where the raw water is not subject to recontamination by surface water runoff and a point downstream before or at the first customer.
   a. 2 log
   b. 2.5 log
   c. 3 log
   d. 4 log

3. Where are the sampling points located for required sampling of organics (except trihalomethanes) in a community water system?
   a. Representative points within the distribution system
   b. 75% at locations representative of population distribution and 25% at the farthest points in the distribution system
   c. Entry points to the distribution system
   d. Entry points to the distribution system and representative points within the distribution system

4. Where are the sampling point(s) located for required sampling of natural radionuclides in a community water system?
   a. Consumer's faucet
   b. Representative points within the distribution system
   c. Each entry point to the system
   d. 75% at locations representative of population distribution and 25% at the farthest points in the distribution system

5. Continuous chlorine residual monitoring is required where the water enters the distribution system under the Surface Water Treatment Rule when the
   a. population served is >3,300 people.
   b. population served is >10,000 people.
   c. number of taps is >1,000.
   d. number of taps is >2,500.

6. A community water system must post the Consumer Confidence Report on a publicly accessible website if it serves more than
   a. 10,000 people.
   b. 25,000 people.
   c. 50,000 people.
   d. 100,000 people.

7. Public water systems that cannot meet the required removal of total organic compounds (TOCs) can comply if the source-water TOC level is ________, calculated quarterly as a running annual average.
   a. <1.0 mg/L
   b. <2.0 mg/L
   c. <2.5 mg/L
   d. <5.0 mg/L

8. The Surface Water Treatment Rule does not specify which of the following treatment technology?
   a. Conventional treatment
   b. Membranes
   c. Slow sand filtration
   d. Chloramine disinfection

9. A regulated organic contaminant concentration that exceeds the MCL
   a. is a Tier 1 violation requiring public notification within 24 hours.
   b. is a Tier 2 violation requiring public notification within 30 days.
   c. is a Tier 3 violation requiring public notification within 12 months.
   d. does not require public notification.

10. Operators must balance satisfying regulations on the amount of certain chlorinated organic compounds with
    a. controlling water hardness.
    b. meeting disinfection limits.
    c. keeping turbidity below 0.1 ntu.
    d. reducing iron levels below the MCL.

11. A water system using conventional filtration treatment does not have to use enhanced coagulation to achieve the total organic carbon (TOC) percent removal if
    a. the total trihalomethanes running annual average is less than or equal to 0.030 mg/L.
    b. the sum of haloacetic acids is less than or equal to 0.040 mg/L.
    c. the treated water quarterly running average TOC is less than 4.0 mg/L.
    d. the treated water quarterly running average TOC is less than 2.0 mg/L.

12. The Stage 1 DBPR established an MCL of ______ for TTHM (total trihalomethanes).
    a. 2.10 mg/L
    b. 0.08 mg/L
    c. 1.05 mg/L
    d. 1.00 mg/L
13. The Safe Drinking Water Act defines a public water system as a supply of piped water for human consumption that has at least ______ service connections, or serves ______ or more persons 60 or more days each year.
    a. 15; 25
    b. 25; 15
    c. 20; 50
    d. 10; 25
14. Under the enhanced coagulation requirements of the DBPR, treatment plants must remove specific percentages of total organic carbon based on their source water total organic carbon and what other factor?
15. The Filter Backwash Recycle Rule applies what sizes of systems?
16. Which rule most directly guards against the presence of *Escherichia coli* in the water supply?

Chapter 4

Water Quality Testing

Water Quality Monitoring

All public water systems monitor water quality to some extent. Small systems with consistently good-quality water from deep wells may need to provide only occasional monitoring. Because surface water is more prone to variations in water quality, systems using surface-water sources are required to monitor their water on a more frequent or continuous basis than systems using groundwater.

Water quality is monitored to meet federal, state, and local requirements and for process control. The contaminants that are monitored under US Environmental Protection Agency (USEPA) requirements are extensive, and public water systems must monitor water quality to ensure proper and economic treatment as well as to comply with regulations.

Sampling

Importance of Sampling

Sampling is a vital part of monitoring the quality of water in a water treatment process, distribution system, and supply source. However, errors occur easily when recording water quality information. Every precaution must be taken to ensure that the sample collected is as representative as is feasible of the water source or process being examined.

Water treatment decisions based on incorrect data may be made if sampling is not correctly performed. Representative analytical results depend on the water treatment plant operator ensuring the following:

- The sample is representative of the water source under consideration.
- The proper sampling techniques are used.
- The samples are protected and preserved until they are analyzed.
- The proper sample containers are used.

Types of Samples

Waterworks operators collect grab samples and composite samples depending on the requirements of the operation or on regulations.

Grab Samples A grab sample is a single water sample collected at any time. Grab samples show the water characteristics at the time the sample was taken. A grab sample may be preferred over a composite sample when

- the water to be sampled does not flow on a continuous basis,
- the water's characteristics are relatively constant, or

grab sample
A single water sample collected at one time from a single point.

- the water is to be analyzed for water quality indicators that may change with time, such as dissolved gases, coliform bacteria, residual chlorine, disinfection by-products, temperature, volatile organics, certain radiological parameters, and pH.

Figures 4-1 and 4-2 illustrate this point. Figure 4-1 shows the changes in surface water **dissolved oxygen (DO)** over a 24-hour period. A grab sample represents the DO level only at the time the sample was taken. DO can change rapidly—for example, because of the growth of algae or plants in the water (diurnal effect). Online process instruments are good examples of instruments that perform grab sample analyses; they analyze a continuous string of grab samples and produce a series of individual analyses that, when plotted, illustrate trends such as those in the figures.

Figure 4-2 shows that levels of total dissolved solids (TDS) in the same water change very little. A grab sample can be representative of the water quality in a stable supply such as a deep well for perhaps a month. TDS levels are a function of the minerals dissolved from rocks and soil as the water passes over or through them and may change only in relation to seasonal runoff patterns. TDS in groundwater (e.g., wells) may also change if certain water-bearing zones in the well become plugged, changing the dilution or zones from which the water is being drawn.

dissolved oxygen (DO)
The oxygen dissolved in water, wastewater, or other liquid, usually expressed in milligrams per liter, parts per million, or percent of saturation.

Composite Samples In many processes, water quality changes with time. A continuous sampler–analyzer provides the most accurate results in these cases. Often the operator is the sampler–analyzer, and continuous analysis could prove costly. Except for tests that cannot wait because of rapid physical, chemical, or biological changes of the sample (such as tests for DO, pH, and temperature), a fair compromise may be reached by taking samples throughout the day at hourly or 2-hour intervals. Each sample should be refrigerated immediately after collection. At the end of 24 hours, each sample is vigorously mixed and a portion of each sample is then withdrawn and mixed with the other samples. The size of the portion is in direct proportion to the flow when the sample was collected (aliquot)

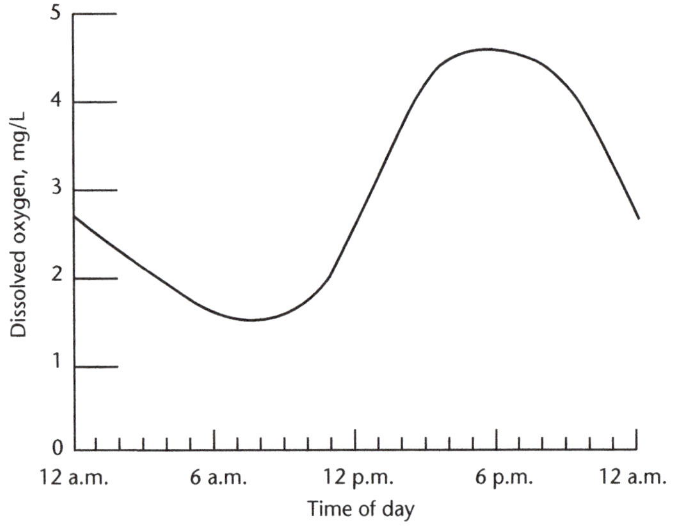

Figure 4-1 Example of hourly changes in dissolved oxygen for a surface water source

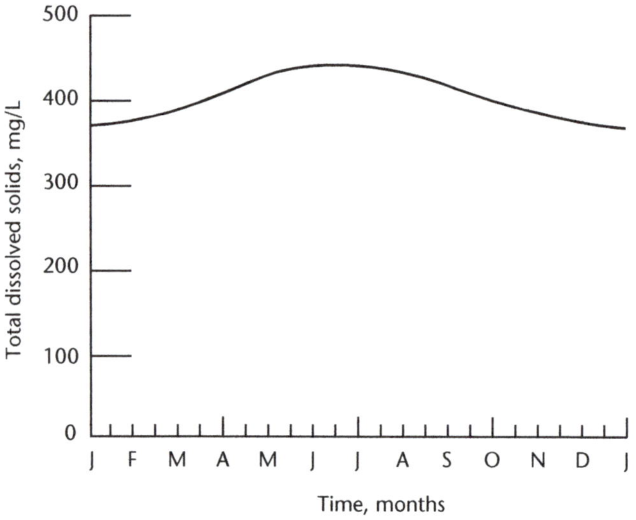

Figure 4-2 Example of monthly changes in total dissolved solids for the surface water source shown in Figure 4-1

and the total size of sample needed for testing. For example, if hourly samples are collected when the flow is 1.2 mgd, use a 12-mL portion of the sample, and when the flow is 1.5 mgd, use a 15-mL portion of the sample. The resulting mixture of portions of samples is a composite sample. *In no instance should a composite sample be collected for bacteriological examination.*

When the samples are taken, they can either be set aside or combined as they are collected. In both cases, they should be stored at a temperature of less than 40°F (4°C) but above freezing until they are analyzed.

Continuous Sampling This type of sampling is used in online or process control sampling devices/instruments. Some of the new regulations call for this type of sampling for the larger systems for chlorine residual under the new Ground Water Rule (GWR) and for surface water filtration or groundwater under the direct influence of surface water (GWUDI) filtration systems. Continuous sampling is also being used in certain circumstances to monitor distribution systems for chlorine levels and for other parameters associated with security monitoring. It is also used by larger systems on the incoming surface water for turbidity, pH, and streaming current measurements for treatment control.

As technology becomes more sophisticated and affordable, this type of monitoring will become more prevalent in the industry. Some systems use this technology to monitor levels of nitrate and other specific ions during the treatment process and fluoride levels of water leaving the treatment facility. Examples of the online instruments are shown in Figures 4-3 and 4-4.

Sampling Point Selection

Careful selection of representative sample points is an important step in developing a sampling procedure that will accurately reflect water quality. The criteria used to select a sample point depend on the type of water sampled and the purpose of the testing. Check with primacy regulations as to *compliance samples* versus *process samples*. Any sample taken from a *compliance sample tap* may have

Figure 4-3 Online chlorine residual analyzer
Courtesy of HACH Inc.

Figure 4-4 Online particle counter
Courtesy of HACH Inc.

composite sample
A series of individual or grab samples taken at different times from the same sampling point and mixed together.

to be reported as a *performance sample* even if it is just being collected for process control. Samples are generally collected from three broad types of areas:

- Raw-water supply
- Treatment plant
- Distribution system

Raw-Water Sample Points The choice of collection points for raw-water samples depends on the type of system being sampled. There are at least three general types of systems:

- Raw-water transmission lines
- Groundwater (wells)
- Rivers, reservoirs, and lakes

Raw-water transmission lines and groundwater sources are sampled directly from the transmission line or well-discharge pipe. After a sampling point has been selected (prior to any chemical addition or treatment), the pipeline is equipped with a small sample valve or tap, often called a *sample cock* (Figure 4-5). The valve must be fully opened before sampling to flush out any standing water and accumulated sediment. The flow may then be adjusted to achieve the optimal flow for the type of sample being collected. For example, a slow, steady stream to prevent aeration is best for analysis of volatile organics or DO.

Most of the physical factors known to promote mixing in surface waters are absent or are much less effective in groundwater systems. Wells usually draw water from a considerable thickness of saturated rock and often from several different strata. These water components are mixed by the turbulent flow of water in the well before they reach the surface and become available for sampling. Most techniques for well sampling and exploration are usable only in unfinished or nonoperating wells. Usually the only means of sampling the water tapped by a well is to collect a pumped sample. The operator is cautioned to remember that well pumps and casings can contribute to sample contamination. If a pump has not run for an extended period of time prior to sampling, the water collected may not be representative of the normal water quality. Good records of static and pumping levels of the wells should be kept to determine if the well is performing as designed and when the sample should be drawn to be representative of the water in the well column.

Rivers To adequately determine the composition of a flowing stream, each sample (or set of samples taken at the same time) must be representative of the entire flow at the sampling point at that instant. The sampling process must be repeated at a frequency sufficient to show changes of water quality that may occur over time in the water passing the sampling point.

Figure 4-5 Sample cock attached to pipeline for sampling

On small or medium-size streams, it is usually possible to find a sampling point at which the composition of the water is presumably uniform at all depths and across the stream. Obtaining representative samples in these streams is relatively simple. For larger streams, more than one sample may be required. A portable conductivity meter is very useful in selecting good sample sites.

Reservoirs and Lakes Water stored in reservoirs and lakes is usually poorly mixed. Thermal stratification and associated depth changes in water composition (such as DO) are among the most frequently observed effects. Single samples can therefore be assumed to represent only the spot of water from which the sample came. Therefore, several samples must be collected at different depths and from different areas of the impoundment to accurately sample reservoirs and lakes. See Figures 4-6 and 4-7.

Treatment Plant Sample Points Treatment plants are sampled to evaluate the treatment efficiency of unit processes or to evaluate operational changes. Selection of in-plant sample points is an important step in developing an overall process control program for a water treatment plant. Samples from the points selected can be tested to determine the efficiency of the various treatment processes. The test results will also help to indicate operational changes that will improve contaminant removal efficiencies or reduce operating costs.

Collection of representative samples in the water treatment plant is similar to sample collection in a stream or river. The operator must ensure that the water sampled is representative of the water passing that sample point. In many water plants, money has been spent to purchase sample pumps and piping only to find that the sample from that point is not representative of the passing water. A sample tap in a stagnant area of a reservoir or on the floor of a process basin serves little purpose in helping the plant operator with control of water quality. The operator is urged to ensure that each and every sample point is located to provide useful and representative data. If the sampling point is improperly located, the operator should make arrangements to move the piping to a better location. Multiple sample locations for the same analysis point may be needed if changes in plant conditions, such as flow, affect the quality of the sample. For example, if using a streaming current detector or other instruments such as an oxidation–reduction meter, flow can change mixing times and affect the validity of the value read on the instrument.

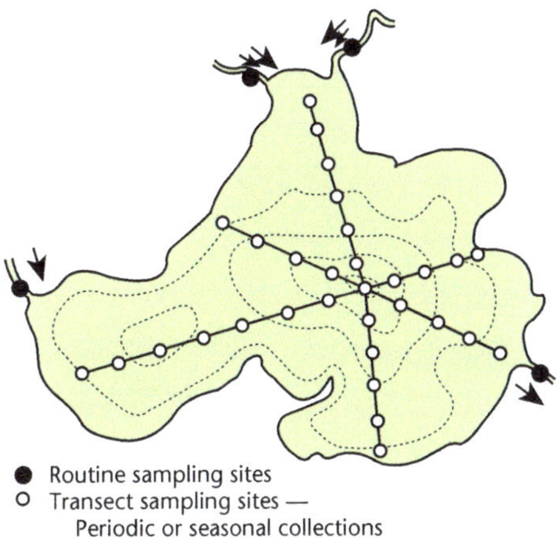

Figure 4-6 Routine and transect sample points in a natural lake

Source: Mackenthun and Ingram. 1967.

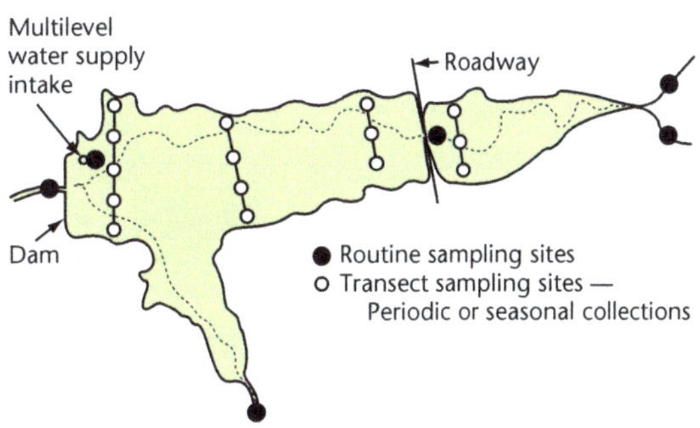

Figure 4-7 Routine and transect sample points in a reservoir

Source: Mackenthun and Ingram. 1967.

Treatment plants vary widely in the kinds of treatment processes used and the configurations of the processes. In general, in-plant sample points are established at every place where, because of a treatment method or group of methods, a measurable change is expected in the treated-water quality. Simply put, if you are adding something to the process, either chemically or mechanically, you should have some way of determining the effect. Other in-plant sampling sites may have to be selected and installed for changes in testing needed by changes in governmental laws and regulations. Figure 4-8 identifies 10 suggested locations where process control samples are routinely collected in a plant employing several different treatment processes. These locations are described as follows:

- Between sample points 1 and 2, test results should show a reduction in algae and the associated tastes and odors (the result of chemical pretreatment), a reduction in sediment load (the result of presedimentation), and a reduction in debris (the result of screening).
- Between points 2 and 3, aeration should cause oxidation of iron and manganese and a significant reduction in undesirable dissolved gases while increasing the oxygen content.
- Between points 3 and 4, the combined effects of coagulation, flocculation, and sedimentation should cause a reduction in turbidity and color.
- Water quality changes between sample points 4 and 5 will allow the operator to monitor the effectiveness of the softening process.
- Sample points 5 and 6 allow monitoring of the efficiency of filtration in removing turbidity and previously oxidized iron and manganese as well as the reduction in pathogenic organisms.
- Sampling at points 6 and 7 will indicate the efficiency of the adsorption process in removing organic chemicals.
- Point 8 is used for the measurement of fluoride concentration to ensure that water entering the distribution system contains the proper level.
- Sampling at point 9 will provide a final check on pH and alkalinity for corrosion control.
- Point 10 is used for monitoring chlorine residual, turbidity, and the presence of coliform bacteria in the finished water.

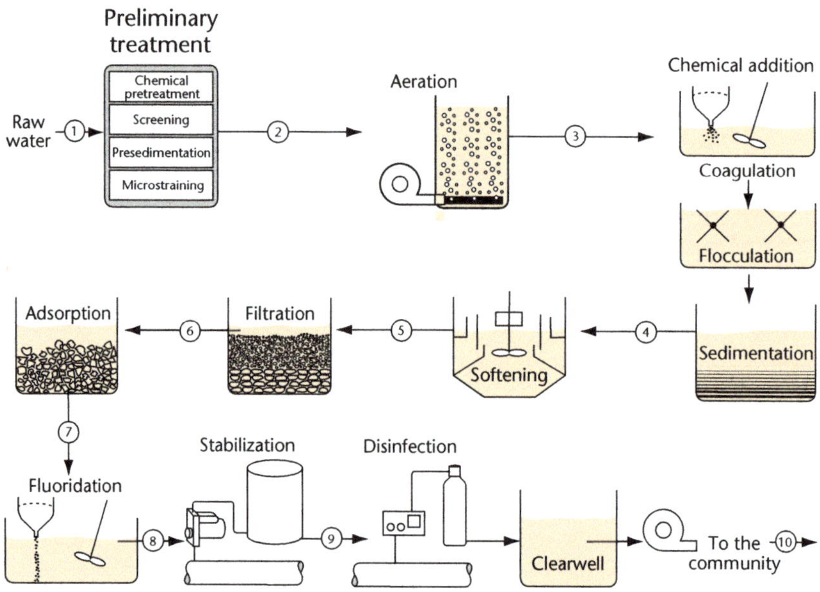

Figure 4-8 Suggested in-plant sample points (indicated by numbered circles)

In the selection of in-plant sample collection points, certain precautions should be kept in mind. Sampling immediately after chemical addition or at points immediately downstream from chemical additions should be avoided because proper mixing and reaction may not have had time to take place. Sampling of combined filter effluent turbidity should take place before post-treatment chemicals are added to avoid measuring added turbidity, such as from lime or phosphates. Operators should always take samples from the main stream of flow, avoiding areas of standing water, algae mats, and floating or settled debris.

Finished-water sample points are normally established downstream of the final treatment process at or just before the point where the water enters the distribution system, such as the point of discharge from the clearwell. For example, turbidity samples required by the National Primary Drinking Water Regulations must be collected before the water enters the distribution system while avoiding an area where added chemicals (e.g., lime or corrosion inhibitors) may affect the results.

Distribution System Sample Points Representative sampling in the distribution system is an indication of system water quality. Results of sampling should show if there are quality changes in the entire system or parts of it, and they may point to the source of a problem (such as tastes and/or odors). Sampling points should be selected, in part, to trace the course from finished-water source (at the well or plant) through the transmission mains, and then through the major and minor piping of the system. A sampling point on a major transmission main, or on an active main directly connected to it, would be representative of the plant effluent water quality.

Sample points in the distribution system are used to determine the quality of water delivered to consumers. In some cases, the distribution system samples may be of significantly different quality than samples of finished water at the point of entry to the distribution system. For example, corrosion in distribution system pipelines can cause increases in water color, turbidity, taste and odor, and physical constituents such as lead and copper. Microbiological growth may also be taking place in the water mains, which degrades water quality. In addition, a cross-connection between the distribution system and a source of contamination can result in chemical or microbiological contamination of the water in the system.

The inlet and outlet of storage tanks are good sampling sites because they give information about the effects of storage. The direction of flow should be noted when sampling; that is, it should be noted whether the tank is filling or emptying at the time of sampling.

Most of the samples collected from the distribution system will be used to test for coliform bacteria and chlorine residual. Others may be used to determine water quality changes. Still others will be used to test for maximum contaminant levels (MCLs) of inorganic and organic contaminants and for compliance with the Lead and Copper Rule, as required by the applicable drinking water standards. Distribution system sampling should always be performed at locations representative of conditions within the system. Radiological samples must be from the source water supply under current regulation. Bacteriological sampling and the associated chlorine residual testing are regulated.

There are two major considerations in determining the number and location of sampling points, other than those required by regulation:

- They must be representative of each different source of water entering the system.
- They must be representative of conditions within the system, such as dead ends, loops, storage facilities, and pressure zones.

The precise location of sampling points depends on the configuration of the distribution system.

In general, samples should not be taken from customers' taps. The exception is for lead and copper testing and for water quality complaint sampling.

Frequency of sampling is based on population served.

WATCH THE VIDEO
Sampling Procedures (www.awwa.org/wsovideoclips)

Collection of Samples

The steps described in the following sections are general sample collection procedures that should be followed regardless of the constituent tested. Special collection procedures required for certain tests are described later in the chapter.

Note that the volume of sample will vary according to the testing procedures used and often is set by regulations. For example, coliform compliance sampling requires a 100-mL grab sample. The laboratory analyst will provide instructions for sample volume.

Only containers designed for water sampling and provided by the laboratory should be used. Mason jars and other recycled containers cannot be trusted to function properly no matter how well they are cleaned, and they are generally not accepted by a laboratory for water analysis. Some laboratories reuse sample containers by washing them under carefully controlled conditions and sterilizing them prior to reuse. In other cases, it has been found more economical to dispose of used bottles and provide only new ones for collection.

When a container with a screw-on lid is used, the lid should be removed and held threads down while the sample is collected in the container. The lid can easily be contaminated if the inside is touched or if it is set facedown or placed in a pocket. A contaminated lid can contaminate the sample, which will necessitate resampling, wasting a great deal of time and expense.

Raw-Water Sample Collection

If no raw-water sample tap is available and the sample must be taken from an open body of water, the following procedures should be used.

On a well supply, if no raw-water sample tap is available, the well should be put to waste with any treatment shut off, the samples collected, the treatment restarted, and the well placed back in service. A clean, wide-mouth sampling bottle should be used for raw-water sampling. The bottle should not be rinsed; this is especially important if the bottle has been pretreated or contains a preservative. The open bottle should be held near its base and plunged neck downward below the surface of the water. The bottle should then be turned until the neck points slightly upward for sampling, with the mouth directed toward any current present. Care must be taken to avoid floating debris and sediment. In a water body with no current, the bottle can be scooped forward to fill the bottle. Once the bottle has been filled, it is retrieved, capped, and labeled.

If the sampler is wading, the sample bottle should be submerged upstream from that person. If a boat is being used for stream sampling, the sample should be taken on the upstream side.

When samples are being taken from a large boat or a bridge, the sample bottle should be placed in a weighted frame that holds the container securely. The opened bottle and holder are then slowly lowered toward the water with a rope or with the handle that comes with certain devices available through water supply equipment catalogs, DO sampling cans, "swing samplers" on poles, long-handled dippers, or weighted bailers. When the bottle or sample device approaches the surface, the unit is dropped quickly into the water. Slack should not be allowed in the rope because the bottle could hit bottom and break, or it could pick up mud and silt. After the bottle is filled, it is pulled in, capped, and labeled. There are also specialized sampling devices to be used as required for specific samples. For example, for DO, the device with the sample bottles is lowered into the water and then the stopper is remotely removed, the sample container is filled, and the stopper is replaced before the unit is removed from the water. This type of device can also be used when sampling at a certain depth to ensure the water is from the zone desired.

Treatment Plant Sample Collection

The procedure used to collect samples from an open tank or basin or in an open channel of moving water is essentially the same as for raw-water sampling. Treatment plants should be equipped with sample taps. These faucets provide a continuous flow of water from various locations in the treatment plant, including raw-water sources. In some plants these taps do not run continuously because of operational constraints, so the operator may have to turn the taps on and run the water for a specific period of time to obtain a representative sample. To collect a sample, the operator or laboratory technician draws the required volume from the sample tap. Figure 4-9 shows a typical bank of sample faucets in a laboratory.

Distribution System Sample Collection

Once the distribution system sample locations have been selected, sample collection consists of a few simple, carefully performed steps. First, the faucet is turned on and set to produce a steady, moderate flow of water (Figure 4-10). If a steady flow cannot be obtained, the tap should not be used. The water is allowed to run long enough to flush any stagnant water from the service line. (The important exception to this procedure is with samples collected for lead and copper analyses; these must be first-draw samples collected immediately after the faucet has been opened.) Depending on the length of the service line, as mentioned before, this process can take from 2 to 5 minutes or longer. The line is usually flushed when the water temperature changes (depending on climate and source, the temperature may increase or decrease) and stabilizes. The sample is then collected

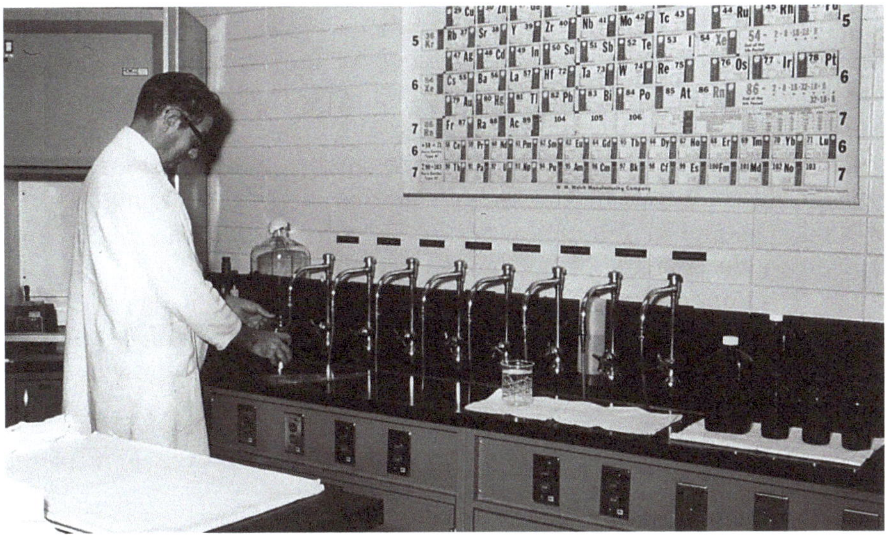

Figure 4-9 Sample faucets in a laboratory

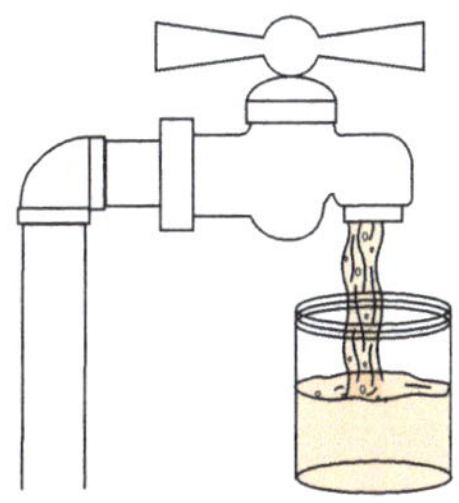

Figure 4-10 Sample faucet should be set to produce a steady, moderate flow

without the flow changing. The sample bottle lid should be held threads-down during sample collection and replaced on the bottle immediately. If the lid must be set down during the sampling process, it should be placed threads-up and protected from splatter or falling matter (rain, for example) that could contaminate the sample. The final step after sampling is to label the bottle.

A once common practice was to flame the outside of a faucet. This procedure is no longer recommended. Experience showed that the flame could not be held on the faucet long enough to kill all the bacteria on the outside of the faucet without potential damage to the faucet. Many faucets are now made partly or entirely of plastic, which will quickly melt if high heat is applied. The current common practice is to dip the faucet into a small container of bleach or swab it with bleach or other approved disinfectant.

Samples should not be collected from sill cocks or other faucets with hose threads unless local regulations require it and the threads can be thoroughly cleaned. Because of the way they are constructed, these faucets do not usually throttle to a smooth flow. Also, if any water splashes up onto the threads and then drains into the sample bottle, it will bring with it contaminants from the outside of the faucet.

Monitoring for Chemical Contaminants

Drinking water may contain contaminants considered a threat to the public. The contaminants of concern may occur naturally in the water, be human-made, or be formed during the water treatment process. The chemicals are broken into four general classes for regulation:

1. Inorganic chemicals
2. Synthetic organic chemicals
3. Volatile organic chemicals
4. Radionuclides

The need to establish regulations for new chemical contaminants has presented USEPA with the problem of creating, adapting, proving, and promulgating *analytical techniques*. The method must be for a specific chemical and not for a "contaminant" defined by a physical description such as a boiling fraction An example of this distinction can be found in kerosene, which is a member of the petroleum distillates; these chemicals boil at between 150°C and 275°C (300–530°F), which results in a mixture of organic chains of 6 to 16 carbon atoms in each compound. Before a requirement to monitor for a contaminant can be imposed, the testing methods must be developed to ensure that an adequate number of laboratories will be available to perform the tests and that they will get consistent, reliable results. Many of the chemicals now being added to the list of regulated contaminants must be analyzed at the parts-per-billion level or in even smaller concentrations.

Faucets selected should be on the lines connected directly to the main. Only cold-water faucets should be used for sample collection. A sampling faucet must not be located too close to a sink bottom. Contaminated water or soil may be present on the exteriors of such faucets, and it is difficult to place a collection bottle beneath them without touching the neck's interior against the faucet's outside surface. In most instances samples should not be taken from the following types of faucets (Figure 4-11):

- Leaking faucets, which allow water to flow out around the stem of the valve and down the outside of the faucet
- Faucets with threads

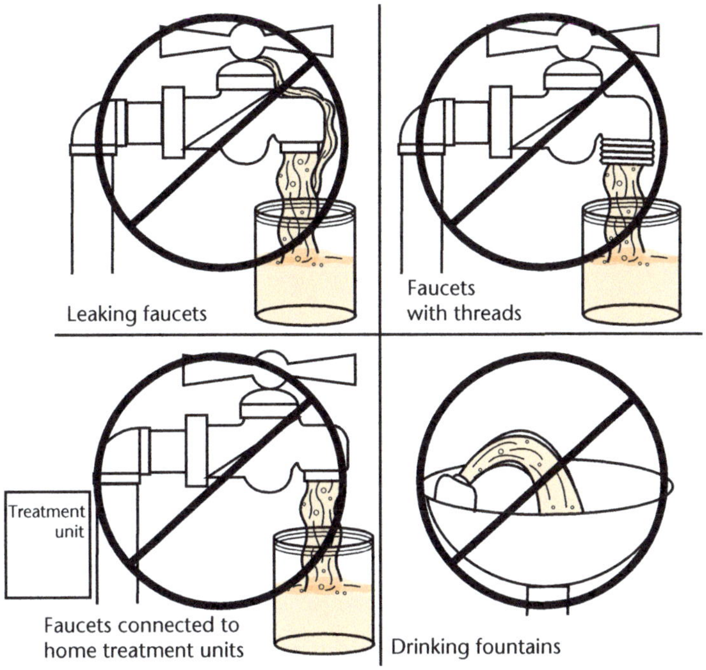

Figure 4-11 Types of faucets that should not be used for sampling

- Faucets connected to home water-treatment units, including water softeners and hot-water tanks
- Faucets that swivel, since the swivel joint may act as a siphon and introduce contamination
- Faucets with single-lever handles that do not guarantee that only the cold-water sample is being selected

Laboratory Certification

Each of the approximately 155,000 public water systems affected by the Safe Drinking Water Act (SDWA) must routinely monitor water quality to determine if the water is adequately protected from regulated microbiological, chemical, and radiological contaminants. It is imperative that the analyses for all of this monitoring be performed by standard methods approved for compliance testing so that the results are comparable for all systems. Consequently, states are required by federal regulations to consider analytical results from water systems only if samples have been analyzed by a certified laboratory. Some exceptions are measurements for turbidity, chlorine residual, temperature, and pH, which may be performed by a person acceptable to the state using approved equipment and methods.

Federal regulations require each state with primary enforcement responsibility to have available laboratory facilities that have been certified by USEPA, with capacities sufficient to process samples for water systems throughout the state. Certified laboratories fall into the following general classes:

- State-operated laboratories
- Water-system laboratories
- Commercial laboratories

In most states, the necessary capacity is provided by a combination of all three types of laboratories. Some laboratories may be certified to perform only one type of analysis; for instance, some laboratories are set up to handle only microbiological analyses. Analyses requiring expensive equipment and highly trained technicians, such as for organic chemical and radiological monitoring, are also generally handled by specialized laboratories.

Consistency among laboratories in analytical results is overseen by USEPA and by state primacy programs for each of the types of analyses for which the laboratories are certified. Periodically, an independent vendor contracted by USEPA provides to each laboratory carefully prepared proficiency testing samples containing a known concentration of a contaminant. The values of the samples are unknown to the laboratory, and its staff must be able to determine the contaminant concentration within an appropriate tolerance to maintain the laboratory's certification. The results determined by the laboratory are submitted to the agency having primacy for laboratory certification for the particular state.

Historically, most states have operated their own laboratories to process water system samples. But the number of samples has increased severalfold in recent years, so it is difficult for the states to continue providing laboratory service with state funding only. Some states have instituted charges to water systems to help fund the laboratory services. Other states process only a certain number of samples from any one water system, and if more are required, commercial laboratories must be used.

Record Keeping and Sample Labeling

Records should be kept for every sample that is collected. A sample identification label or tag should be filled out at the time of collection. Each label or tag should include at least the following information:

- Water utility name
- Water system's public water system identification number
- Date on which the sample was collected
- Time when the sample was collected
- Location where sample was collected
- Type of sample—grab or composite
- Tests to be run
- Name of person sampling
- Preservatives used
- Bottle number

The samples provided to laboratories should always be clearly labeled. The information on the label should also be entered on a record-keeping form that is maintained as a permanent part of the water system's records and placed on the chain-of-custody forms submitted to the laboratory. Each laboratory may have its own forms that request the required information for compliance with regulations.

Sample Preservation, Storage, and Transportation

Samples cannot always be tested immediately after they are taken. Ensuring that the level of the constituent remains unchanged until testing is performed requires careful attention to techniques of sample preservation, storage, and transportation. It is also extremely important that records be kept of the chain of custody of samples collected for SDWA compliance.

Preservation and Storage

After a sample has been collected, its quality may change because of chemical and/or biological activity in the water. Some characteristics (alkalinity, pH, dissolved gases, disinfectant residuals, temperature, and odor) can change quickly and quite significantly, and so samples to be analyzed for these parameters should not be stored under any conditions. The tests for disinfectant residuals, pH and temperature must be completed in the field at the time of collection. Other parameters, such as pesticides and radium, change more slowly and much less noticeably, and these samples can usually be stored for considerable lengths of time if necessary.

To extend the storage time of samples requiring chemical analysis, sample-preservation techniques have been developed that slow the chemical or biological activity in the sample. These techniques allow the sample to be transported to the laboratory and tested before significant changes occur.

Sample preservation usually involves the following steps:

- Refrigeration and/or
- Chemical preservatives

For some samples storage time can be prolonged by keeping samples refrigerated until the analysis is performed. In some cases, it is recommended that samples be transported or shipped to the laboratory in a portable cooler containing an ice pack.

Often the laboratory provides bottles for specific analyses with the preservative already added. To avoid loss of preservative during sampling, it is important not to allow these containers to overflow as they are filled. These containers must also be kept out of the reach of children, because the preservative material could be harmful to a child who opens a container. Inorganic metals, for example, are preserved by concentrated nitric acid to reduce the pH to less than 2. If preservatives are to be added by the sampler, specific instructions on the procedures should be obtained from the laboratory that will perform the analyses.

Most laboratories do not maintain a full staff on weekends, so they generally request that samples with short holding times, such as bacteriological samples, be collected and shipped early in the week. If a sample arrives on a weekend and cannot be processed, the delay will probably exceed the required holding time and the sample will be rejected. However, most laboratories accept emergency samples on weekends.

Samples that must be submitted within a specified compliance period should generally be collected and sent to the laboratory early in the compliance period. Some of the problems that can require resampling are described in the following list:

- The sample is frozen or broken during shipment.
- The sample is lost or delayed in shipment and arrives at the laboratory after the specified holding time has elapsed.
- The laboratory makes an error in processing the sample.
- The laboratory analysis is inconsistent or shows a concentration greater than the MCL, and another sample to confirm the results is required.
- The sample was not properly preserved or was too warm, or no preservative was present.
- The sample container was not the proper container for the required sample, size, volume, or material.

Sampling early in the compliance period ensures that time is available for one or more resamplings, if necessary, before the end of the period. If resampling has not been completed before the end of a compliance period, a water system is usually deemed out of compliance and will be instructed by the state to provide public notification.

Transportation

If samples arrive at a laboratory past the specified holding time following collection, the laboratory must reject the samples. New sample bottles must then be shipped to the water system and another set of samples will have to be collected and shipped back.

The mail is usually the best and easiest method of shipment, except for microbiological or certain radiological samples that require delivery within about a 24-hour period. If regular mail service fails to deliver samples reliably within the required time period, overnight shipping services or package delivery services may be tried. In some cases, changing to a laboratory at a different location may improve delivery time. Some water system operators who are located near a laboratory have found it best just to drive the samples directly to the laboratory or arrange with the laboratory for pickup of the samples as part of the analysis price. Depending on distance and availability of personnel, a bonded courier service may be used.

If samples are shipped, it is important to make sure the bottle caps are tight to prevent leakage. Systems that have had bottle caps loosen during shipment have found that wrapping the lids with electrical or packing tape is an easy method of further securing them. Samples must be packed in a sturdy container with enough cushioning material to prevent breakage. The box should be marked to indicate which end is up, that the contents are fragile, that they must not be allowed to freeze, and that priority should be given to the shipment.

Chain of Custody

As more and more parameters are added to the list of regulated and unregulated contaminants, and with the MCLs and MCLGs in the micrograms per liter range, the practice of good quality assurance and quality control (QA/QC) procedures becomes very important. One essential part of QA/QC is maintaining a written record of the sample's history of SDWA compliance, from the time of collection to the time of analysis and subsequent disposal. This record, called the **chain of custody**, is important if the analyses are ever challenged and need to be defended. Chain-of-custody requirements vary by state, so water system operators should be sure that the requirements for their state are being met.

Field Log Sheet

One method of establishing the chain-of-custody record is to use a daily field log sheet, which should contain the following information:

- Date on which the samples were collected
- Name of the sampler
- List of all the samples collected by the sampler on this date
- List of all the sample locations for this date
- Time of day when each sample was collected
- Comments concerning any unusual situations
- Signature of the individual receiving the samples from the sampler

chain of custody
A written record of the sample's history of SDWA compliance, from the time of collection to the time of analysis and subsequent disposal.

- Date and time when the samples were received by the laboratory
- Location or identification of the laboratory

This log sheet states that the samples were in the custody of the sampler until they were turned over to the shipper. The laboratory record then follows the history of the sample to disposal.

Sampler's Liability

If the results of an analysis of a specific sample are ever questioned, the sampler will be asked to verify that the sample was in his or her custody until it was turned over or sent to the laboratory. The sampler will be asked to verify that the sample was collected, stored, and transported using proper procedures and that no other person could have in any way altered the concentrations of any contaminant(s) present.

Sampler's Responsibility

The sampler has the basic responsibility to ensure that the sample is properly collected, labeled, stored, and transported to the laboratory. The sample collector must be able to testify that the sample was under his or her custody at all times. The sample collector is also responsible for knowing and performing the proper sampling routine for each type of analysis required, including preservation.

Microbiological Test Methods

Coliform Analyses

The detection of coliform bacteria in a water sample by any of the four analytical techniques is a warning of possible contamination. One positive test does not conclusively prove contamination, however, and additional tests must be conducted. Samples are often contaminated by improper sampling technique, improperly sterilized bottles, and laboratory error. Regulatory agencies recognize this fact, and drinking water regulations require check or repeat sampling after findings that show a positive test for coliform in a sample. Drinking water regulations and maximum contaminant levels for coliform bacteria are discussed in Chapter 3.

Sampling

Sterile containers must be used for all samples collected for bacteriological analysis. The same sampling procedures should be used for coliform analysis and heterotrophic plate count analysis (see Table 4-1). Samples are collected in sterile bottles (usually 100 mL) containing sodium thiosulfate ($Na_2S_2O_3$) to remove any chlorine residual.

coliform bacteria
A group of bacteria predominantly inhabiting the intestines of humans or animals but also occasionally found elsewhere. Presence of the bacteria in water is used as an indication of fecal contamination (contamination by human or animal wastes).

Test Methods

The MTF and P–A tests are designed on the principle that coliform bacteria produce gas from the fermentation of lactose within 24–48 hours when incubated at 35°C (95°F). Although the bacteria themselves cannot be seen, their presence is signified by the gas that is formed and trapped in an inverted vial in the fermentation tube.

Other testing methods that will be discussed include the MMO–MUG technique, the membrane filter method, and the heterotrophic plate count method.

Table 4-1 Total coliform sampling requirements, according to population served

Population Served	Minimum Number of Routine Samples per Month*	Population Served	Minimum Number of Routine Samples per Month*
15 to 1,000†	1‡	59,001 to 70,000	70
1,001 to 2,500	2	70,001 to 83,000	80
2,501 to 3,300	3	83,001 to 96,000	90
3,301 to 4,100	4	96,001 to 130,000	100
4,101 to 4,900	5	130,001 to 220,000	120
4,901 to 5,800	6	220,001 to 320,000	150
5,801 to 6,700	7	320,001 to 450,000	180
6,701 to 7,600	8	450,001 to 600,000	210
7,601 to 8,500	9	600,001 to 780,000	240
8,501 to 12,900	10	780,001 to 970,000	270
12,901 to 17,200	15	970,001 to 1,230,000	300
17,201 to 21,500	20	1,230,001 to 1,520,000	330
21,501 to 25,000	25	1,520,001 to 1,850,000	360
25,001 to 33,000	30	1,850,001 to 2,270,000	390
33,001 to 41,000	40	2,270,001 to 3,020,000	420
41,001 to 50,000	50	3,020,001 to 3,960,000	450
50,001 to 59,000	60	3,960,001 or more	480

Source: *Water Quality and Treatment*. 1999.

*In lieu of the frequency specified in this table, a noncommunity water system using groundwater and serving 1,000 persons or fewer may monitor at a lesser frequency specified by the state until a sanitary survey is conducted and the state reviews the results. Thereafter, noncommunity water systems using groundwater and serving 1,000 persons or fewer must monitor in each calendar quarter during which the system provides water to the public, unless the state determines that some other frequency is more appropriate and notifies the system in writing. Five years after promulgation of the Total Coliform Rule (TCR), noncommunity water systems using groundwater and serving 1,000 persons or fewer must monitor at least once per year.

†Includes public water systems that have at least 15 service connections but serve fewer than 25 persons.

‡For a community water system serving 25 to 1,000 persons, the state may reduce this sampling frequency if a sanitary survey conducted in the last 5 years indicates that the water system is supplied solely by a protected groundwater source and is free of sanitary defects. However, in no case may the state reduce the sampling frequency to less than once per quarter.

multiple-tube fermentation (MTF) method
A laboratory method used for coliform testing that uses a nutrient broth placed in culture tubes. Gas production indicates the presence of coliform bacteria.

presumptive test
The first major step in the multiple-tube fermentation test. The step presumes (indicates) the presence of coliform bacteria on the basis of gas production in nutrient broth after incubation.

Multiple-Tube Fermentation Method

The multiple-tube fermentation (MTF) method (Figure 4-12) progresses through three distinct steps:

1. Presumptive test
2. Confirmed test
3. Completed test

Presumptive Test The presumptive test is the first step of the analysis. Samples are normally pipetted into tubes containing a culture medium (lauryl tryptose broth [LTB]) with inverted filed vials containing the media in the tubes. The samples are incubated for 24 hours and then checked to see if gas has formed in the inner vial and cloudiness has developed in the broth. If neither is the case, they are incubated for 24 hours more and checked again.

If coliform bacteria are present in the water, the gas they produce will begin to form a bubble in each inverted vial within the 48-hour period; this is called a

Figure 4-12 Typical multiple-tube fermentation setup
Source: Opflow.

positive sample or reported as presence. If no gas forms, the sample is called negative or reported as absence. If gas is produced after either the 24-hour or the 48-hour incubation period, the sample must undergo the confirmed test.

Confirmed Test The confirmed test is more selective for coliform bacteria. This test increases the likelihood that positive results obtained in the presumptive test are caused by coliform bacteria and not other kinds of bacteria. Cultures from the positive samples in the presumptive test are transferred to brilliant green lactose bile (BGB) broth and incubated. If no gas has been produced after 48 hours of incubation, the test is negative and no coliform bacteria are present. If gas is produced, the test is positive, indicating the presence of coliform bacteria.

Bacteriological testing of most public water supplies stops after the confirmed test. This is the minimum testing that all samples must undergo when the MTF method is used. To check its procedures, the laboratory should conduct the completed test on at least 10 percent of the positive tubes from the confirmed test.

Completed Test The completed test is used to definitely establish the presence of coliform bacteria for QC purposes. A sample from the positive confirmed test is placed on an eosin methylene blue (EMB) agar plate and incubated. A coliform colony will form on each EMB plate.

A small portion of the coliform colony is transferred to a growth medium and incubated for 18 to 24 hours. A second portion is transferred to an LTB and incubated for 24–48 hours. The completed test is positive if (1) gas is produced in the LTB and (2) red-stained, non-spore-forming, rod-shaped bacteria are found. If no gas is produced in the LTB or if red-stained, chainlike cocci or blue-stained, rod-shaped bacteria are found on the agar, the test is negative.

Figure 4-13 provides a summary of the MTF method. (The bacteria are not visible to the human eye. A microscope is needed to read the plates and determine the bacteria type.)

Presence–Absence Test

The presence–absence (P–A) test is a simple modification of the MTF method. It is intended for use on routine samples collected from a distribution system or water treatment plant. A 100-mL portion of the sample is inoculated into a 250-mL milk

positive sample or presence
In reference to the multiple-tube fermentation or membrane filter test, any sample that contained coliform bacteria.

negative sample or absence
When referring to the multiple-tube fermentation or membrane filter test, any sample that does not contain coliform bacteria.

confirmed test
The second major step of the multiple-tube fermentation method. This test confirms that positive results from the presumptive test are due to coliform bacteria.

completed test
The third major step of the multiple-tube fermentation method. This test confirms that positive results from the presumptive test are due to coliform bacteria.

presence–absence (P–A) method
An approved bacteriological procedure for detecting the presence or absence of total coliforms.

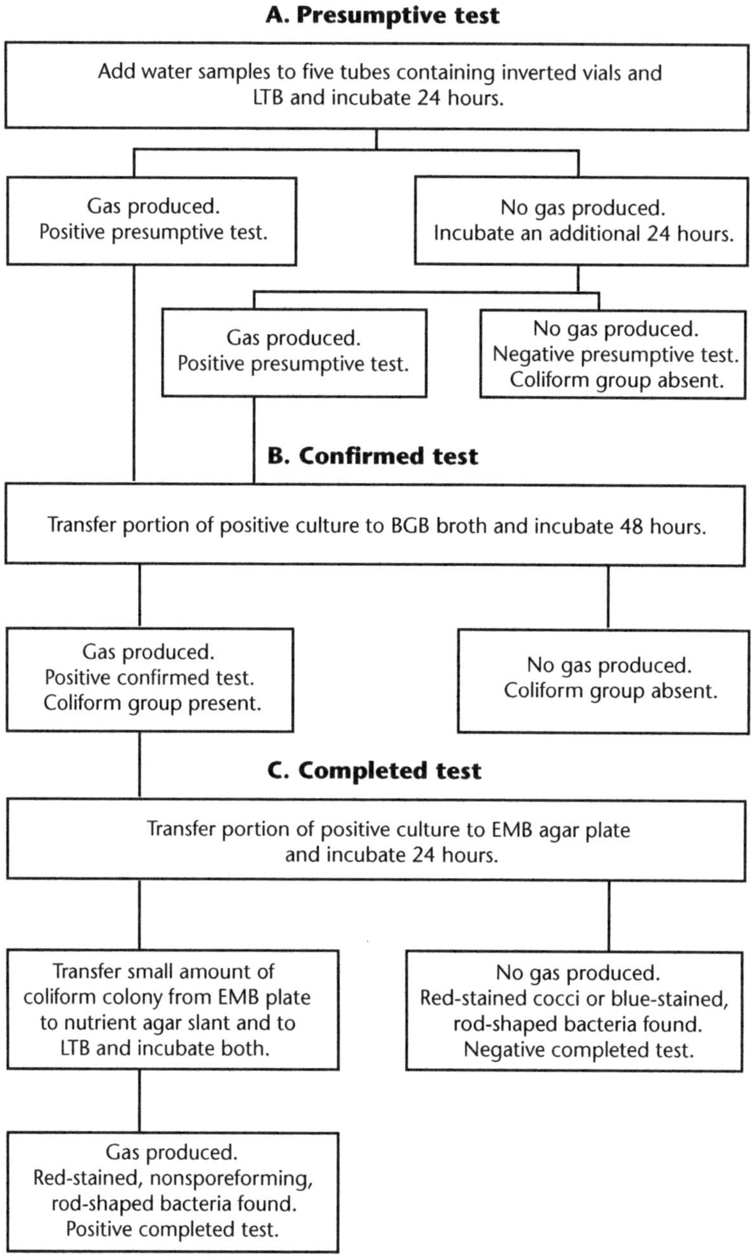

Figure 4-13 Summary of the multiple-tube fermentation method

dilution bottle containing special P–A media and a small inverted tube. The sample is then incubated at 35°C (95°F) for 24 and 48 hours.

The presence of total coliforms is indicated by the purple P–A medium turning yellow (indicating acid production) and by the formation of gas in the medium. All yellow and gas-producing samples from this presumptive stage must then be confirmed as described for the MTF-confirmed step using BGB tubes. Gas production indicates the presence of total coliforms and must be reported as a positive sample (presence) in the monthly report to the primacy agency.

Samples confirmed for total coliforms must also be analyzed for either fecal coliforms or *Escherichia coli*. A check or repeat sample must also be collected and analyzed. A check/repeat sample that is positive can result in an acute violation of the Total Coliform Rule (TCR) and must be reported to the primacy agency within 24 hours after results become known.

Fecal Coliform Procedure If the MTF or P–A method is being used, as the presumptive positive samples are being inoculated into the BGB broth, 0.1 mL of the presumptive broth is also transferred into an EC broth tube. (The actual name of the broth is EC and is a test for *E. coli*.) If the membrane filter method is used, bacterial growth is transferred into an EC tube. This tube is then incubated for 24 hours in a water bath at 44.5°C (112°F). The presence of gas in the tube confirms the presence of fecal coliforms.

E. coli *Procedure* The presence of *E. coli* can be determined using the MUG test discussed in the next section. A 0.l-mL portion of the presumptive media or a swab is used to transfer a sample from a membrane filter into an EC–MUG tube. A tube that fluoresces under a long-wave ultraviolet (UV) light is confirmation for E. coli.

MMO–MUG Technique

The MMO–MUG technique was approved by USEPA shortly after promulgation of the TCR. MMO and MUG are acronyms for the constituents in the medium used in the tests. MMO represents minimal media with ONPG (ONPG stands for *ortho*-nitrophenyl-beta-D-galactopyranoside). *E. coli* produce a specific enzyme that reacts with ONPG to give a yellow color. MUG stands for 4-methylumbelliferyl-beta-D-glucuronide. Only *E. coli* produce an enzyme that reacts with MUG. Therefore, a medium containing MMO and MUG can be used to identify both total coliforms and *E. coli* in a single-sample inoculation.

Two procedures may be used. In the *ten-tube procedure*, ten tubes are purchased with the medium already in them. A 10-mL portion of sample is transferred into each tube and incubated at 35°C (95°F) for 24 hours. In the *P–A procedure*, the medium is purchased in vials. The medium is transferred into a bottle containing 100 mL of sample, is mixed, and is incubated as in the ten-tube procedure. If total coliforms are present in either procedure, the medium will turn yellow. If *E. coli* are present, the medium will also fluoresce blue under a UV light.

Membrane Filter Method

The membrane filter method of coliform testing begins with the filtering of 100 mL of sample under a vacuum through a membrane filter. The filter is then placed in a sterile container/petri dish (Figure 4-14) and incubated in contact with a selective culture medium.

A coliform bacteria colony will develop at each point on the filter where a viable coliform bacterium was left during filtration. After a 24-hour incubation period, the number of colonies is counted (Figure 4-15).

A typical coliform colony on M-Endo media is pink to dark red with a distinctive green metallic surface sheen. All organisms producing such colonies within 24 hours are considered presumptive coliforms. For confirmation, representative colonies are inoculated into LTB and BGB broth.

USEPA has revised the TCR, and the new rule uses *E. coli* as the indicator for fecal contamination. Total coliform is used as an indicator that there may be a problem in the system and triggers a system investigation.

Heterotrophic Plate Count Procedure

The heterotrophic plate count (HPC) procedure is a way to estimate the population of bacteria in water. The test determines the total number of bacteria in a sample that grows under specific conditions in a selected medium.

The HPC is performed by placing diluted water samples on plate-count agar. The samples are incubated for 48–72 hours. Bacteria occur singly, in pairs, in

MMO–MUG method
An approved bacteriological procedure for the detection of total coliforms and *E. coli*.

membrane filter method
A laboratory method used for coliform testing. The procedure uses an ultrathin filter with a uniform pore size smaller than bacteria—less than a micron. After water is forced through the filter, the filter is incubated in a special media that promotes the growth of coliform bacteria. Bacterial colonies with a green-gold sheen indicate the presence of coliform bacteria.

heterotrophic plate count (HPC)
A laboratory procedure for estimating the total bacterial count in a water sample.

Figure 4-14 Placement of membrane on pad soaked with culture medium

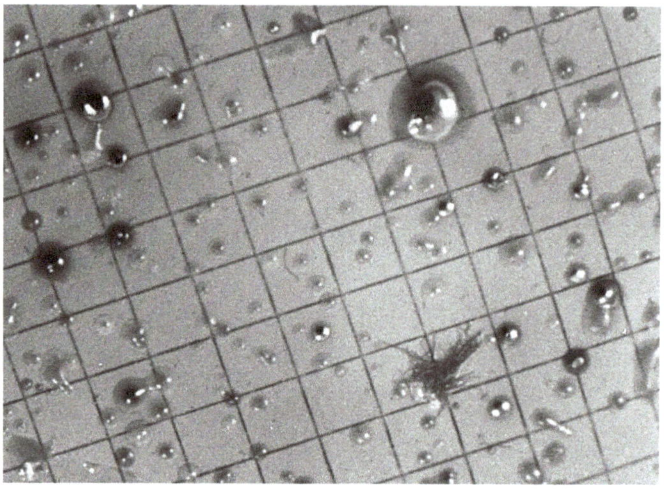

Figure 4-15 Membrane filter after incubation with positive growth colonies

chains, and in clusters. The bacteria colonies that grow on the agar are counted using colony-counting equipment. Detailed procedures are described in the latest edition of *Standard Methods*. These procedures must be closely followed to provide reliable data for water QC measurements. Properly treated water should have an HPC of less than 500 colonies per milliliter. Higher counts indicate an operational problem that should be investigated.

Alternate Methods

Other methods for coliform and *E. coli* detection are being developed using a combination of enzymes, ß-glucuronidase, and ß-galactosidase in combination with ONPG and MUG. In tests that are positive for coliform, a yellow substance is produced that fluoresces at 366 nm UV after an incubation period of 24 ± 2 hours at 35.0°C ± 0.5°C (~95°F). There are various methods of determination, from just adding the water sample to a test bottle containing the media and incubating, to pouring the water sample into a tray containing the detection media with volumetric cells for counting. The method has been approved for use by USEPA but may need to be confirmed with the area's drinking water primacy agency.

Chemical Test Methods

Physical and Aggregate Properties of Water

Physical testing of drinking water measures the physical properties of the water as distinguished from chemical or biological contaminants. Several physical tests are either required by regulations or necessary for control of the water treatment processes. Water systems using surface water sources perform physical tests frequently because of changing water quality. Groundwater quality generally fluctuates slowly or infrequently; as such, most physical, chemical, and microbiological testing of groundwater is performed periodically but not as frequently as testing of surface water sources. Tests ensure that optimum treatment is provided and that the finished water meets state and federal standards.

Acidity

The definition of acidity is base neutralizing power. There are two types of acidity: phenolphthalein acidity (carbon dioxide acidity), which is prevalent above pH 8.3, and mineral acid acidity, which tends to be in the lower pH ranges. Phosphoric acid, hydrochloric acid, and sulfuric acid are all mineral acids. Acidity and alkalinity exist in the water at the normal pH levels found in water since the existence of basic molecules such as negative ions does not exclude the existence of the acid type of molecule positive ions.

Methods of Measurement Mineral acids are measured by titration to a pH of 4.3 using methyl orange as an indicator (orange on the alkaline side to salmon pink on the acid side). Mineral acidity plus acidity caused by weak acids (e.g., carbonic acid) is measured by titration of the sample to the phenolphthalein end point (pH 8.3). This measurement is called *total acidity* or *phenolphthalein acidity*. Phenolphthalein is colorless at a pH less than 8.3 and pink or red above pH 8.3. To a theoretical chemist, a pH of 7 is considered neutral. To a water chemist, though, who wants to know how much free or combined carbon dioxide is present and how much total alkalinity is present, a pH of 7 has less meaning. The water chemist is interested in what makes up the pH and how much buffering exists.

Significance Acidity plays a more important role in industrial water applications, especially in neutralization reactions. In drinking water, acids are associated with corrosive environments. Chemical reaction rates are influenced by acidity. Acids are used in the regeneration of ion exchange systems. They are also used to lower the pH of water treated by nanofiltration and reverse osmosis membranes.

Alkalinity

The definition of alkalinity is acid neutralizing power (see Table 4-2).

Types Phenolphthalein alkalinity (P alkalinity) has a pH end point of 8.3. Methyl orange or total alkalinity (MO alkalinity) has a pH end point of 4.3 and is directly related to the amount of hydroxide (OH^-), carbonate (CO_3^-), or bicarbonate (HCO_3^-) alkalinity present.

More specifically, when water samples have a pH above 10, hydroxide anions (OH^-) are the primary constituent of the pH-causing ions. Titration with strong acid is complete at the phenolphthalein end point (8.3). Hydroxide alkalinity is equal to phenolphthalein alkalinity.

Table 4-2 Alkalinity relationships

Result of Titration	Hydroxide Alkalinity	Carbonate Alkalinity	Bicarbonate Alkalinity
P* = 0	0	0	MO
P < ½ MO†	0	2P	MO – 2P
P = ½ MO	0	2P	0
P > ½ MO	2P – MO	2(MO – P)	0
P – MO	MO	0	0

*P = phenolphthalein

†MO = methyl orange

acidity
Base neutralizing power.

alkalinity
Acid neutralizing power.

In water samples with a pH of 8.3 or higher, carbonate ions are the primary alkalinity component. Titration to phenolphthalein end point is exactly one-half of total titration to pH 4.3. Carbonate alkalinity equals total alkalinity.

If only hydroxide-carbonate is present, water samples will have a high pH, usually well above 10. Titration from pH 8.3 to 4.3 represents one-half of carbonate alkalinity.

If only carbonate-bicarbonate is present, water samples will have a pH greater than 8.3 and less than 11. Titration to pH 8.3 represents one-half of carbonate alkalinity.

If only bicarbonate is present, water samples will have a pH of 8.3 or less (usually less). Bicarbonate alkalinity equals total alkalinity.

Color

Color in water may result from the presence of minerals, inorganic chemicals, metals (e.g., iron and manganese), the decomposition of organic matter from soils, aquatic organisms, and vegetation. A problem in surface waters, some groundwaters containing iron or manganese and dissolved organic matter can also have significant color levels. Color in water is classified as either true color or apparent color. True color is defined as the color of water from which turbidity has been removed. Apparent color includes true color and color caused by suspended matter.

Color in the yellow to brown range is determined by a visual or spectrophotometric comparison of the sample with either a known colored chemical solution or a calibrated color disk.

The unit of measurement is the color unit. Color units are the estimated color of a diluted sample times 50, divided by the milliliters of sample taken for dilution. For finished waters, some pristine surface water supplies, or groundwater, the sample normally would not have to be diluted and could be estimated directly without any division. Thus, a 50-mL sample with a low color of 1 would have a result of 1 color unit. Color is reported in whole numbers from 1 to 500. Sample pH is always reported with the color units.

Other colors in the water, such as blue or red, have other causes and are not assigned a numerical value. They may be listed as to their intensity or shade (e.g., light blue, aqua, pink).

Significance Color in drinking water should be removed to produce a pleasing, acceptable appearance. The color of a drinking water affects consumers' acceptance. Consumers expect a colorless product and will reject colored water; they may change to another source of water even if the other source is less safe. Color may also indicate high levels of organic compounds, which may produce high levels of trihalomethanes (THMs) and other DBPs on contact with chlorine or other disinfectants.

Color data from raw- and finished-water sample points comprise one indicator of the efficiency of the treatment plant's processes.

Sampling Samples collected should represent raw water, finished water, and distribution system water, and should include water pH at the point of sampling. Sampling in surface water systems is especially important.

Methods of Determination The platinum–cobalt method is the preferred method for color analysis. It is useful for measuring color derived from naturally occurring materials, but it is not applicable to color measurement of waters containing highly colored industrial wastes.

color
A physical characteristic of water. Color is most commonly tan or brown due to oxidized iron, but contaminants may cause other colors, such as green or blue.

true color
The color of water from which turbidity has been removed.

apparent color
A characteristic of water that includes true color and color caused by suspended matter.

platinum–cobalt method
A procedure used to determine the amount of color in water.

Conductivity

Conductivity (specific conductance) is a measure of the ability of an aqueous solution to carry an electrical current. This ability depends on the presence of ions; on their total concentration, mobility, and valence; and on the temperature of measurement. Solutions of most inorganic compounds are relatively good conductors. Conversely, molecules of organic compounds that do not dissociate in aqueous solution conduct a current very poorly, if at all.

The units for conductivity are the inverse of unit of resistance (ohm), or (1/ohm-m), or mho (pronounced like "Moe") per centimeter. In Système International (SI) units, conductivity is reported as millisiemens per meter (mS/m). To report results in SI units, divide mho/cm by 10.

Conductivity (specific conductance) is a measure of the ionic strength of a solution. It is a required water quality parameter of the Lead and Copper Rule. The conductivity of potable waters in the United States ranges from 50 to 1,500 µmho/cm or (S) (siemens).

Significance Conductivity is a general parameter that assists the analyst in evaluating many aspects of water quality. In the laboratory, conductivity measurements are used to

- establish the degree of mineralization of a water to assess the effect of the total concentration of ions, which is particularly relevant to corrosion rates;
- evaluate variations in the concentration of dissolved minerals in a water source;
- estimate the concentrations of TDS in water supplied to a distribution system or from points in the system; and
- approximate the milliequivalents per liter of either cations or anions in a water sample

Sampling A 50-mL aliquot of sample for conductivity measurements may be taken from an unpreserved sample collected from a regularly used bacteriological sample point or from a sample collected at the entry point to the distribution system. The sample should be preserved by cooling it to 39°F (4°C) until it is analyzed.

The sample should also be protected from exposure to the atmosphere, since the water could desorb or adsorb gases (e.g., ammonia and carbon dioxide) that could dissolve in an ionic state. This development would affect conductivity.

Methods of Determination Laboratories analyzing a sample for conductivity may not have to be certified, but a method approved by USEPA must be used if the results are to be reported to the state for the Lead and Copper Rule.

An approved method uses a self-contained conductivity instrument with conductivity cells containing either platinum or nonplatinum electrodes. Procedures for running the test may be found in *Standard Methods* or *Methods of Chemical Analyses of Water and Wastes* from USEPA. Conductivity measurements may be taken in the field.

Hardness

Hardness is a measure of the concentration of calcium and magnesium salts, which are generally present in water as bicarbonate salts. Water hardness is derived largely from the water contacting soil and rock formations. Hard waters usually occur where the topsoil is thick and contains minerals and metals that will dissolve and where limestone formations are present. Soft waters occur where soil is sandy, the topsoil layer is thin, and limestone formations are sparse or absent. Another mineral in the soil, gypsum or calcium sulfate ($CaSO_4$), can also contribute to hardness.

Significance Hard and soft waters are both satisfactory for human consumption. However, consumers may object to hard water because it causes scale to form in household plumbing fixtures and on cooking utensils. Hardness is also a problem for industrial and commercial users because of scale buildup in boilers and other equipment.

According to the National Research Council,(National Academy of Sciences) water with <75 mg/L as calcium carbonate ($CaCO_3$) of hardness is considered soft water. Other organizations say that water below about 100 mg/L of $CaCO_3$ is soft, between 100 and about 175 mg/L of $CaCO_3$ is moderately hard, and over 175 mg/L $CaCO_3$ is considered hard. Water that is most satisfactory for household use contains about 75 to 100 mg/L as $CaCO_3$. Waters with 300 mg/L as $CaCO_3$ are generally considered too hard. When water is softened in a water treatment plant, it is either partially softened or blended to result in a hardness concentration between 75 and 150 mg/L as $CaCO_3$. Waters containing calcium sulfate hardness at these same levels can cause gastric irritation in some individuals until they become acclimated. USEPA has a secondary standard of 250 mg/L of sulfate as $CaSO_4$ in drinking water with a TDS of 500 mg/L.

Very soft waters found in some sections of the United States have hardness concentrations of 30 mg/L as $CaCO_3$ or less. These waters are generally corrosive and are sometimes treated to increase hardness.

Sampling If a water treatment plant softens water, hardness analyses of the finished water should be conducted daily to determine whether the desired degree of softening has been achieved. Analyses should be conducted on samples collected immediately after softening and before the water enters a clearwell.

Hardness determinations should also be performed on raw-water samples whenever weather conditions (e.g., spring rains) affect the supply. This sampling will reveal any variation in the hardness of the raw water and will provide advance information for chemical dosage changes that may be necessary for softening. Even if softening is not practiced, hardness determinations should still be made periodically as a general water quality measurement.

At least 100 mL of sample should be collected in either glass or plastic bottles. Samples should be cooled to 39°F (4°C) and acidified with 0.5 mL/100 mL to a pH of <2 with nitric acid unless they are going to be analyzed immediately. These acidified and cool samples may be stored for 28 days before analysis.

Methods of Determination The ethylenediaminetetraacetic acid (EDTA) titrimetric method is the preferred method of analysis. It consists of sequestering (tying up) the calcium and magnesium ions by titrating with an EDTA solution. The sample is titrated in the presence of an indicator. The initial solution is red, and it changes to blue when all the ions have been sequestered.

Taste and Odor

Tastes and odors in water are difficult to measure. They are caused by a variety of substances, including organic matter, dissolved gases, and industrial wastes. Odors in water supplies are most frequently caused by algae or decaying organic matter. The intensity and offensiveness of odors vary with the type of organic matter.

Odors are generally classified as aromatic, fishy, grassy, musty, septic, or medicinal. Industrial wastes, such as phenolic or oil waste, are also responsible for some odors in surface waters.

The human sense of smell is much more sensitive than is the sense of taste, so odor tests are most commonly run in water treatment plants. The taste test, which classifies tastes as sweet, sour, bitter, and salty, can be run only on water known to be safe for drinking; thus its usefulness is limited.

Significance An odor test can be used to evaluate how well a water treatment plant removes taste- and odor-causing organic materials. The test can also be used to trace the origins of contamination in source water. For example, an odor-causing industrial waste discharge might be occurring upstream from a water treatment plant. Samples can be collected at intervals upstream until the problem-causing area has been reached. The odors should become stronger closer to the discharge and should not be evident in samples collected upstream from the discharge. This technique is time consuming; however, it can be conducted by water plant personnel, and extensive laboratory facilities are not necessary. An odor test can also be used to detect problems in the distribution system. For example, odors occur in dead-end water mains having a significant bacteriological buildup. A definite chlorine odor can indicate the loss of free chlorine caused by stagnation, slime buildup, and/or anaerobic conditions.

The threshold odor number (TON) is designed to help monitor all types of odors, independent of source. The TON cannot, however, be used to indicate the concentration of the odor-producing substance, because some substances produce strong odors at low concentrations. For example, some chemical wastes, such as phenol in chlorinated water, have been detected by the threshold odor test at a 0.001-mg/L concentration. Other odor-producing substances, such as detergents, may not be detected until the concentration is as high as 2.5 mg/L.

An odor with a TON of 3 might be detected by a consumer whose attention is called to it, but it probably would not be noticed otherwise. If an odor appears gradually, consumers will adapt to it, and it will be noticed less than if it appears suddenly. Finished-water quality with a TON above 5 will begin to draw complaints from consumers. When a TON of 3 or more is detected in a finished-water supply, quick action should be taken to solve the problem.

Sampling Water supplies with seasonal or recurring taste and odor problems should be analyzed regularly, and, as problems occur, corrective action should be taken. The tests may be time consuming, so it is not generally possible to conduct more than a few tests per day. Water samples should be taken from raw and finished waters. At least 1,000 mL of sample should be collected for an odor analysis. Samples should be collected in clean bottles that have not been used for any samples that might leave a taste or an odor. The bottles should be washed with detergent and rinsed with distilled water and then odor-free water. Glass sample bottles should be used; plastic containers may add some odor of their own or an odor from substances that were previously in the container.

Aeration and mixing of the sample should be kept to a minimum before testing because air strips or oxidizes odor-producing compounds. An air space should be left at the top of the bottle so the sample can be thoroughly shaken before testing.

Odor tests should be run as soon as possible after collection. If the sample must be stored, it should be tightly capped and placed in an odor-free refrigerator. The sample should be analyzed no later than 24 hours after collection.

threshold odor number (TON)
A number indicating the greatest dilution of a water sample (using odor-free water) that still yields a noticeable odor.

Odor-free water can be generated in the plant or laboratory with an activated carbon filter purchased and installed on a cold-water tap. Be sure to follow the manufacturer's instructions for the use of the filter.

Methods of Determination Two standard methods of quantifying taste and odor in water supplies are in use: the threshold odor test and the flavor profile analysis.

Threshold Odor Test Most tastes and odors are extremely complex, and the best way to detect them is with the human sense of smell. A series of sample dilutions are prepared and placed in bottles for observers to test. Each bottle contains 200 mL of liquid consisting of a mixture of sample water and odor-free distilled water.

The bottles are arranged so that the observer smells the most dilute samples first and then from a bottle of completely odor-free water as a reference. Then the next strongest sample is smelled, and again the odor-free water, and the process is continued until an odor is first detected. The TON may be calculated as follows:

$$\frac{V_s + V_d}{V_s} = \text{TON}$$

where:

V_s = volume of sample, in mL
V_d = volume of dilution water, in mL

The lowest obtainable TON is 1. If no odor is detected in an undiluted sample, the TON is reported as "no odor observed" and no number is assigned. If an odor is first detected in a bottle that has 100 mL of sample diluted to 200 mL with distilled water, the TON is 2. TONs corresponding to various dilutions are shown in Table 4-3.

The threshold odor test is not precise and is based on human judgment. The ability to detect odors varies among individuals, and if very accurate results are desired, a panel of five or more persons is recommended to overcome this variability. Persons performing odor tests should not have colds or allergies that would affect their sense of smell. They also should be nonsmokers and should not use perfumes or aftershaves, which tend to dull the sense of smell. Plant operators should make odor observations, but because they work in the environment where the odor may exist, their sense of smell can become desensitized to an odor

Table 4-3 Threshold odor number corresponding to various dilutions

Volume of Sample (mL) Diluted With Odor-Free Water to 200 mL	Threshold Odor Number
0.8	256
1.6	128
3.1	64
6.3	32
12.5	16
25	8
50	4
100	2
200	1

(anosmia). If they step out of the environment for a short period, their sensitivity may return. All tests must, of course, be conducted in an odor-free atmosphere.

Flavor Profile Analysis The flavor profile analysis (FPA) was approved by the Standard Methods Committee in 1990 and first appeared as a proposed standard method in the 18th edition of *Standard Methods*.

The FPA differs from the TON in that the samples are not diluted and each individual odorant in the sample is evaluated and numerically rated. FPA can be applied to both taste- and odor-causing compounds.

In the procedure, a panel consisting of four to six members and a panel coordinator conduct each round of testing. Panel members must be selected for their desire to participate, their tested ability to accurately taste and smell samples, and their ability to interact with other panel members. Dominant personalities are not desirable.

The panel members gather in an odor-free room. The panel coordinator prepares and presents the samples to the panel members, and they independently write down the taste or odor attributes they have observed. The coordinator writes each observation on a blackboard or flip chart and leads a discussion to reach consensus among the panel members.

Panelists are trained in the proper methods of tasting and sniffing samples and are taught to identify and rate the attributes of both tastes and odors. They must also be trained in how to prepare for a round of testing—i.e., no aftershave or perfume; no smoking, gum chewing, or consuming of spicy food or drink 1 hour prior to the test; no colds or allergies.

Standards and references for tastes and odors are commercially available to assist in training panelists and reaching consensus. For example, to help panel members identify the four taste attributes, standards can be made from sucrose for sweetness, citric acid for sourness, sodium chloride for saltiness, and coffee for bitterness. Natural material is also available for panel use; for example, geranium leaves may be used to assist the panel in coming to a consensus on what constitutes a geranium odor in a water sample.

Panel members should follow all the safety precautions mentioned in the method since this is the only laboratory test in which the sample is consumed, or at least placed directly into the body.

Temperature

Temperature is measured on either the Fahrenheit (°F) or the Celsius (°C) scale. The freezing point of water is 32°F or 0°C; the boiling point is 212°F or 100°C. Because temperature is a factor in computing the Langelier saturation index, it is one of the water quality parameters required by the Lead and Copper Rule.

Significance Water temperature determines, in part, how efficiently certain unit processes operate in the treatment plant. The rate at which chemicals dissolve and react is somewhat dependent on temperature. Cold water generally requires more chemicals for efficient coagulation and flocculation to take place. Water with a higher temperature may result in a higher chlorine demand because of increased reactivity and also because there is usually an increased level of organic matter, such as algae, in the raw water. For surface water plants in the northern regions, the dropping of the plant effluent water temperature to about the 39°F (4°C) point can indicate the start of the cold-weather main breaks.

Sampling Temperature readings must be taken on-site, either directly from the water or from samples immediately after collection. Immediate readings are necessary because the water temperature will begin to change once the sample is taken.

Methods of Determination A laboratory thermometer is used for temperature analysis. The thermometer is left in the water long enough to get a constant reading, and the measured temperature is expressed to the nearest degree or less depending on the thermometer's accuracy. Digital battery-operated thermometers are currently available. They have the advantage over glass bulb–type thermometers of being easier to read and less prone to breakage.

Total Dissolved Solids

Total dissolved solids (TDS), also referred to as *total filterable residue*, in natural waters consists mainly of carbonates, bicarbonates, chloride, sulfate, calcium, magnesium, sodium, and potassium. Dissolved metals, dissolved organic matter, and other substances also account for a small portion of the dissolved residue in water.

Significance Dissolved solids in drinking water tend to change the water's physical and chemical nature. Distilled or deionized water has a flat taste; most consumers prefer water that contains some dissolved solids. Different salts in solution may interact and cause effects that each salt alone would not cause. The presence of some dissolved compounds or ions (such as arsenic and mercury) can be harmful in water even where the total solids concentration is relatively low.

It is generally agreed that the TDS concentration of palatable water should not exceed 500 mg/L. Lime softening and ion exchange facilities both significantly reduce the quantity of TDS in finished water.

Many communities in the United States use waters containing 2,000 mg/L or more TDS because better-quality water is not available. These waters tend to be unpalatable, may not quench thirst, and can have a laxative effect on new or transient users. However, no lasting harmful effects have been reported from such waters. Waters containing more than 4,000 mg/L TDS are considered unfit for human consumption.

Raw-water source samples can be dipped from just below the surface of the water in the area of the intake structure. They may be collected in clean wide-mouth glass or plastic 1-L containers. Filter effluent samples should be collected from an effluent sample tap or drain line. The sample should be stored in a cooler away from sunlight for transportation to the laboratory.

Sampling Total dissolved solids are typically analyzed in raw water and finished water, although they may be collected from any point at which the concentration of dissolved solids may be of concern. Samples for total dissolved solids should be collected in a clean plastic or glass container, which has been rinsed with the sample prior to filling the bottle. At least 200 mL of sample should be collected. Although it is preferable to analyze for total dissolved solids as soon as possible after sample collection, samples may be held for up to 7 days if cooled to 4°C.

Methods of Determination Total dissolved solids are typically determined gravimetrically. A well-mixed sample is filtered through a standard glass fiber filter. A known volume (typically 100 mL) of the filtered water is evaporated to dryness in a pre-weighed dish and dried to a constant weight at 180°C. The increase in dish weight represents the TDS, which is expressed in mg/L. The values obtained through this analysis may differ from theoretical values determined by chemical analysis. Since the gravimetric total dissolved solids analysis procedure can be time-consuming, a sample's approximate total dissolved solids concentration may also be estimated based on its conductivity. Conductivity is a relatively simple measurement that can be performed in the lab or the field using an electrode and a calibrated conductivity meter.

Turbidity

Turbidity is an optical property caused by particles suspended in water. These particles cause light rays to be scattered and absorbed rather than transmitted in a straight line, making the water appear cloudy. Turbidity is the measurement of the clarity of a water sample. Waters showing very little light scattering produce low-turbidity measurements; those with a great deal of light scattering have high turbidity. The suspended particles causing turbidity include organic and inorganic matter and plankton.

Turbidity should not be confused with suspended solids. Turbidity expresses how much light is scattered by the sample. Suspended solids measurements express the weight of suspended material in the sample. In most cases, turbidity cannot be correlated to suspended solids concentration.

Interferences with turbidity measurement include color, high turbidity (raw muddy water requires dilution and calculations to obtain a reliable reading), and gas bubbles. Scratched sample cells, condensation, poor calibration, and stray light also interfere.

Significance Turbidity is expressed in nephelometric turbidity units (ntu), and a reading in excess of 5 ntu is noticeable to consumers. Turbidity is significant in water supplies because it creates a potential public health hazard, unpleasant appearance, and operational difficulties. The most important of these is the potential public health hazard. The effectiveness of chlorine or other disinfectants depends on the disinfectant making contact with the pathogenic organisms in the water. Suspended particles in turbid water can shelter microorganisms from the disinfectant and allow them to still be viable when they reach the customers. Turbid water may also contain particles of organic matter that can react with chlorine to form THMs or other DBPs.

Turbidity analyses are also used to evaluate in-plant operations. Turbidity measurements taken after settling and before filtration reflect the performance of the coagulation, flocculation, and sedimentation processes. A rise in turbidity after settling indicates that the coagulant application should be changed and/or that operational corrections must be made.

Settled water before filtration should have a turbidity of less than 10 ntu. If water with high turbidity reaches the filter, it will cause high filter head loss and shorten filter runs. Changes in raw-water turbidity usually require that the coagulant dose be changed. Any significant change in turbidity within the unit process should be an immediate warning that operational adjustments are necessary. The operator should set up parameters that define a significant change for a specific unit process and the type of water being treated. For example, if the normal turbidity level doubles (e.g., the raw water goes from 3 to 6 ntu for no apparent reason), operational adjustment is necessary.

Turbidity analyses are also used to monitor finished-water quality for compliance with state and federal drinking water standards.

Sampling Turbidity analyses are usually conducted on samples collected from raw water, sedimentation basin effluent, filter effluent, and finished water. Figure 4-16 shows some typical turbidity sampling points.

At least 100 mL of sample should be collected in a clean glass or polyethylene container. Samples should be shaken and analyzed immediately after collection because the level of turbidity can change if the sample is stored. If it is not possible to run a turbidity test immediately, cool the sample to <4°C and analyze within 48 hours.

turbidity
A physical characteristic of water that makes the water appear cloudy. The condition is caused by the presence of suspended matter.

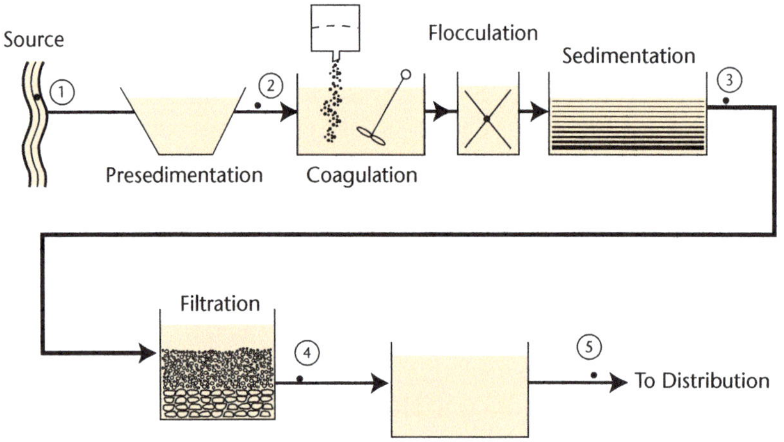

1. Turbidity of raw water entering the plant.
2. Turbidity reduction by presedimentation; helps the operator determine coagulant dose.
3. Turbidity removal by coagulation, flocculation, and sedimentation processes; assists the operator in monitoring the efficiency of the process.
4. Turbidity after filtration; continuous monitoring of turbidity for each filter monitors for turbidity break-through, which is one of the indicators of the need for filter backwashing.
5. Turbidity of all treated water leaving the plant; monitors compliance with drinking water regulations for maximum allowable effluent turbidity.

Figure 4-16 Typical turbidity sampling points

All filter plants should keep a continuous record of finished-water turbidity. Continuous-reading turbidimeters with recorders installed on the filter-effluent piping will continuously determine, report, and record the quality of the filter effluent. The turbidimeter signal can sound alarms to indicate the need to shut down an improperly operating filter. This alarm system increases the reliability of the filter operation and is especially important in assuring the safe operation of pressure filters and high-rate (4–6 gpm/ft^2 [2.7–4.1 mm/sec]) filter plants. It should be noted that the continuous-reading turbidimeter may not match the laboratory meter exactly because of differences between the instruments and the methods used. This would be similar to the differences between two cells used to analyze one sample for turbidity; slight variations in the manufacturing of the equipment can vary the result.

Methods of Determination The nephelometric turbidimeter measures the scattering of light in nephelometric turbidity units. USEPA drinking water regulations specify the use of a nephelometric turbidimeter for all required monitoring (Figure 4-17).

nephelometer
An instrument that determines turbidity by measuring the amount of light scattered by turbidity in a water sample. It is the only instrument approved by the US Environmental Protection Agency to measure turbidity in treated drinking water.

Analysis is quick and easy with the nephelometric method. Nephelometry is useful for in-plant monitoring, and results can be compared from plant to plant, which is an advantage to operators seeking performance information from other facilities.

Under the requirements of the SWTR, water systems serving a population in excess of 500 must perform turbidity monitoring of filtered water at least every 4 hours that the plant is in operation. A system may substitute continuous turbidity monitoring if the equipment is validated on a regular basis using a procedure approved by the state. The Enhanced Surface Water Treatment Rule (ESWTR) requires continuous monitoring of the individual filters in a surface water treatment plant and calls for the turbidity for each filter to be recorded at a minimum of every 15 minutes.

Figure 4-17
Nephelometric turbidimeter with flow-through chamber
Courtesy of Hach Company.

Inorganic Chemicals

An inorganic chemical, substance, compound, or contaminant is one not derived from hydrocarbons. An inorganic contaminant is an inorganic substance regulated by USEPA. Inorganic contaminants contained in this list are antimony, asbestos, barium, beryllium, cadmium, chromium (total), copper, cyanide, fluoride, lead, mercury nitrate, nitrite, selenium, thallium, and arsenic. Interestingly, nickel, which was once regulated by USEPA was remanded; monitoring is still required even though there is currently no MCL.

In drinking water, other inorganic compounds are of interest. These include the divalent cations, calcium, magnesium, iron, and manganese.

Carbon Dioxide

Carbon dioxide is a colorless, odorless, noncombustible gas that is found in all natural waters. Carbon dioxide in surface waters can originate from the atmosphere, but most comes from biological oxidation of organic matter. Biological oxidation is also the primary source of carbon dioxide in groundwater.

Significance Consuming excess carbon dioxide in water has not been found to have adverse health effects. In fact, carbon dioxide is present in commercial carbonated beverages in concentrations far greater than those found in natural waters. However, carbon dioxide in water forms carbonic acid, which can cause corrosion problems. In addition, carbon dioxide values must be known to calculate proper lime dosages when softening water. If recarbonation is used following lime softening, carbon dioxide values must be determined to control the process. In groundwater, a high level of carbon dioxide reduces the pH and increases the dissolution rate of metals and minerals from the surrounding earth, iron, manganese, and so on.

Sampling Carbon dioxide analyses should be run on raw and finished water. Special precautions must be taken during collection and handling of the sample if the titrimetric method is being used. Exposure to the air must be kept to a minimum. Field determination of free carbon dioxide immediately after sampling is advisable. If field determination is impossible, the sample should be kept cool and the analysis completed as soon as possible.

Samples may be collected in glass or plastic bottles. At least 100 mL of sample should be collected. The bottle should be filled to the top with no air space left, and no preservatives should be added. A piece of tubing from the faucet to the bottom of the collection vessel should be used to minimize aeration of the sample from splashing or bubbling. Fill the container with several volume changes while the tubing is submersed. Slowly remove the tubing from the sample container with the water still flowing to ensure that the container remains full with no air space.

Methods of Determination The amount of carbon dioxide in water may be determined by using the nomographic method or the titrimetric method. To use the nomographic method, the pH, bicarbonate alkalinity, temperature, and TDS must be known. Results are most accurate when the pH and alkalinity are analyzed immediately after sample collection. The titration method may be performed potentiometrically or with a phenolphthalein indicator.

Chlorine Residual and Demand

Chlorine is usually added to source water as the water enters the treatment plant (prechlorination) and again just before it leaves the plant (postchlorination). In the plant, chlorine is also often added at intermediate points during the treatment process. Postchlorination is primarily administered to provide an excess of chlorine for continued disinfection in the distribution system. Tests of chlorine levels in the plant and throughout the distribution system are necessary to determine that chlorine dosage levels are adequate and to monitor water quality.

Significance Destruction of pathogenic organisms by chlorine is directly related to contact time and the concentration of the chlorine. High chlorine doses with short contact periods will provide essentially the same results as low doses with long contact periods. Chlorination also oxidizes substances such as iron, manganese, and organic compounds, making their removal from the water easier.

Successful chlorination requires that enough chlorine be added to complete the disinfection or oxidation process. However, chlorine must not be added in amounts that are wasteful, creating unnecessarily high operational costs. Determining effective and efficient chlorine dosage levels is the responsibility of the plant operator.

Chlorine Residual There are two types of chlorine residual: combined residual and free available residual. The process by which these are formed is illustrated in Figure 4-18.

The first amount of chlorine (e.g., 1 mg/L) that is added to raw water is used in oxidizing reducing compounds such as iron and manganese (from point 1 to point 2 in the figure). The chlorine oxidizes the iron and manganese and in the process is used up, meaning no residual forms and no disinfection occurs.

If the initial chlorine dosage used is higher (e.g., 2.5 mg/L), the reaction will go to point 3. Between points 2 and 3, the chlorine reacts with the organic substances and the ammonia in the water, forming chloroorganics and chloramines. These two products are called combined chlorine residual. This is a chlorine residual that—because it is combined with other chemicals, normally ammonia compounds, in the water—has lost some of its disinfecting strength. Compared with free chlorine, combined chlorine residual is a less effective disinfecting agent. If it is not properly controlled, it may cause tastes and odors characteristic of water in a swimming pool.

As the chlorine dosage is increased further (point 3 to point 4), the chloramines and some of the chloroorganics are destroyed. This process reduces the combined chlorine residual until, at point 4, the combined residual reaches its lowest point. Point 4 is called the breakpoint. At the breakpoint, the chlorine residual changes from combined to free available.

nomographic method
A method that uses a graph or other diagram to solve formulas and equations.

titrimetric method
Any laboratory procedure that uses titration to determine the concentration of a constituent in water.

combined chlorine residual
The chlorine residual produced by the reaction of chlorine with substances in the water. Because the chlorine is "combined" it is not as effective a disinfectant as free chlorine residual. In water treatment, this usually refers to compounds formed by the combination of chlorine and ammonia.

breakpoint
The point at which the chlorine dosage has satisfied the chlorine demand.

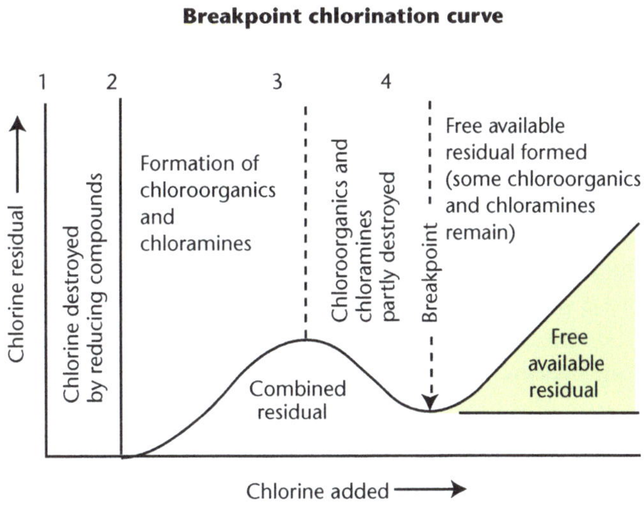

Figure 4-18 Formation of combined chlorine residual and free available chlorine residual

As the initial chlorine dosage is increased still further (beyond 4 mg/L in this example), free available chlorine residual is formed—free in the sense that it has not reacted with anything and available in the sense that it can and will react if necessary. In terms of disinfecting power, free available residual is 25 times more powerful than combined residual, and it will not produce the characteristic swimming-pool odor that combined residuals do. Because free available chlorine residual forms only after the breakpoint, the process is called **breakpoint chlorination**.

The free available chlorine residual at the consumer's tap should be at least 0.2 mg/L or at a level specified by the state. This level helps ensure that the water is free from harmful bacteria. However, higher levels may be necessary to control special problems, such as iron bacteria. If maintaining a free chlorine residual in a distribution system becomes difficult, several possible problems are indicated. Stagnant water in dead ends or storage tanks, biological growths, contamination of mains during main-break repair, and contamination caused by cross-connections all cause dissipation of free chlorine residuals. Further, a drop in chlorine residual in the distribution system may indicate inadequacies in the treatment process itself.

Chlorine Demand Test results for chlorine residual can be combined with operating data regarding the amount of chlorine added at the plant to yield information on chlorine demand. **Chlorine demand** is a measurement of how much chlorine must be added to the water to achieve breakpoint chlorination or whatever free chlorine residual is desired.

The most significant reason for analyzing a water supply's chlorine demand is to determine the proper dosage. However, changes in chlorine demand can also indicate water quality changes. For example, if a water supply suddenly requires more chlorine to maintain a residual (that is, if the water exhibits a higher chlorine demand), then the chlorine is oxidizing some contaminants that previously were not present in the water supply.

When chlorine demand increases, two steps are necessary. First, the chlorine dose must be increased to meet the higher demand. Second, the reason for the increased demand should be investigated. A sudden increase in chlorine demand frequently occurs in surface water because of seasonal water quality changes. Chlorine demand in groundwater should not change substantially because the quality of groundwater is usually very stable.

breakpoint chlorination
The addition of chlorine to water until the chlorine demand has been satisfied and free chlorine residual is available for disinfection.

chlorine demand
The quantity of chlorine consumed by reaction with substances in water.

Disinfection By-products

The disinfection by-products (DBPs) currently regulated are the THMs and the sum of five haloacetic acids. Continuing studies and research have revealed that chlorine (and all other alternate disinfectants) reacts with organic compound precursors in the water to form many different kinds of organic compounds. In current thinking, many of these compounds are considered to be potentially toxic and are suspected of being carcinogenic. Haloacetic acids (HAA5), halonitriles, haloaldehydes, and chlorophenols are just a few of the organic compounds associated with chlorine disinfection. Thus, chlorination has its good and its bad points; the water plant operator must know how to adequately disinfect the water without producing undesirable levels of DBPs.

Sampling Chlorine residual sampling is done at the treatment plant and at the consumer's faucet.

Treatment Plant Sampling In-plant sampling of chlorine residual determines whether sufficient chlorine has been added to the water before it leaves the treatment plant. This is the only way to be sure that finished water leaving the plant contains the desired chlorine residual. Obtaining representative samples is the most critical part of in-plant chlorine sampling.

In some instances, a sample must be collected at a point near the location of chlorine addition. In this case, analyses will probably show disproportionately high chlorine residual. To obtain data that approximate the actual chlorine residual in a particular basin, the sample should be held for a time period equal to the basin detention time or, in any case, at least 10 minutes. Under requirements of the Surface Water Treatment Rule (SWTR), surface water systems serving populations larger than 3,300 are required to provide continuous chlorine residual monitoring where the water enters the distribution system.

Distribution System Sampling For systems that chlorinate, sampling for chlorine residual from the distribution system is done to determine whether consumers are receiving water that is of good quality. In other words, if there is no chlorine residual, the operator will have to determine the cause of the reduced residual and take corrective measures to restore the chlorine residual to the area to ensure the safety of the water supply. If the water samples are positive for bacteria and chlorine residual is analyzed at the time of sample collection, the information regarding the level of chlorine residual may aid in determining whether the contamination was in the water sample or whether the equipment used for the analysis was contaminated. A strong chlorine residual in the sample collected may indicate that the water was not contaminated.

If analysis is made in the field, only about 10 mL of sample is required. Current regulations require that chlorine residual samples taken for compliance samples must be analyzed immediately or at the most within 15 minutes of collection. If samples must be taken to a laboratory, a 100-mL sample should be collected. Analysis should be completed as soon as possible after collection. There is no recommended preservation for chlorine samples; chlorine is unstable in water and residual chlorine will continue to diminish with time, and so immediate field testing is preferred. Chlorine analyses performed after the 15-minute window are not considered acceptable results for compliance testing for regulatory purposes.

Agitation or aeration of the sample should be avoided because it can cause reduction of the sample's chlorine concentration. Chlorine will also be destroyed and subsequent analysis will be erroneously low if samples are exposed to sunlight. The same sample bottle should never be used for both chlorine residual and coliform analyses. Bottles used for coliform analysis contain a chemical (sodium thiosulfate) that neutralizes the chlorine residual.

Methods of Determination The *N,N*-diethyl-*p*-phenylenediamine (DPD) test kit is the simplest and quickest way to test for residual chlorine. The test takes approximately 5 minutes to complete. The old orthotolidine method has been eliminated as an acceptable method and should not be used. It is not as accurate as the DPD method, particularly for measuring free chlorine residual.

Another technique, used primarily in laboratories because of its accuracy, is amperometric titration. The method is unaffected by sample color or turbidity, which can interfere with colorimetric determinations. However, performance of amperometric titration requires greater skill and care than does the DPD method. Because of the equipment and sample volumes required, the amperometric method is normally not used as a field test outside of the treatment plant.

The chlorine demand can be determined by treating a series of water samples with known but varying chlorine dosages. After an appropriate contact time, the chlorine residual of each sample is determined. This procedure indicates which dosage satisfied the demand and provided the desired residual.

Dissolved Oxygen

Dissolved oxygen (DO) in water is not considered a contaminant. Either an excess or a lack of DO does, however, help create unfavorable conditions. Generally, a lack of DO in natural waters creates the most problems, specifically an increase in tastes and odors as a result of anaerobic decomposition.

The amount of DO in water is a function of the water's temperature and salinity. Cold water contains more DO, and saline water contains less DO. Natural waters are seldom in equilibrium (exactly saturated with DO). Temperature changes as well as chemical and biological activities all use or release oxygen, causing the amount of DO in water to change continually.

Significance Nominal levels of DO in municipal water supplies are generally not a problem. DO has no adverse health effects and actually increases water's palatability. Most consumers prefer water that has a DO content near the saturation point. However, a concentration this high is detrimental to metal pipes because oxygen helps accelerate corrosion.

Introducing oxygen into water can be a method of treatment for purposes such as oxidizing iron and manganese into forms that will precipitate out of the water. DO also has the ability to remove excess carbon dioxide. It degrades some organic compounds that cause taste and odor problems, provided the contact time is long enough.

Where additional DO is desired for water treatment, some form of aeration is used. Aeration processes are discussed in Chapter 9.

DO data from raw-water storage-reservoir samples can also be used to indicate the general quality of the water. On the basis of these data, operators may be able to make treatment changes or alter the way the reservoir releases are made to prevent taste, odor, and other problems.

Sampling DO analyses should be conducted routinely on raw-water samples from surface water sources, particularly if storage reservoirs are being used. Treated-water samples should also be analyzed routinely if aeration is used as a treatment process; otherwise the tests may be conducted on a weekly basis for general quality data.

DO should be determined on-site if the electrode method (see next section) is used. If the modified Winkler test (see next section) is being used, the sample must be collected in a glass bottle and "fixed" (treated with a chemical additive to retard change) on-site. The sample should be stored in the dark at the temperature of the collection water, or the water should be sealed and kept at a temperature of 50–68°F (10–20°C) until the analysis can be performed.

Methods of Determination The **electrode method** and the **modified Winkler method** (also called the *iodometric method*) are preferred for DO measurements. Because the electrode method is not as sensitive to interferences as is the modified Winkler test, it is excellent for analyzing DO in polluted waters, highly colored waters, and strong waste effluents. Drinking waters and supply reservoirs have few interferences that cause problems with the modified Winkler procedure.

Inorganic Metals

The effects of metal in water are varied. Table 3-1 in Chapter 3 provides information on inorganic contaminants including metals covered under the National Primary Drinking Water Regulations, the MCLG, MCL, or treatment technique. Information is provided about potential health effects from ingestion of water, sources, and treatment technology information.

Three types of metals are discussed in this section. Dissolved metals are metals in an unacidified sample that pass through a 0.45-μm membrane. Suspended metals are metals in an unacidified sample retained by a 0.45-μm membrane. Total metals are the sum of suspended and dissolved metals or the concentration determined in an unfiltered digested sample.

Sampling and Sample Preservation Before collecting a sample, it must be decided what fraction is to be analyzed (dissolved, suspended, total, or acid extractable). This decision will determine in part whether the sample is acidified with or without filtration and the type of digestion required.

Serious errors may be introduced during sampling and storage because of contamination from the sampling device, failure to remove residues of previous samples from the sample container, and loss of metals by adsorption on or precipitation in the sample container caused by failure to acidify the sample properly.

Sample Containers Because of expense, the preferred sample container is made of polypropylene or linear polyethylene with a polyethylene cap. Borosilicate glass containers also may be used, but soft glass containers should be avoided for samples containing metals in the microgram-per-liter range. Samples should be stored for determination of silver in light-absorbing containers. Only containers and filters that have been acid rinsed should be used.

Preservation The samples should be preserved immediately after sampling by acidifying with concentrated nitric acid (HNO_3) to pH <2. The samples should be filtered for dissolved metals before preserving. Usually 1.5 mL concentrated HNO_3 added to a liter sample (or 3 mL 1 + 1 HNO_3 added to a liter sample) is sufficient for short-term preservation. For samples with high buffer capacity, the amount of acid should be increased (5 mL may be required for some alkaline or highly buffered samples). The sample should be checked to ensure the pH of the sample has been reduced to below 2. Commercially available high-purity acid should be used.

After acidifying the sample, it is preferable to store it in a refrigerator at approximately 39°F (4°C). Under these conditions, samples with metal concentrations of several milligrams per liter are stable for up to 6 months (except mercury, which is stable up to 5 weeks).

electrode method
Any analytical procedure that uses an electrode connected to a millivoltmeter to measure the concentration of a constituent in water.

modified Winkler method
A modification of the standard Winkler (iodometric) method that uses an alkali-iodide-azide reagent to make the procedure less subject to interferences.

Methods of Determination The presence of metals is determined using both colorimetric and instrumental methods. Instrumental methods include atomic absorption spectrometry, which includes flame, electrothermal (furnace), hydride, and cold-vapor techniques; flame photometry; inductively coupled plasma emission spectrometry; inductively coupled plasma mass spectrometry; and anodic stripping voltametry.

Sources of Contamination Introducing contaminating metals from containers, distilled water, or membrane filters should be avoided. Some plastic caps or cap liners may introduce metal contamination; for example, zinc has been found in black Bakelite-type screw caps as well as in many rubber and plastic products, and cadmium has been found in plastic pipette tips. Lead may be a contaminant in urban air and dust depending on the industry, construction, and demolition taking place in areas where lead may have been used in the past. It used to be found in gasoline, so the by-products of gasoline engine use would have contained lead, which may have been deposited on a variety of surfaces including the soil in the area.

Take care not to introduce metals into samples during preliminary treatment. During pretreatment, contact with rubber, metal-based paints, cigarette smoke, paper tissues, and all metal products including those made of stainless steel, galvanized metal, and brass should be avoided. Conventional fume hoods can contribute to sample contamination, particularly during acid digestion in open containers if the solutions boil and splatter their contents to adjoining sample containers. Vessels should be covered with watch glasses and spouts turned away from incoming air to reduce airborne contamination. Plastic pipette tips often are contaminated with copper, iron, zinc, and cadmium. Colored plastics, which can contain metals, should not be used. Certified metal-free plastic containers and pipette tips should be used when possible. Glass should not be used if analyzing for aluminum or silica.

Metal-free water should be used for all operations. Reagent-grade acids used for preservation, extraction, and digestion should be pure. If excessive metal concentrations are found, ultrapure acids should be used. Blanks should be processed through all digestion and filtration steps, and blank results should be evaluated relative to corresponding sample results. Either corrections should be applied to sample results or other corrective actions should be taken as necessary or appropriate.

Inorganic Nonmetallic Compounds

Inorganic nonmetallic contaminants include cyanide, fluoride, nitrate, and nitrite. The analytical methods used for determination include wet-chemical techniques and instrumental methods such as ion chromatography.

Nitrate can occur in trace quantities and in high concentrations in surface water and groundwaters depending on the type of soil and the land uses in the watershed for surface water or the recharge zones for wells. Nitrite is an intermediate product formed when ammonia is oxidized to nitrate and when nitrate is reduced. Cyanide is not normally found in either surface water or groundwater. Fluoride, iron, and manganese are discussed in detail in the following sections.

Fluoride Fluoride is found naturally in many waters. It is also added to many water systems to reduce tooth decay.

Significance Research has demonstrated that drinking water containing a proper amount of fluoride results in reduction in tooth decay during the years of children's tooth formation (from birth to between the ages of 12 and 15 years). This is assuming that the children are actually drinking the tap water in the appropriate amounts to provide them with the correct dose for their age and body mass.

Fluoride concentrations in drinking water that are optimum for reducing tooth decay are based on average air temperature. Depending on air temperature, the regulatory agencies set the levels the treatment plants that feed fluoride are allowed to introduce. Because more water is consumed in warmer climates, fluoride concentrations should be lower in these areas. Excessive fluoride concentrations can cause teeth to become stained or mottled. This problem generally occurs only where natural fluoride concentrations exceed 2.0 mg/L (the secondary MCL).

Close control of fluoride concentrations is necessary to ensure the maximum benefit of fluoridation with an adequate margin of safety. A reduction of only 0.3 mg/L below the optimum concentration can drastically reduce the dental benefits of fluoride. Specific guidelines from the state should be obtained concerning recommended concentrations for a given water supply if fluoride is being added to the water.

Sampling Fluoride samples should be collected from raw and finished water in polyethylene bottles. Raw-water samples are necessary because the total amount of fluoride reaching the consumer is equal to the fluoride concentration in the raw water plus the fluoride added at the plant. Although the fluoride level in most source water is fairly stable, it can vary somewhat and so should be periodically analyzed. The amount of fluoride to be added to the raw water is calculated by subtracting the raw-water concentration from the desired treated-water concentration.

Finished-water samples are tested to ensure that the fluoride feeders are operating correctly and the final fluoride concentration is at the desired level. Nearly all regulatory agencies require that samples of the plant effluent be tested daily.

Samples collected for fluoride analysis may be held for 7 days before analysis. They should be stored in a refrigerator at 39°F (4°C) with no preservatives added.

Methods of Determination Two methods for fluoride analysis are commonly used: the **SPADNS method** and the electrode method. (SPADNS stands for sodium 2-[parasulfophenylazo]-1.8-dihydroxy-3,6-naphthalene disulfonate.) The electrode method requires a selective ion fluoride electrode connected to a pH meter with a millivolt scale or to a meter having a direct concentration scale for fluoride. With either method, interferences may require a distillation step prior to the test. The water to be tested should be checked for interferences that might be present.

Iron Iron occurs naturally in rocks and soils and is one of the most abundant of all elements. It exists in two forms: ferrous (Fe^{+2}) and ferric (Fe^{+3}). Ferrous iron is found in well waters and in waters with a low level of DO. Under anaerobic conditions, waters can have significant dissolved-iron concentrations.

Dissolved iron in water is derived naturally from soils and rocks. It may also result from the corrosive action of water on unprotected iron or steel mains, steel well casings, and tanks. Surface waters may also occasionally contain appreciable amounts of iron that originates from industrial wastes or from acid runoff from mining operations.

SPADNS method
A colorimetric procedure used to determine the concentration of fluoride ion in water. SPADNS (sodium 2-[parasulfophenylazo]-1.8-dihydroxy-3,6-naphthalene disulfonate) is the chemical reagent used in the test.

Significance Iron at levels normally found in properly treated drinking water does not present a health issue. Even though iron poisoning can take place, the normal dosage would be in the range of 20 mg/kg or above for symptoms of iron toxicity to develop. Water quality limits on allowable concentrations of iron in water supplies are based on the problem of discoloration and undesirable taste and of the iron staining porcelain plumbing fixtures. Iron concentrations above 0.3 mg/L can cause undesirable "red water." Concentrations at or above this level in finished water indicate that steps should be taken to remove iron.

Iron also provides a nutrient source for some bacteria that grow in distribution systems and wells. Iron bacteria, such as *Gallionella*, cause red water, tastes and odors, clogged pipes, and pump failure.

Whenever tests show increased iron concentrations between the water plant and the consumer's tap, corrosion and/or iron bacteria may be present, and corrective action should be taken. If the water is corrosive, pH adjustment might be considered first. If the problem is caused by bacteria, flushing of the mains, shock chlorination (temporarily high concentrations), or increased everyday chlorination may prove effective.

Sampling Samples should be taken from raw and finished water. The samples should be collected in glass or plastic bottles and may be stored as long as 6 months before analysis. At least 100 mL of sample should be collected. Samples should be preserved with approximately 0.5 mL of concentrated nitric acid per 100-mL sample to lower the pH to less than 2.

Methods of Determination Iron concentration may be determined by the phenanthroline method or the atomic absorption spectrophotometric (AA) method. The phenanthroline method is simple and reliable. It is a colorimetric test and can be run with a spectrophotometer or filter photometer. The AA method, used by large laboratories, is very accurate and is particularly advantageous when large numbers of samples must be tested. Another method allowed in the regulations is Inductively Coupled Plasma (ICP) analysis.

Manganese Manganese, a metal, creates problems in a water supply similar to those created by iron. Manganese occurs naturally in ores but not in a pure state. It exists in soils primarily as manganese dioxide. It is found both in the manganous divalent form (Mn^{+2}) and in the quadrivalent form (Mn^{+4}). Manganese is much less abundant in nature than is iron; therefore, it is found less often in water supplies and is often present at lower concentrations. It is also more difficult than iron is to oxidize or cause to precipitate because manganous solutions are more stable than are ferrous solutions.

The most common forms of manganese—oxides, carbonates, and hydroxides—are only slightly soluble. Consequently, manganese concentrations in surface waters seldom exceed 1.0 mg/L. In groundwaters subject to anaerobic or reducing conditions, manganese concentrations, like iron concentrations, can become very high.

Significance Consumption of manganese when it is present in properly treated drinking water has shown no known harmful effects on humans. Water quality limits on allowable concentrations of manganese have been based on aesthetic problems rather than health concerns. Manganese does not usually discolor water, but it stains clothes and bathroom fixtures black. Staining problems can begin at 0.05 mg/L, a much lower concentration than for iron.

Raw-water and finished-water analyses will indicate whether manganese removal is necessary or whether the desired removal has been achieved in the treatment plant. Increases in manganese concentration in the distribution system are not generally experienced, except that a rapid flow change in the distribution system may result in some deposits breaking loose and entering consumers' water. This problem is best controlled by flushing the lines in areas where the problem occurs.

atomic absorption spectrophotometer (AA)
A spectrophotometer used to determine the concentration of metals in water and other types of samples.

Sampling Samples for manganese analysis should be taken from raw and finished water. The samples should be collected in glass or plastic bottles and may be stored as long as 6 months before analysis. At least 100 mL of sample should be collected. Samples should be preserved with concentrated nitric acid. Approximately 0.5 mL of concentrated nitric acid per 100-mL sample should be added to lower the pH to less than 2.

Methods of Determination The AA method or ICP mass spectroscopy (ICPMS) analysis are the preferred methods of determination. Also approved for use are colorimetric methods, which are more readily available and economical for the treatment plant operator.

pH

pH is a measure of the hydrogen ion concentration present in water, or it can be stated as the logarithm of the reciprocal of the hydrogen ion concentration, which is the same as the negative $\log_{10}$ of the hydrogen ion concentration in water.

$$pH = \log_{10} 1/[H^+]$$
$$= -\log_{10}[H^+]$$

Expression of pH Because pH is a logarithmic function, small changes in the measured pH mean large changes in the hydrogen ion concentration. The usual method for reporting pH results is to present the high and low pH along with an average pH or median value depending on what the reporting agency requests.

Importance of pH pH controls many chemical reactions, including coagulation, disinfection, water softening, corrosion, biochemical reactions, and ammonia removal. It can be affected by many of the treatment processes in the plant, including ion exchange, corrosion control, disinfectant chemicals, aeration, and coagulation, to name a few. It also indicates to the design engineer what construction materials to use. A question often asked in water treatment is, "How many pH meters should I install and where do I put them?"

Sampling Samples for pH analysis should be collected in clean plastic or glass containers and analyzed within 15 minutes of sample collection. Since pH may impact the performance and operation of many water treatment plant processes, it is typically analyzed at multiple points throughout the water treatment plant.

Methods of Determination In the lab and the field, sample pH is typically measured using a calibrated pH meter and probe. Buffers are used to calibrate pH meters, using buffer solutions that bracket the pH range of the samples being analyzed. To measure samples, a pH probe is immersed in a gently stirred sample until the pH reading stabilizes. Care should be taken to minimize interferences in pH calibration and measurement; for example, carbon dioxide, which can form carbonic acid in solution and depress pH. The probes used to measure pH require regular cleaning and maintenance, which should be performed according to the manufacturer's instructions. Online, continuously monitoring pH analyzers are also commonly used in water treatment plants for data trending and process control.

pH
A measure of water's acidity or alkalinity. A scale of 0 to 14 is used, with 0 being extremely acidic and 14 being extremely alkaline.

Organic Contaminants

No single analytical method is capable of measuring all of the organic substances in a water sample. However, available analytical methods can be grouped into two categories, general and specific.

General Analytical Methods

Threshold odor tests, flavor profiles, and color determinations, discussed earlier in the chapter, have been used in the water utility industry for many years to obtain general measures of the levels of natural organic compounds in water. Two other methods used occasionally in monitoring water quality are UV light absorbance and fluorescence. These tests are used in some plants for control of organic compound removal processes because the measurement can be made quickly and easily.

Another test commonly used to determine the overall content of organic compounds in water is the measurement of **total organic carbon (TOC)**. The typical concentration of TOC in water sources ranges from less than 0.5 mg/L to more than 10 mg/L. Highly colored water may have a TOC concentration of more than 30 mg/L.

Total organic halogen (TOX) is another measurement that is being used increasingly because it is specific to halogenated organic compounds. The presence of TOX in a sample is an indication of the presence of either synthetic organic compounds or DBPs. Only fairly sophisticated laboratories are currently capable of carrying out the procedures for TOC and TOX determinations.

Specific Analytical Methods

The list of organic compounds that have been identified in drinking water samples has grown from approximately 200 in 1975 to thousands today, and it is constantly lengthening. In many cases, of course, a compound may have been identified only in isolated samples or at extremely low concentrations. However, the growth in the list is primarily because of steadily increasing improvements in analytical methods.

Gas chromatography (Figure 4-19) requires very specialized equipment, detailed procedures, and trained operators, but in general the three steps are as follows:

1. Extraction and concentration of the organic compounds in the sample
2. Separation of the extracted organic compounds in a gas chromatograph
3. Detection of individual compounds

Extraction and Concentration Organic substances are first extracted from a water sample. One method uses an organic solvent such as methylene chloride or pentane. This process is called liquid–liquid extraction. Another method strips the organic compounds out of the sample using an inert gas such as nitrogen or helium. This process is called purge-and-trap analysis.

Separation The complex solution must then be separated into its individual organic components. This process is carried out with a gas chromatograph or a high-performance liquid chromatograph. Chromatographs have a column of long, thin tubing through which individual organic compounds are driven off the sample as the temperature is elevated. Thus, these processes may be viewed as sophisticated distillation or separation functions.

Detection

As the chromatograph separates the organic compounds by the temperature at which they are vaporized, they travel to a detector. Several types of detectors are available, each with certain advantages and disadvantages. The types in general use include flame ionization, electron capture, electrolytic conductivity, photoionization, and mass spectrometry. An organic compound is identified by comparing

total organic carbon (TOC)
The results of a general analysis performed on a water sample to determine the total organic content of the water.

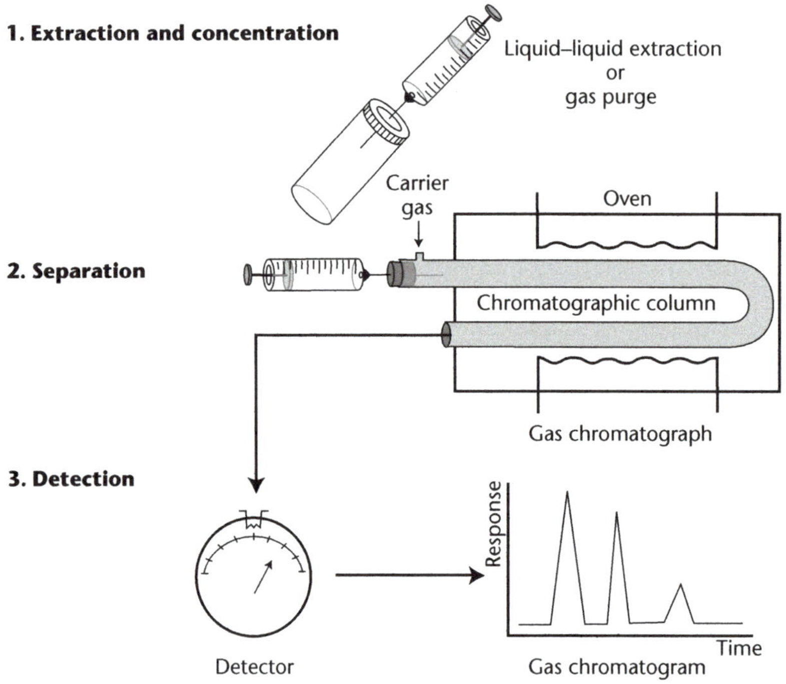

Figure 4-19 Steps in gas chromatography

the signal the detector obtains (shown graphically in a gas chromatograph [GC] generated by the detector) with known standards for the compound. This process is generally aided by a computer connected to the equipment.

Figure 4-20 shows a chart produced by a GC, showing the presence of THMs in a water sample.

Sampling for Organic Compounds

The location of sampling points in a water distribution system is very important; certain points should definitely not be used. The following are among typical locations to be avoided:

- Public restrooms should not be used as sampling locations because the deodorizer commonly used in restrooms contains an organic chemical that may be in the air in sufficient concentration to contaminate the water sample.
- Gasoline service stations should be avoided because of the prevalence of petroleum products that could be in the air or that could have gotten on the sampling faucet.
- Any location where there are unusual odors, such as a freshly painted room, or where there is a smell from cleaning materials, should be avoided.
- A location where a pump or piping has recently been installed or repaired, especially if organic solvents have been used for cleaning and degreasing, should be avoided because of the possibility that organic solvents may have been used in the plumbing.
- Other unsuitable locations are those where solvents may be present in the atmosphere such as paint or hardware stores, barber and beauty shops (hair spray, etc.), and dry-cleaning establishments.

gas chromatography (GC)
A technique used to measure the concentration of organic compounds in water.

Ultraclean glass vials having lids with polytetrafluoroethylene (PTFE; trade name Teflon) liners are used for collecting volatile organic compound samples. The samples must be collected so that there is zero headspace in the vial; in other words, there must be no bubble of air in the vial after it is filled. If any air remains in the vial, a portion of the more volatile organic compounds will come out of

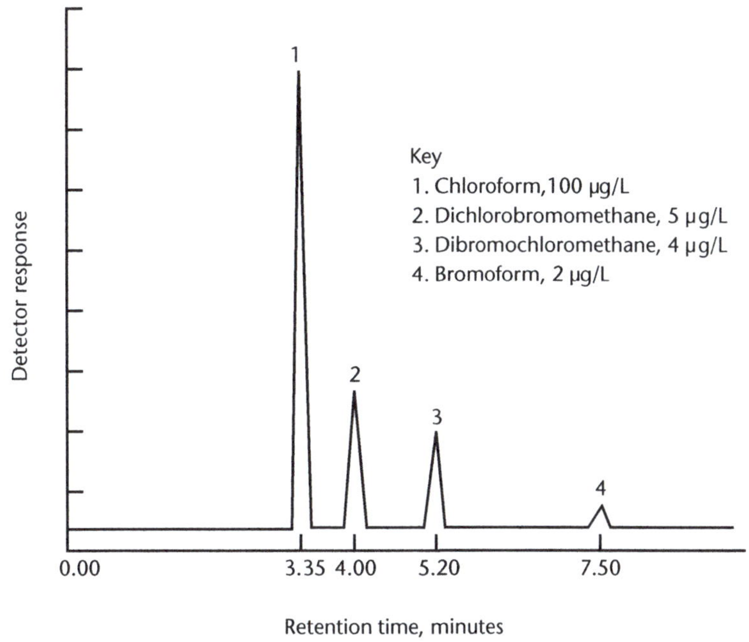

Figure 4-20 Sample readout from a gas chromatograph

solution and into the air space, which will cause inaccurate analysis of the water sample. Trip blanks and field blanks should also be incorporated into the normal sampling practices.

Each sample container must be completely labeled. A general rule is that the description of the sampling site must be complete enough so that a person unfamiliar with the initial sampling could return and collect a repeat sample from the same location if necessary.

Coagulant Effectiveness

The removal of suspended solids from surface water is necessary both to make the water aesthetically pleasing to customers and to assist in the elimination of pathogenic organisms. The SWTR, the TCR, and the Disinfectants and Disinfection By-products Rule (DBPR) require more complete removal of turbidity and dissolved organics than was previously practiced by most water systems. This requirement in turn demands more efficient coagulation, flocculation, and sedimentation. Effective coagulation is also a tool in removing organic chemical precursors from the raw water.

Significance

Coagulation and flocculation involve the addition of chemical coagulants such as aluminum sulfate, ferric chloride, or polyelectrolytes to raw water to hasten the settling of suspended matter. Plant operation is most efficient when the lowest turbidity is obtained in finished water with the lowest cost for coagulant chemicals. Several laboratory tests can provide the information necessary to accomplish this goal. These tests allow operators to select optimal chemical dosages in the laboratory rather than using trial and error in the plant.

The tests can also be used to check the adequacy of flash mixing or flocculation mixing in the plant. Test results may indicate when to improve or modify flash mixers and flocculation basins to obtain more efficient operation.

When coagulants are evaluated, the goal is to identify the one coagulant (or combination of coagulant and coagulant aids) that will produce low turbidity with the least expensive dose of chemicals. Chemical prices must also be evaluated, because a low dose of an expensive chemical may be more cost effective than a high dose of an inexpensive chemical. One must also consider the cost and ease of disposal of the coagulant and the contaminants removed in the process.

Sampling

Where the samples are collected for analysis depends on the procedure being used and the information desired from the test. The tests should be conducted whenever there is a significant change in water quality or when other conditions may require a change in coagulant dose.

Methods of Determination

The following methods are commonly used to determine optimum coagulant effectiveness:

- Jar test
- Zeta potential detector
- Streaming current detector (SCD)
- Particle counting

Of these methods, the **jar test** is the most readily available to most operators and has been commonly used for many years. There is no standard procedure for conducting the jar test, nor is there standard test equipment that must be used. A typical procedure for conducting a jar test is provided in American Water Works Association (AWWA) *Manual M12, Simplified Procedures for Water Examination.*

Briefly, the procedure is to collect a sample of the raw water and add progressively larger doses of the coagulant chemical being tested to several jars of the sample. The test should be run as quickly as possible after the sample has been collected because a change in water temperature can have a significant effect on the results. Actual plant conditions of flash-mixing speed and time, flocculation time, and settling time are then simulated in the jars.

Visual observation of the floc and turbidity readings of the settled water in the jars provide the data necessary to make determinations. The types of equipment available for performing jar tests ranges from simple hand-shaken jars to computer-controlled testing devices. A motor-driven jar test device is pictured in Figure 4-21.

Figure 4-22 illustrates typical jar test results from a water supply in which alum is used as a coagulant. The data show that the benefit of the coagulant decreases with doses larger than 35 mg/L for turbidity removal. In other words, beyond that dose a very large increase in the amount of chemical is required to

jar test
A laboratory procedure for evaluating coagulation, flocculation, and sedimentation processes. Used to estimate the proper coagulant dosage.

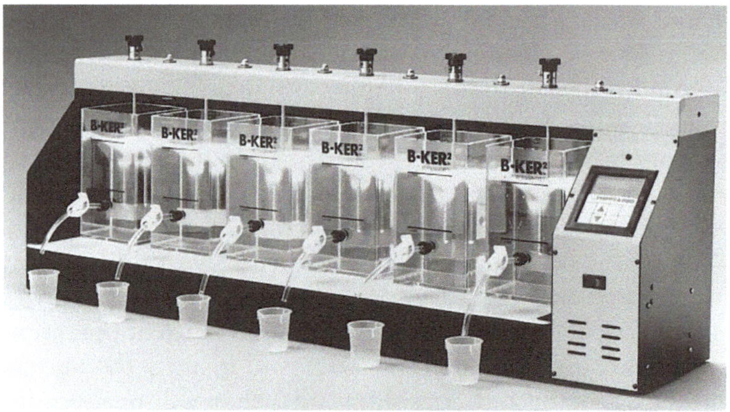

Figure 4-21 Jar test apparatus

Courtesy of Phipps & Bird.

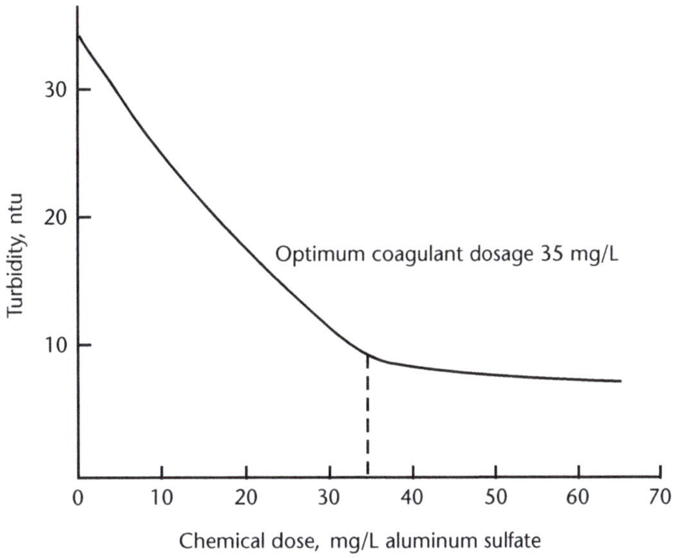

Figure 4-22 Typical jar test results

produce a small increase in turbidity removal. Therefore, 35 mg/L should be considered the optimal dose for alum alone for the water tested for turbidity removal. Additional coagulant may be required to obtain the necessary TOC removal as required by the DBPR.

Study Questions

1. Which type of sample should always be collected for determining the presence of coliform bacteria?
   a. Time composite
   b. Grab sample
   c. Proportional
   d. Composite

2. Which chemical is used to remove residual chlorine from water?
   a. $Na_2S_2O_3$
   b. Na_2SiO_3
   c. Na_2SiF_6
   d. NaOCl

3. Water that is to be analyzed for inorganic metals should be acidified with
   a. dilute hydrochloric acid.
   b. concentrated hydrochloric acid.
   c. dilute nitric acid.
   d. concentrated nitric acid.

4. A typical coliform colony in the membrane filter method has the following characteristics:
   a. Blue with lustrous surface sheen
   b. Pink to dark red with green metallic surface sheen
   c. Pink or yellow with lustrous to metallic surface sheen, depending on species
   d. Yellow with silver metallic to lustrous surface sheen

5. Which of the following is the first level of total dissolved solids that is considered unfit for human consumption?
   a. 1,000 mg/L
   b. 2,500 mg/L
   c. 4,000 mg/L
   d. 5,000 mg/L

6. Which type of special photometer is used to analyze for heavy metals in water?
   a. Electrophotometer
   b. Magnetophotometer
   c. Spectrophotometer
   d. Atomic absorption spectrophotometer

7. In the multiple-tube fermentation method, which type of culture media is used in the confirmed test?
   a. Lactose agar
   b. Green lactose bile broth
   c. Eosin methylene red agar
   d. Glucose bile tubes

8. All positive samples in the presence–absence test must be confirmed using
   a. brilliant green bile tubes.
   b. eosin methylene red agar.
   c. lauryl tryptose broth.
   d. eosin 12-amino acid broth.

9. What is pH_s?
   a. pH of a solution.
   b. The Langlier Index
   c. pH saturation
   d. Salt pH

10. _____ is a measure of the hydrogen ion concentration present in water.
    a. Turbidity
    b. pH
    c. Gas chromatography
    d. Covalence

11. What steps are involved in the multiple-tube fermentation method?

12. Conductivity (specific conductance) is a measure of the ability of an aqueous solution to do what?

13. Describe the characteristics of carbon dioxide.

14. In regard to the testing of coagulant effectiveness, which method is the most readily available to most operators and has been commonly used for many years?

Chapter 5
Corrosion Control

Many water systems must apply special chemical treatment because their source water either causes damaging corrosion or deposits scale on pipelines and plumbing fixtures. The treatment process for controlling these problems is known as stabilization. Many more systems must provide corrosion control treatment under new federal and state regulations enacted to protect the public from the health dangers of lead and copper, and other trace metals in drinking water.

Water system operators are cautioned that complicated interactions often occur in the control of water corrosion and scaling. A seemingly simple change to improve one characteristic may have an adverse effect on some other water characteristic or treatment process. This chapter offers an overview of the need for corrosion and scaling control and their related processes. However, because of all the complex variables involved, it is not intended to be a guide for the best treatment method for any particular system. It is best to get professional guidance and state approval before beginning any new stabilization treatment.

stabilization
The water treatment process intended to reduce the corrosive or scale-forming tendencies of water.

Purposes of Corrosion and Scaling Control

Corrosion and scaling are controlled for the following reasons:

- Protect public health
- Improve water quality
- Extend the life of plumbing equipment
- Meet federal and state regulations

corrosion
The gradual deterioration or destruction of a substance or material by chemical action. The action proceeds inward from the surface.

scaling
Metal deposits left in pipelines and plumbing fixtures.

Protecting Public Health

Corrosive water can leach toxic metals from distribution and household plumbing systems. Lead and copper are the metals most likely to be a problem because they are commonly used in plumbing systems.

In addition, corrosion of cast-iron mains can cause the formation of iron deposits, called tubercles, in the mains. These deposits can protect bacteria and other microorganisms from chlorine, allowing them to grow and thrive. Changes in water velocity or pressure can then cause the microorganisms to be released, creating a potential for disease outbreaks. Some bacteria shielded by the tubercles can also accelerate the corrosion process.

tubercles
A knob of rust formed on the interior of cast-iron pipes as a result of corrosion.

Table 5-1 Estimated effect of scale on boiler fuel consumption

Scale Thickness, in.	mm	Fuel Consumption, % increase
1/50	0.5	7
1/16	1.6	18
1/18	3.2	39

Improving Water Quality

Corrosive water attacking metal pipes can cause taste, odor, and color problems in a water system. Red-water problems occur when iron is dissolved from cast-iron mains by corrosive water. The iron will stain a customer's plumbing fixtures and laundry and make the water's appearance unappealing for drinking and bathing. The dissolved iron also acts as a food source for a group of microorganisms called iron bacteria, which can cause serious taste-and-odor problems. Corrosion of copper pipes can cause a metallic taste, as well as blue-green stains on plumbing fixtures and laundry.

Extending the Life of Plumbing Equipment

Unstable water can also result in significant costs to water systems and customers. Aggressive water can significantly reduce the life of valves, unprotected metal, and asbestos–cement (A–C) pipe. It can also shorten the service life and performance of plumbing fixtures and hot water heaters.

Buildup of corrosion products (a process known as *tuberculation*) or uncontrolled scale deposits can seriously reduce pipeline capacity and increase resistance to flow. This impaired flow in turn reduces distribution system efficiency and increases pumping costs. If scale deposits or tuberculation go unchecked, pipes can become completely plugged, requiring expensive repair or replacement. Scaling can also increase the cost of operation of hot water heaters by increasing their fuel consumption, as shown in Table 5-1.

Meeting Federal and State Regulations

As detailed in Chapter 3 and later in this chapter, the Lead and Copper Rule enacted by the US Environmental Protection Agency (USEPA) in 1991 requires water systems to check if their water is corrosive enough to cause lead and copper corrosion products to appear in customers' water at levels exceeding the new action level. If the level is exceeded, the system is required to take action to reduce the corrosivity of the water.

iron bacteria
Bacteria that use dissolved iron as an energy source. They can create serious problems in a water system because they form large, slimy masses that clog well screens, pumps, and other equipment.

rust
Oxidized iron.

corrosive or aggressive
Tending to deteriorate material, such as pipe, through electrochemical processes.

Water System Corrosion

Corrosion can be broadly defined as the wearing away or deterioration of a material because of chemical reactions with its environment. The most familiar example is the formation of rust (oxidized iron) when an iron or steel surface is exposed to moisture. Corrosion is usually distinguished from erosion, which is the wearing away of material caused by physical causes, such as abrasion. Water that promotes corrosion is known as corrosive or aggressive water.

In water treatment operations, corrosion can occur to some extent with almost any metal that is exposed to water. Whether corrosion of a material will be extensive enough to cause problems depends on several related factors, such as the type of material involved, the chemical and biological characteristics of the water, and the electrical characteristics of the material and its environment.

The relationships among these factors, as well as the process of corrosion itself, are quite complex. As a result, it is difficult to make general statements about what combinations of water and equipment will or will not have corrosion problems. The discussions in this chapter cover only basic principles; the operator faced with persistent corrosion in a given installation may require the assistance of corrosion-control specialists.

Chemistry of Corrosion

The chemical reactions that occur in the corrosion of metals are similar to those that occur in an automobile battery. In fact, corrosion generates an electrical current that flows through the metal being corroded. The chemical and electrical reactions that occur during concentration cell corrosion of iron pipes are illustrated in Figure 5-1.

WATCH THE VIDEO
Corrosion Control—Anodes and Cathodes (www.awwa.org/wsovideoclips)

concentration cell corrosion
A form of localized corrosion that can form deep pits and tubercles.

As shown in Figure 5-1A, minor impurities and variations (present in all metal pipes) have caused one spot on the pipe to act as an electrical anode in relation to another spot that is acting as an electrical cathode. At the anode, atoms of iron (Fe^{+2}) are breaking away from the pipe and going into solution in the water. As each atom breaks away, it ionizes by losing two electrons, which travel through the pipe to the cathode.

In Figure 5-1B, it is shown that chemical reactions within the water balance the electrical and chemical reactions at the anode and cathode. Many of the water molecules (H_2O) have dissociated into H^+ and OH^- ions. This is a normal

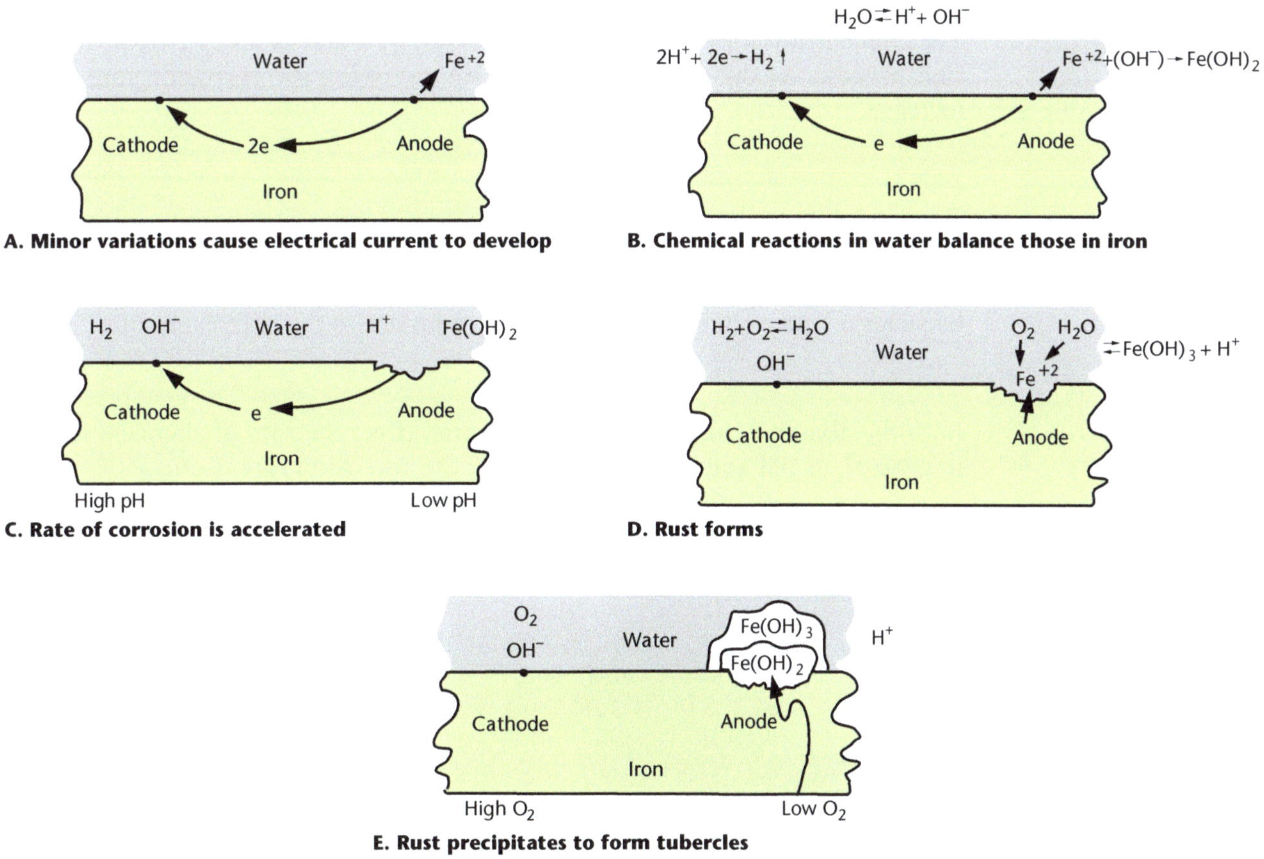

Figure 5-1 Chemical and electrical reactions that occur during corrosion of iron pipe

condition, even with totally pure water. The Fe^{+2} released at the anode combines with two OH^- ions from dissociated water molecules to form $Fe(OH)_2$, ferrous hydroxide. Similarly, two H^+ ions from the dissociation of the water molecules near the cathode pick up the two electrons originally lost by the iron atom then bond together as H_2, hydrogen gas.

The formation of $Fe(OH)_2$ leaves an excess of H^+ near the anode, and the formation of the H_2 leaves an excess of OH^- near the cathode. This change in the normal distribution of H^+ and OH^- accelerates the rate of corrosion and causes increased pitting in the anode area (the concentration cell), as shown in Figure 5-1C.

If the water contains dissolved oxygen (O_2)—most surface water does—then $Fe(OH)_3$, ferric hydroxide, will form (Figure 5-1D). Ferric hydroxide is common iron rust. The rust precipitates, forming tubercles (Figure 5-1E). The existence of tubercles further concentrates the corrosion, increasing both pitting at the anode and growth of the tubercle. Tubercles can grow into large nodules (Figure 5-2), significantly reducing the carrying capacity of a pipe. During rapid pressure or velocity changes, some of the $Fe(OH)_3$ can be carried away, causing "red water."

Factors Affecting Corrosion

The rate of corrosion depends on many site-specific conditions, such as the characteristics of the water and pipe material. Therefore, there are no established guidelines that determine the rate at which a pipe will be corroded.

Chemical reactions play a critical role in determining the rate of corrosion at both the cathode and the anode. Any factor that influences these reactions will also influence the corrosion rate.

Dissolved Oxygen

The concentration of dissolved oxygen (DO) in water is a key part of the corrosion process. As the concentration of DO increases, the corrosion rate will also increase.

Total Dissolved Solids

The total dissolved solids (TDS) concentration is important because electrical flow is necessary for the corrosion of metal to occur. Pure water is a poor conductor of electricity because it contains very few ions. But as the TDS is increased, water becomes a better conductor, which in turn increases the corrosion rate.

Alkalinity and pH

Both the alkalinity and the pH of the water affect the rate of chemical reactions. In general, as pH and alkalinity increase, the corrosion rate decreases.

Figure 5-2
Tuberculated pipe
Courtesy of Girard Industries.

Temperature

Because chemical reactions occur more quickly at higher temperatures, an increase in water temperature usually increases the corrosion rate.

Flow Velocity

The velocity of water flowing past a piece of metal can also affect the corrosion rate, depending on the nature of the water. If the water is corrosive, higher flow velocities cause turbulent conditions that bring DO to the corroding surface more rapidly, which increases the corrosion rate. However, if chemicals are being added to stabilize the water, the higher velocities will decrease the corrosion rate, allowing the chemicals (such as $CaCO_3$) to deposit on the pipe walls more quickly.

Type of Metal

Metals that easily give up electrons will corrode easily. Table 5-2 lists metals commonly used in water systems, with those at the top being most likely to corrode. This listing is called the galvanic series of metals. Where dissimilar metals are electrically connected and immersed in a common flow of water, the metal highest in the galvanic series will immediately become the anode, the other metal will become the cathode, and corrosion will occur. This process is termed galvanic corrosion (as opposed to concentration cell corrosion).

The rate of galvanic corrosion will depend largely on how widely separated the metals are in the galvanic series. Widely separated metals will exhibit extremely rapid corrosion of the anode metal (the highest in the series), and the cathode metal will be protected from corrosion. A common example of galvanic corrosion occurs when a brass corporation cock is tapped into a cast-iron main and attached to a copper service line. The copper will be protected at the expense of the brass and cast iron.

Table 5-2 Galvanic series for metals used in water systems

Corroded End (Anode)	**MOST ACTIVE**
Magnesium	+
Magnesium alloys	↑
Zinc	
Aluminum	
Cadmium	
Mild steel	
Wrought (black) iron	
Cast iron	Corrosion Potential
Lead–tin solders	
Lead	
Tin	
Brass	
Copper	↓
Stainless steel	–
Protected End (Cathode)	**LEAST ACTIVE**

galvanic series
A listing of metals and alloys according to their corrosion potential.

galvanic corrosion
A form of localized corrosion caused by the connection of dissimilar metals in an electrolyte, such as water.

Electrical Current

If electrical current is passed through any corrodible metal, corrosion will be accelerated. The two causes of electrical current in water mains are improperly grounded household electrical systems and electric railway systems.

Bacteria

Certain types of bacteria can accelerate the corrosion process because they produce carbon dioxide (CO_2) and hydrogen sulfide (H_2S) during their life cycles, which can increase the corrosion rate. These bacteria can also produce slime, which will entrap precipitating iron compounds and increase red-water problems and the amount of tuberculation.

Two groups of bacteria cause the most problems. Iron bacteria, such as *Gallionella* and *Crenothrix*, can form considerable amounts of slime on pipe walls, particularly if the water contains enough dissolved iron to allow them to survive. The iron present in the water can be naturally occurring in the source water or can be caused by corrosion of the pipe material. Beneath this slime layer, CO_2 production by the bacteria can significantly lower the pH, which will speed up the corrosion rate.

The periodic sloughing of these slime accumulations can cause other major problems, such as tastes and odors. The slimes can also prevent the effective deposition of a protective calcium carbonate ($CaCO_3$) layer by enmeshing it within the slime layer. As this layer sloughs away, it carries away the $CaCO_3$ and leaves the pipe surface bare.

Sulfate-reducing bacteria, such as *Desulfovibrio desulfuricans*, can accelerate corrosion when sulfate (SO_4^{-2}) is present in the water. They reduce the SO_4 under anaerobic conditions, which occur under the slime layer where oxygen is depleted by other bacteria. The products formed are iron sulfide (FeS) and hydrogen sulfide (H_2S), which causes obnoxious odors and black-colored water. The CO_2 formed can also lower the pH of the water.

All of these factors interact with each other and with metal pipes, tanks, and various equipment in water plants, the distribution system, and the customer's plumbing. As the factors change, the corrosion rate will also change. The only factors over which the operator has significant control are pH, alkalinity, and the bacteriological content of the water. The techniques used to control these parameters are discussed later in this chapter.

Types of Corrosion

Corrosion in water systems can be divided into two broad classes: localized and uniform. Localized corrosion, the most common type in water systems, attacks metal surfaces unevenly. It is usually a more serious problem than uniform corrosion because it leads to a more rapid failure of the metal. Two types of corrosion that produce pitting are galvanic corrosion and concentration cell corrosion (discussed previously).

Uniform corrosion takes place at an equal rate over the entire surface. It usually occurs where waters having very low pH and low alkalinity act on unprotected surfaces.

Scale Formation

The formation of mild scale on the interior of pipes can protect the pipe from corrosion by separating the corrodible pipe material from the water. However, uncontrolled scale deposits can significantly reduce the carrying capacity of a

Figure 5-3 Scaling of pipe
Courtesy of Johnson Controls.

distribution system. Figure 5-3 shows a pipe that is almost completely blocked by scale.

Chemistry of Scale Formation

Scale is formed when the divalent metallic cations associated with hardness, primarily magnesium and calcium, combine with other minerals dissolved in the water and precipitate to coat pipe walls. The most common form of scale is calcium carbonate ($CaCO_3$). Other scale-forming compounds include magnesium carbonate ($MgCO_3$), calcium sulfate ($CaSO_4$), and magnesium chloride ($MgCl_2$).

Factors Affecting Scale Formation

Water can hold only so much of any given chemical in solution. If more is added, it will precipitate instead of dissolve. The point at which no more of the chemical can be dissolved is called the saturation point. The saturation point varies with other characteristics of the water, including pH, temperature, and TDS.

The saturation point of calcium carbonate ($CaCO_3$) depends primarily on the pH of the water. For example, if water with a certain temperature and TDS concentration can maintain 500 mg/L of $CaCO_3$ in solution at pH 7, then the same water will hold only 14 mg/L of $CaCO_3$ in solution if the pH is raised to 9.4.

Temperature also affects the saturation point, although not as dramatically as pH. The solubility of $CaCO_3$ in water decreases as temperature increases. The most common example is when the higher temperature in hot water heaters and boilers causes scale to precipitate out of the water and build up on pipe and tank walls. Because the presence of other minerals in the water affects the solubility of $CaCO_3$, the TDS concentration must be known in order to determine the $CaCO_3$ saturation point. As the TDS concentration increases, the solubility of $CaCO_3$ increases.

WATCH THE VIDEO
Lime Softening—Scaling (www.awwa.org/wsovideoclips)

Corrosion and Scaling Control Methods

The following are the basic methods used for stabilizing water to protect against the problems of corrosion or scaling:

- Adjustment of pH and alkalinity
- Formation of a calcium carbonate coating
- Use of corrosion inhibitors and sequestering agents

The selection of which method or methods are finally used on any water system depends on both the chemical characteristics of the raw water and the effects of other treatment processes being used.

saturation point
The point at which a solution can no longer dissolve any more of a particular chemical. Precipitation of the chemical will occur beyond this point.

The type of source water, the number of sources, and the hydraulics and flow patterns of the system can also have a bearing on the choice of corrosion-control measures that should be taken. Systems that have multiple sources with different chemistry may have particularly complex problems. The pH of the water is sometimes critical in the use of chemical corrosion-control measures; some measures work properly only within a narrow pH range. The principal treatment techniques for corrosion and scaling control are summarized in Table 5-3.

Table 5-3 Summary of treatment techniques for controlling corrosion and scaling

Treatment	Application	Effectiveness	Comments or Problems
To Prevent Corrosion			
Lime alone or lime with sodium carbonate or sodium bicarbonate	Increase pH Increase hardness Increase alkalinity	Most effective in water with low pH and hardness Excellent protection for copper, lead, and asbestos–cement pipe in stabilized waters Good protection for galvanized and steel pipe	May be best overall treatment approach Oversaturation may cause calcium deposits
Sodium hydroxide	Increase pH	Most effective in waters with sufficient hardness and alkalinity to stabilize water May provide adequate protection against lead corrosion in low-alkalinity, soft waters	Should not be used to stabilize waters without the presence of adequate alkalinity and hardness May cause tuberculation in iron pipes at pH 7.5–9.0
Sodium hydroxide and sodium carbonate or sodium bicarbonate	Increase pH Adjust alkalinity	Most effective in water with low pH and sufficient hardness Excellent protection for lead corrosion in soft waters at pH 8.3	Combination of high alkalinity and hardness with low pH is more effective than combination of high pH with low hardness and alkalinity
Inhibition with phosphates (primarily sodium zinc phosphate and zinc orthophosphate)	Form a protective film on pipe surfaces	Effective at pH levels above 7.0 Good protection for asbestos–cement pipe Addition of lime may increase effectiveness of treatment for copper, steel, lead, and asbestos	May cause leaching of lead in stagnant waters May encourage the growth of algae and microorganisms May cause red water if extensive tuberculation is present May not be effective at low pH levels
Inhibition with silicates	Form a protective film on pipe surfaces	Most effective in waters having low hardness and pH below 8.4 Good protection for copper, galvanized, and steel pipe	May increase the potential of pitting in copper and steel pipes May not be compatible with some industrial processes
To Prevent Scale Formation			
Carbon dioxide or sulfuric acid	Decrease pH Decrease alkalinity	Effective with high-pH, high-alkalinity water such as lime-softened water	Overfeeding can cause low pH and corrosion
Sequestering with phosphates (primarily sodium hexametaphosphate and tetrasodium polyphosphate)	Sequester scale-forming ions	Effective in controlling scale formation from lime-softened waters and iron in the source water	Can loosen existing deposits and cause red-water complaints Compounds lose sequestering ability in hot water heaters, causing precipitation of calcium carbonate or iron

Adjustment of pH and Alkalinity

In general, soft waters that have a pH of less than 7 and are slightly buffered will be corrosive to lead and copper. Water that has too much alkalinity can also be quite corrosive. Water that is nominally corrosive naturally can also have the corrosivity increased by the addition of other water treatment chemicals. For instance, gaseous chlorine will reduce pH levels.

Usually, a moderate increase in pH and alkalinity levels can reduce corrosion, and a decrease can prevent scale formation. The formation of a protective film on the interior of lead and copper pipes is also usually aided by increasing pH.

Lime is normally used to increase both pH and alkalinity because it is less expensive than other chemicals having the same effect. Soda ash (sodium carbonate) can be added along with the lime to further increase the alkalinity. Sodium bicarbonate is sometimes used instead of sodium carbonate because it will also increase alkalinity without as much of an increase in pH. The increased alkalinity buffers the water against pH changes in the distribution system. This has proven particularly effective in controlling corrosion of lead and copper service pipes. Instead of lime, caustic soda with soda ash or sodium bicarbonate can be used to increase pH and alkalinity.

Lime-softened water can cause severe scale problems if it is not stabilized (see Chapter 6). Stabilization after softening is accomplished by the addition of carbon dioxide (a process called *recarbonation*) or sulfuric acid. Both chemicals lower the pH so that calcium carbonate will not precipitate in the distribution system.

Formation of a Calcium Carbonate Coating

Because corrosion attacks the surface of a pipe, a protective coating on the pipe surface can inhibit corrosion. A coating of cement, plastic, or asphaltic material is commonly applied to the interior of pipes and tanks and on metal equipment to protect them from corrosion. Although these coatings form an effective barrier against corrosion, very aggressive water can sometimes attack the coatings. In addition, if there are any breaks in the coating, corrosion will be particularly severe at these points.

As a result, many systems apply an additional protective coating by controlling the chemistry of the water. A common protective-coating technique is to adjust the pH of the water to a level just above the saturation point of calcium carbonate. When this level is maintained, calcium carbonate will precipitate and form a protective layer on the pipe walls. This process must be closely controlled. A pH that is too low may result in corrosion, and a pH that is too high may result in excessive precipitation, which will cause a clogging of service pipes and a restriction of flow in the distribution system. Lime, soda ash, sodium bicarbonate, or sodium hydroxide can be used to raise the pH level. Lime is often used because it also adds needed calcium (hardness) and alkalinity.

Use of Corrosion Inhibitors and Sequestering Agents

Some waters do not contain enough calcium or alkalinity to make the formation of calcium carbonate coatings economical. Water obtained primarily from snowmelt is an example, having alkalinity and calcium concentrations as low as 2 mg/L as $CaCO_3$. In this event, other chemical compounds can be used to form protective coatings.

The most common compounds are polyphosphates and silicates. The chemical reactions by which these compounds combine with corrosion products to form a protective layer are not completely understood; however, the chemicals have proven successful in many water systems.

Some polyphosphates can also be used as sequestering agents to prevent scale formation. These compounds sequester, or chemically tie up, the scale-forming ions of calcium and magnesium so that they cannot react to form scale. Because these compounds remain in solution, they are eventually ingested by consumers. Therefore, any sequestering agent selected must be suitable for use in drinking water.

Polyphosphates also sequester iron, whether it is dissolved in water from the source or from corrosion of the system. This prevents the precipitation of the iron compounds, so red water will not result. However, this effect does not prevent corrosion; it merely prevents the corrosion by-products from being noticed.

Corrosion and Scaling Control Facilities

Whatever method or methods are used to control corrosion and scaling, the facilities required for the process consist primarily of selected chemicals and the equipment required to store, handle, and feed these chemicals. A wide range of chemicals is available for stabilization. Some of these are primarily intended for industrial or other nonpotable applications. Therefore, all chemicals used to stabilize drinking water must be approved for such use. Chemicals must also comply with NSF International and American Water Works Association standards to ensure quality. Table 5-4 lists the chemicals commonly used for stabilizing potable water.

Chemical Storage and Handling

Chemicals should be purchased in quantities that will maintain a 30-day minimum supply at all times. This practice guards against interruptions of service because of temporary shortages and other unforeseen events. In determining the quantity of chemical to be stored at any one time, the operator should consider storage space available, discounts for purchasing large quantities, and length

Table 5-4 Chemicals commonly used for stabilization of potable water

Treatment Method	Chemical Name	Chemical Formula
Increase pH and alkalinity	Unslaked lime (quicklime)	CaO
	Slaked lime (hydrated lime)	$Ca(OH)_2$
	Sodium bicarbonate	$NaHCO_3$
	Sodium carbonate (soda ash)	Na_2CO_3
	Sodium hydroxide (caustic soda)	$NaOH$
Decrease pH and alkalinity	Carbon dioxide	CO_2
	Sulfuric acid	H_2SO_4
Form protective coatings	Sodium silicate (water glass)	$Na_2O(SiO_2)n$*
	Sodium hexametaphosphate (sodium polyphosphate, glassy)	$(NaPO_3)n \bullet Na_2O$†
	Sodium zinc phosphate	$(MPO3)n \bullet M_2O$‡
	Zinc orthophosphate	$Zn_3(PO_4)_2$
Act as sequestering agents	Sodium hexametaphosphate (sodium polyphosphate, glassy)	$(NaPO_3)n \bullet Na_2O$†
	Tetrasodium pyrophosphate	$Na_4P_2O_7 \bullet 10H_2O$

*Typically $n = 3$.

†Typically $n = 14$.

‡M = Na and/or ½ Zn, typically $n = 5$.

of time the chemical can be held without losing potency or caking so that it becomes unusable.

Lime

Lime is available in either unslaked or slaked form. Unslaked lime, also called quicklime or calcium oxide (CaO), is available in a variety of sizes from a powder to pebble form. Although available in bags, it is considerably more economical and easier to handle in bulk. As a result, quicklime is normally used only by treatment plants that use large quantities. It should not be stored for more than three months because it can deteriorate. Quicklime is noncorrosive in the dry form and can be stored in steel or concrete bins. A minimum of two bins should be available to provide for maintenance and ease in unloading. The bulk deliveries are transferred from the railcars or trucks to the bins by mechanical or pneumatic conveyors.

For treatment plants that do not use large quantities of lime, hydrated (or slaked) lime—calcium hydroxide, $Ca(OH)_2$—is more cost-effective. Hydrated lime is a finely divided powder available in 50- or 100-lb (23- or 45-kg) bags and in bulk shipments. The bulk form is unloaded and stored like quicklime and should be used within three months. The bagged hydrated lime can be stored for up to one year without serious deterioration. Both types should be stored in a dry, well-ventilated area. This is particularly important for quicklime because moisture will start the slaking process, generating a tremendous amount of heat, which could start a fire if combustible materials are nearby.

Soda Ash

Soda ash (sodium carbonate, Na_2CO_3) is a white, alkaline chemical. It is available as granules and powder in bulk or in 100-lb (45-kg) bags. Because soda ash absorbs moisture and will readily cake, it should be stored in a dry, well-ventilated area. Dry soda ash is not corrosive, so it can be stored in steel or concrete bins. Sodium carbonate should not be stored near products containing acid. If acid contacts the chemical, large quantities of carbon dioxide can be released, creating a safety problem.

Sodium Bicarbonate

Sodium bicarbonate ($NaHCO_3$) is also commonly known as *baking soda*. It is a white, alkaline chemical available in either powdered or granular form. It can be purchased in 100-lb (45-kg) bags or in barrels up to 400 lb (180 kg). It must be stored in a cool, dry, well-ventilated area because it decomposes rapidly as the temperature nears 100°F (38°C).

As with sodium carbonate, sodium bicarbonate should not be stored near products containing acid. Corrosion-resistant materials must be used for storing and transporting sodium bicarbonate solution.

Sodium Hydroxide

Sodium hydroxide (NaOH) is also commonly called *caustic soda*. It is available as a liquid or in dry form in flakes or lumps. Regardless of the form, it must be handled carefully because of the many hazards involved. As a liquid, caustic soda is available in solution concentrations of 50 percent or 73 percent NaOH. The 50 percent solution will begin to crystallize if its temperature drops below 54°F (12°C). The 73 percent solution will crystallize as its temperature drops below 145°F (63°C). Special storage facilities are therefore required to maintain

quicklime
Another name for calcium oxide (CaO), which is used in water softening and stabilization.

the temperature of the solution high enough to prevent crystallization. This can be accomplished by further dilution or by installing heaters in the storage tanks.

The manufacturer's recommendations should be followed closely to prevent problems. The storage tanks should be constructed of nickel–cadmium or nickel–alloy steel, lined with caustic-resistant material, such as rubber or polyvinyl chloride (PVC). Tanks made of PVC or fiberglass can also be used.

If the dry form of caustic soda is used, it is normally dissolved immediately on delivery in a storage tank. The dilution concentration is at a point where crystallization will not be a problem at local temperatures.

With either the dry or the liquid form, the addition of water for dilution generates a considerable amount of heat. Therefore, the rate of dilution must be carefully controlled so that boiling or splattering of the solution does not occur. A source of flushing water should be readily available in case of spills.

Sulfuric Acid

Sulfuric acid (H_2SO_4) is a corrosive, dense, oily liquid available in strengths containing 62 percent, 78 percent, or 93 percent H_2SO_4. It can be purchased in 55- or 110-gal (210- or 420-L) barrels or in bulk in tank cars or trucks. The acid must be stored in corrosion-resistant tanks.

If diluted, the acid must always be added to the water (not the water to the acid). If the water is added to the acid, there will be splattering and rapid generation of heat.

Sodium Silicate

Sodium silicate is also frequently called *water glass*. It is an opaque, syrupy, alkaline liquid, available in barrels and in bulk. The barrels can be stored, or the liquid can be added immediately to storage tanks. Bulk deliveries are unloaded directly into storage tanks. The tanks can be constructed of steel or plastic because sodium silicate is not corrosive.

Carbon Dioxide

Carbon dioxide (CO_2) is a colorless, odorless gas that can be generated on-site or purchased in bulk in a liquified form. On-site generation involves producing the CO_2 in a furnace or in an underwater burner.

Liquified CO_2 is stored in insulated, refrigerated pressure tanks ranging in capacity from 6,000 to 100,000 lb (2,700–45,000 kg). The liquified CO_2 must be kept at about 0°F (–18°C) and 300 psig (2,070 kPa [gauge]) so that it will remain in liquid form. Carbon dioxide vapor forms above the liquid surface and is withdrawn and piped to the application point. Each tank has a built-in vaporizer, which helps maintain a constant vapor supply. The storage tank should be located as close as possible to the application point.

Phosphate Compounds

A number of phosphate compounds can be used for stabilization. Three of the most common are sodium hexametaphosphate (sodium polyphosphate, glassy), sodium zinc phosphate, and zinc orthophosphate. Sodium hexametaphosphate is available in bags or drums in a form that looks like broken glass. The other two compounds are generally available in dry form in 50-lb (23-kg) bags and in liquid form in drums or tank trucks. The concentrated liquid solutions are slightly acidic, so they should be stored in corrosion-resistant tanks constructed of fiberglass, PVC, or stainless steel.

Chemical Feed Equipment

Feeding Lime

The type of equipment needed for lime feeding depends on the type of lime being used. If slaked lime is used, a gravimetric or volumetric dry feeder adds the lime to a solution chamber, where water is added to form a slurry (Figure 5-4). When unslaked lime is used, the dry feeder adds lime to a slaker, where the lime and water are mixed together to slake the lime and form a slurry ($CaO + H_2O \rightarrow Ca(OH)_2$). This slurry is often called milk of lime. Slakers are described in more detail in Chapter 6. The hoppers or bins supplying the dry feeders must be equipped with agitators or vibrators, because lime does not flow freely and will pack and "bridge" during storage.

Regardless of the type of lime used, the slurry must be added to the water. It is best to minimize the distance from the feeders to the application point because lime slurry will cake on any surface. Open troughs or flexible feed lines should be used to carry the slurry to the application point because they must frequently be cleaned and flushed. Solution feeders used with the slurry should be designed for easy cleaning.

Feeding Soda Ash and Sodium Bicarbonate

Soda ash and sodium bicarbonate can be fed with gravimetric or volumetric dry feeders. Agitators should be provided on the hoppers to prevent the chemical from bridging, which will block the flow.

Soda ash solutions can be fed using conventional solution feeders and lines. Because soda ash goes into solution slowly, it is necessary to provide larger than usual dissolving chambers to provide proper detention time, as well as good mixers to achieve thorough mixing. Sodium bicarbonate solutions are caustic. Therefore, metering pumps, dissolving chambers, and pipes should be constructed or lined with caustic-resistant materials, such as stainless steel or PVC.

slurry
A thin mixture of water and any insoluble material, such as activated carbon.

milk of lime
The lime slurry formed when water is mixed with calcium hydroxide.

Feeding Sodium Hydroxide

Sodium hydroxide solutions can be fed by metering pumps designed for handling corrosive and caustic solutions. All piping should be of caustic-resistant material, such as PVC. Valves and fittings should not contain any copper, brass, bronze, or aluminum because these materials will deteriorate rapidly.

Figure 5-4 Lime-feeding equipment

Feeding Sulfuric Acid

Sulfuric acid can be fed directly from the shipping container by a corrosion-resistant metering pump and piping. The diluted acid solution is even more corrosive than the pure acid and must be handled carefully.

Feeding Sodium Silicate

Sodium silicate can be fed directly from the shipping container or from a day tank, by using a conventional metering pump and piping.

Feeding Carbon Dioxide

If carbon dioxide is generated on-site, the feeding system consists of the generation equipment and related piping. A typical generation system consists of a combustion unit located beneath the surface of the water in a reaction basin. The carbon dioxide is produced and mixed by the same unit.

Another common system uses gas drawn from a carbon dioxide storage tank. The equipment necessary is shown in Figure 5-5. The gas can be fed to a recarbonation basin or directly to a pipeline for pH control.

Feeding Phosphate Compounds

Phosphate compounds purchased in liquid form are often fed directly from the shipping container; they can also be diluted in a day tank. If dry compounds are used, they are normally dissolved in the storage tanks by placing them in stainless-steel baskets suspended in the tanks. Metering pumps must be designed to handle corrosive materials, and piping, valves, and fittings should be made of PVC or stainless steel.

Feeding With Limestone Contactors

A limestone bed contactor is a treatment device in which water flows through a container packed with crushed limestone and thereby dissolves calcium carbonate. The dissolving of the calcium carbonate increases the pH and alkalinity of the water. This treatment has been shown to reduce the corrosion of lead and copper in water systems having nominally corrosive water.

The contactors are simple, are low cost, and require minimal maintenance, making them especially suitable for corrosion control in small water systems.

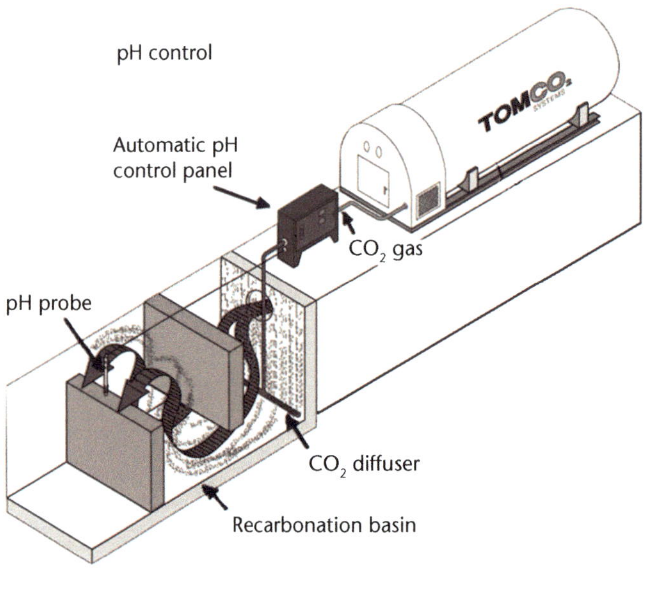

Figure 5-5 Liquid CO_2 recarbonation system

Courtesy of TOMCO Equipment Co.

Regulations

On June 7, 1991, USEPA enacted the Lead and Copper Rule (LCR), a regulation designed to reduce exposure of the public to excessive lead and copper in drinking water. The rule requires all community and nontransient, noncommunity water systems to set up special programs to monitor for lead and copper in drinking water at customer taps. Where excessive amounts are found, the system must provide special treatment to reduce the corrosivity of the water; in addition, it may have to provide a special customer education program.

The sampling protocol for the LCR requires that utilities do a fair amount of work to identify customer locations that are most likely to have piping systems that contribute lead and copper to the water, and to obtain "first-draw" samples from taps in these locations under tightly controlled conditions. This is an ongoing program for most utilities.

It is relatively rare for there to be an appreciable amount of either lead or copper in raw groundwater or surface water. The principal source of excessive lead and copper in customers' water is from corrosion of materials in household plumbing systems. When water is in contact with lead or copper for a period of time, some of the metal is dissolved. The more aggressive the water and the longer the contact time, the higher the lead and copper level will usually be.

Details of the required monitoring, reporting, action levels, and other provisions of the rule are included in Chapter 3.

Operation of the Control Processes

Some water systems already practicing corrosion control treatment may find that the procedures are not sufficient to meet the requirements of the LCR. Many additional systems will find that stabilization is necessary to meet regulations or for other reasons. The first step for the operators of these systems is to review the processes most appropriate for their system.

Selecting a Corrosion or Scaling Control Process

A water system selecting a corrosion-control process should consider the following system dynamics.

Distribution System Conditions

A prime objective of a control program is to reduce corrosion and scaling in the distribution system. The effects of corrosion or scaling can be evaluated to some extent by examining (1) records of main breaks and leaks, (2) information on how well older valves operate, and (3) information on any known problems of reduced flow rates in mains.

Whenever possible, pieces of pipe, valves, and pipe wall sections removed when pressure taps are made should be tagged with their removal location and date and stored for future reference. Inspection of these pieces can often help when the condition of the distribution system piping is being evaluated. Likewise, pieces of water service pipe can be inspected for signs of corrosion or scaling.

Home Plumbing and Customer Complaints

Customer complaints should be recorded based on type of complaint, location, and date. These records can often identify parts of the system or times of the year when particularly serious problems occur. For example, complaints of red water, dirty water, and loss of pressure provide useful insights. Local plumbers are also usually a good source of information on conditions and problems with home plumbing that may be related to excessive corrosion or scale formation. Examples are premature replacement of hot water heaters and replacement of plumbing clogged with scale and precipitates.

Meeting Regulation Requirements

Systems that fail to meet the lead and copper action levels required by the LCR *must* take corrective action as directed by the state. Most systems will probably need only slight adjustment in treatment, but some systems will have to provide extensive treatment changes in order to comply with the requirements. The steps taken to meet these requirements will probably also reduce general corrosion in the distribution system, as well as customer complaints related to water corrosivity.

Water Quality Data

Water quality data for the source water and for several distribution system samples should be examined. These data should cover the following:

- pH
- Alkalinity
- Conductivity
- Temperature
- Iron
- Hardness
- Chloride
- Sulfate
- Total dissolved solids
- Fluoride
- Color
- Zinc
- Copper
- Lead
- Cadmium
- Dissolved oxygen
- Carbon dioxide

Data for the raw water should be compared with data for samples taken from points within the system. If water from the system shows a significant increase in constituents such as iron, copper, zinc, lead, or cadmium, then the water is probably causing corrosion.

Water quality data can be used to calculate stability indexes or indicators. There are many indexes available to engineers and operators. Two indexes of significance are the Langelier saturation index (LSI) and the calcium carbonate precipitation potential (CCPP). Both indexes predict the tendency of a water to either precipitate or to dissolve calcium carbonate. Neither is a direct measure of the potential of the water to be corrosive or scaling.

As discussed in Chapter 1, the LSI is expressed as the difference between the actual system pH and the theoretical pH at which the water is saturated with calcium carbonate.

$$LSI = pH_{actual} - pH_s$$

where:

pH_{actual} = measured pH of the water

pH_s = theoretical pH at which the water will be saturated with calcium carbonate

The value of pH_s is calculated using a formula that takes into account calcium ion concentration, alkalinity, pH, temperature, and TDS.

If the LSI equals zero, the water is considered stable. If the LSI is a positive value, calcium carbonate will precipitate, and the water has scale-forming tendencies. If the LSI is a negative value, calcium carbonate will be dissolved, and the water has corrosive tendencies.

The equation used to determine CCPP index is

$$CCPP = 50{,}045 \times [(Alk)I - (Alk)eq]$$

where:

CCPP = calcium carbonate precipitation potential, in mg/L as $CaCO_3$

50,045 = a unit conversion factor

(Alk)I = measured alkalinity of the water

(Alk)eq = total alkalinity that the water would have in equilibrium, measured as $CaCO_3$

Determining the Best Corrosion Control Treatment for a System

System operators can follow these general suggestions when determining the best type of corrosion control treatment to use on a water system:

- Do not initiate treatment changes without first checking with the office of the state drinking water program. The officials have experience with many different systems and can provide valuable advice on treatment methods and possibly prevent a costly mistake.
- If a corrosion inhibitor is to be used, obtain the recommendations of the proposed supplier to ensure there is no conflict in water chemistries that might cause turbidity or reduce effectiveness of the chemical.
- Do not make treatment changes without first checking with industries, hospitals, other customers, and the local waste treatment authority to ensure the change in water chemistry will not adversely affect them.
- Check with other water systems that use the same water source or have similar water quality to see if they have initiated corrosion control measures. If so, obtain information on their effectiveness. In particular, small systems may be able to use the experience of larger systems that must comply with the new requirement sooner.
    - Unless the effects of a treatment change are known for sure, do not experiment on a whole-system basis. If the results of a proposed treatment change are unknown, run pipe loop experiments under controlled conditions for a time to evaluate the effectiveness and possible side effects. Although pipe loop tests can at times be misleading, they can be of help in determining the best treatment method. If a portion of the distribution system can be isolated and treated separately, it may be possible to run a controlled evaluation on that section.

- Consider the advantages and disadvantages of handling and feeding various chemicals, and the advantages and disadvantages of liquid and dry chemicals. Important considerations include the following:
  - Chemical storage space and facilities required
  - Chemical shelf life
  - Ease or problems in handling chemicals
  - Chemical feed equipment required
  - Hazards or special handling requirements
  - Any extra advantages of a specific chemical, such as improvements to water quality in other ways
  - The cost of chemical treatment per unit of water supplied to the system

Adjusting pH and Alkalinity

Table 5-5 shows the alkalinity adjustments possible with chemicals commonly used to adjust pH or alkalinity. The dosage in a given system depends on the stabilization goals identified as appropriate for that system's water.

Lime

The most common chemical used for pH and alkalinity adjustment is lime because it is readily available, inexpensive compared with other chemicals, and relatively easy to feed. However, lime systems can create many maintenance problems if not operated properly. These problems generally relate to scaling caused by the lime slurry. Because of the high pH of the slurry and the low solubility of lime in water, calcium carbonate ($CaCO_3$) will precipitate on anything the slurry touches. Solution chambers, piping, and pumps must be cleaned frequently to prevent clogging and equipment damage.

The lime slurry is usually added to the filtered-water conduits or into the clearwell. The length of lime slurry-feeding lines should be minimized by locating the feeder as close as possible to the application point. Pumps can be eliminated if the slurry can flow by gravity into a reaction chamber. However, for low rates of lime application, a suitable metering pump should be used. The parts of the pump that contact the slurry must be cleaned routinely to maintain accurate feed rates.

Caustic Soda

Caustic soda can be used to increase both pH and alkalinity. The sodium hydroxide solution is fed by chemical metering pumps into a filtered-water conduit or the clearwell. Because sodium hydroxide is a strong caustic chemical, the pumps, tanks, and feed lines should be routinely inspected for leaks to prevent safety hazards and damage to equipment. Any leaks should be repaired immediately and spills cleaned up promptly.

Other Chemicals

Soda ash or sodium bicarbonate is often fed in combination with lime for additional alkalinity. Neither chemical poses difficult operation or maintenance

Table 5-5 Alkalinity adjustment by chemical treatment

Chemical, 1 mg/L	Alkalinity Change, as $CaCO_3$
Hydrated lime	Increase 1.35 mg/L
Soda ash	Increase 0.94 mg/L
Caustic soda	Increase 1.23 mg/L

problems. Carbon dioxide is used primarily for recarbonation of lime-softened water. Sulfuric acid is used to lower pH and requires the same operating precautions as caustic soda.

Creating Protective Coatings

Lime Coating

Lime, alone or in combination with soda ash or sodium bicarbonate, can be added to precipitate a $CaCO_3$ scale on the pipe walls in the distribution system. This coating will protect pipes from corrosion. If the process is not properly controlled, excessive scale can result, which will increase head loss in the distribution system and clog home plumbing systems.

For best results, the protective coating should be dense and provide uniform coverage. An acceptable coating can be produced by treated water that meets all of the following characteristics:

- Alkalinity and calcium concentrations are both maintained at a minimum of 40 mg/L as $CaCO_3$.
- The water is slightly oversaturated with $CaCO_3$ (4–10 mg/L over the saturation concentration).
- The pH is held within the range of 6.8–7.3.

In water that has a widely fluctuating pH, stabilization should never be used to try to dissolve previously deposited $CaCO_3$ or to maintain a certain thickness. Such attempts will result only in increased corrosion or scaling problems.

Polyphosphate Coating

A protective coating can be formed on pipe walls by polyphosphates. Sodium zinc phosphate and zinc orthophosphate are typically used for this purpose. The dosage will depend on local conditions but will generally range between 0.5 and 3 mg/L. If sodium hexametaphosphate is used, the dosage can be as high as 10 mg/L. For all of the compounds, a somewhat higher dose is applied during the first few weeks to help spread the chemical throughout the system.

Polyphosphate solution is usually transferred from the bulk storage or mixing tank to a day tank sized to store about one day's feeding requirement. Phosphate is a bacterial nutrient, and bacterial growth in the storage tanks can be a problem. A small amount of chlorine added to the storage tanks will inhibit bacterial growth. The solution is pumped from the day tank and added to the water either just after filtration or in the clearwell through a diffuser.

Sodium silicate can also be used to form a protective film; it is fed in much the same way as the polyphosphate compounds. The initial dosage is typically 15–30 mg/L, which is then reduced to 5–10 mg/L as a continuous dosage.

Operational Control

Once a stabilization program is started, monitoring and record keeping are essential to control the process and determine if it is effective.

Water Quality Analyses

The water quality analyses must include those parameters needed to control the stabilization process. In most cases, these will be at least the pH and total alkalinity. However, the data needed to calculate the LSI (or one of the other indexes) should also be collected. These data include the concentrations of the calcium ions and the TDS (total filterable residue), as well as the water temperature.

In-Plant Monitoring The pH of the treated water entering the distribution system should be monitored by a continuously recording pH meter. The pH must be kept within the proper range at all times to prevent excessive corrosion or scaling. The total alkalinity concentration should be determined at least every 8 hours, or more frequently if the quality of the source water is rapidly changing. The LSI (or other index) of the raw and treated water should be calculated daily.

To calculate the chemical dosage necessary for proper pH and alkalinity adjustment, the pH and total alkalinity of the water entering the stabilization process must also be monitored. Samples should be tested at least every 8 hours. It is important that the samples be taken just before the application point of the stabilization chemicals. The raw water entering the plant cannot provide meaningful data because other water treatment chemicals (such as alum, chlorine, and fluoride adjustment chemicals) will change the pH.

Distribution System Monitoring Water quality analyses should periodically be conducted on samples from points in the distribution system to help determine if the stabilization process is effective. These analyses should take into consideration the materials in the distribution and plumbing systems, particularly if corrosion is a problem.

Table 5-6 lists some of the analysis parameters that should be included for various types of pipe materials. For example, the pH and alkalinity are important characteristics to measure if A–C pipe or cement-lined pipe is in use. If leaching of the cement is occurring because of aggressive water, the pH and alkalinity will increase through the distribution system (because cement contains lime). For lead and galvanized pipe, lead and cadmium levels should also be tested because these metals are toxic to humans and may be present as a result of corrosion. Samples from household taps should also be taken as required by state and federal regulations.

Pipe and Coupon Testing

A useful method for determining what is happening in the distribution system is to inspect pipe specimens taken from the system. This inspection allows the operator to determine the extent of corrosion or scaling. A number of procedures can be used to determine the corrosion or scaling rate based on pipe specimens. One procedure involves calculating the depth of pits in the pipe or the pipe section's loss of weight. In addition, the scale on the pipe can be analyzed to help determine why a pipe is being protected or corroded or why excessive scale is being deposited.

Table 5-6 Important water characteristics based on water-main materials

Water-Main Material	Water Characteristic
Ductile and cast iron	Color, conductivity, dissolved oxygen, iron, manganese, pH (alkalinity and calcium if main is cement-lined)
Steel	Color, conductivity, dissolved oxygen, iron, manganese, pH
Concrete cylinder	Alkalinity, calcium, conductivity, pH
Asbestos–cement or cement-lined	Alkalinity, asbestos fibers, calcium, conductivity, pH
Lead and galvanized steel	Alkalinity, cadmium, color, conductivity, dissolved oxygen, iron, lead, pH, zinc

Another valuable technique is the coupon test, which measures the effects of the water on a small section of metal (the coupon) inserted in a water line. After a minimum of 120 days, the inserts are removed, cleaned, weighed, and examined. The weight loss or gain of the coupon can provide an indication of the corrosion or scaling rate. Coupons cannot provide the day-to-day information needed to adjust the chemical feed rates. However, they can be used along with water quality data to provide valuable information for long-term stability control.

Common Operating Problems

Scaling or corrosion of plant equipment is the most common in-plant operating problem. Problems within the distribution system include excessive scaling, persistence of red-water problems, and failure to meet LCR requirements.

Excessive Scaling

The chief cause of excessive scaling is poor control of the stabilization process. This is certainly the case when lime (or lime with soda ash or sodium bicarbonate) is being added to form a $CaCO_3$ coating. If the pH is not kept within a narrow range near the saturation pH, excessive $CaCO_3$ can precipitate and cause clogging of household plumbing.

The same problem can occur if the pH of lime-softened water is not adjusted down to near the saturation pH, either by recarbonation or by addition of sulfuric acid. Properly monitoring the treated water and making necessary adjustments to the chemical feeders can prevent the scaling from becoming a problem.

A similar scaling problem can result when lime is added in conjunction with alum. This is often done to ensure that the pH is in the proper range (6.0–7.8) for effective alum coagulation. However, if too much lime is added and the pH goes above 7.8, the alum will remain in solution and pass into the distribution system, where it will then precipitate and form a scale. A water system can prevent this problem by monitoring the pH and closely controlling the amount of lime fed.

Persistence of Red-Water Problems

A frustrating problem for a water treatment operator is the persistence of red water and other corrosion-related problems, even when corrosion control treatment is being practiced. These problems can occur for a number of reasons, but in most cases they are related to poor flow velocity, tuberculation on the pipe surface, or the presence of iron bacteria.

Flow Velocity The velocity of water flowing in the distribution system plays an important role in corrosion control. There must be enough velocity to carry the chemicals throughout the system and bring them into contact with the pipe surfaces. Where the velocity is low, in dead ends and in areas where water use is low, the stabilization chemicals will not form a protective coating or react with the corrosion by-products. A regular main-flushing program should concentrate on such problem areas to prevent the buildup of corrosion by-products until the dead ends and other problem areas can be eliminated.

Tuberculation If tuberculation is already present in mains, the rough, uneven surface will prevent a uniform protective coating from forming. As velocity and pressure changes occur, the tubercles can break away, which will also destroy the protective layer. The addition of polyphosphates can actually increase the

coupon test
A method of determining the rate of corrosion or scale formation by placing metal strips (coupons) of a known weight in the pipe and examining them for corrosion after a period of time.

problems because these chemicals can loosen the deposits and cause them to break away from the pipe. A better approach is to clean the mains and then start the stabilization process.

Iron Bacteria Iron bacteria can interfere with stabilization. The slimy covering these bacteria produce can prevent the stabilization chemicals from reaching the pipe surface. If an attempt is made to lay down a protective layer, that layer sticks to the slime and is destroyed as the slime sloughs off during pressure and velocity changes. The pipes must be cleaned to remove these slime deposits. In addition, proper disinfection to prevent the bacteria from thriving in the pipe must be performed. Once these measures are accomplished, corrosion control can be effective.

Interference With Other Treatment Processes

There are many complex reactions in water chemistry, and the total effect of making a seemingly simple adjustment is not always predictable. A treatment change concerning one parameter so as to meet LCR requirements may have side effects that adversely affect other water treatment processes or the finished water quality. The following are some examples:

- Some corrosion control methods could raise total trihalomethane levels or promote the formation of other undesirable disinfection by-products.
- An increase in pH may make it more difficult to meet disinfection $C \times T$ values (the product of residual disinfectant concentration and contact time) required by the Surface Water Treatment Rule.
- A corrosion control chemical containing sodium may cause a system to exceed the sodium standard in those states that have established a primary standard for sodium.
- An inhibitor chemical may require simultaneous changes in other water parameters, such as the pH, for it to perform properly.
- The addition of a chemical containing phosphate to the water could promote biological growth in an open reservoir.
- A reaction may occur between inhibitors and other minerals in the water to cause a suspended precipitate or undesirable deposits on pipe walls.
- Increasing pH before coagulation and filtration can, under some conditions, adversely affect these processes.
- Some chemicals applied for scaling, iron, or manganese control may result in *higher* lead or copper levels.
- Under certain conditions, changing the water chemistry to reduce corrosion of lead and copper piping could create a problem with other piping. For example, it might cause red water.

Effects on Customer Uses

The effect of a change in water chemistry on how a customer uses the water must also be considered. Industrial users should be consulted before any treatment changes are made to see if the proposed change will affect their processes. In particular, a change in pH or the addition of zinc, phosphate, or silicate could have a major impact on industries that incorporate water into their product—for example, pharmaceutical, beverage, food-packaging, and food-processing firms. Hospitals should also be notified in case a change will affect dialysis equipment or their other water uses. In addition, compounds that precipitate in the distribution system can clog point-of-use treatment devices in homes and in industries.

Effect on Waste Treatment Facilities

Most of the water from a water system eventually ends up at a wastewater treatment plant. Consideration must therefore be given to the effects of any changes in water treatment on wastewater treatment facilities. For instance, the addition of a product containing zinc or phosphate to drinking water could cause a waste treatment plant to exceed its effluent standard for these substances.

Safety Precautions

In operating corrosion or scaling control processes, the operator is exposed to caustic or acidic chemicals and solutions. When dry chemicals, such as lime, soda ash, and sodium bicarbonate, are handled, care must be taken to minimize dust. Chemical dusts are irritating to the respiratory system, skin, and eyes. Dust collectors should be installed on storage hoppers and cleaned regularly. A vacuum cleaner should be used routinely to clean dust around the feeders and hoppers. When handling dry chemicals, operators should wear proper protective clothing, including a close-fitting respirator and tight-fitting safety glasses with side shields.

In hot weather, when workers are perspiring, chemical burns become more of a problem, especially where quicklime dust is present. A long-sleeved shirt with buttoned collar and pants with the pant legs covering the tops of the shoes or boots are recommended. Gloves and headgear should also be worn. A protective cream can also be placed on exposed skin near the face, neck, and wrists. The operator should shower immediately after handling dusty, dry chemicals. Dusty clothing should be laundered before it is worn again.

Proper storage of chemicals can prevent many safety problems by minimizing dust and reducing opportunities for dangerous chemical reactions. The chemicals should be stored in dry areas that have adequate ventilation. Each chemical should be stored in its own area, not together with other chemicals and equipment. If bags are used, they should be stored on pallets and as close to the feeder hopper as possible.

The chemical solutions made from the dry chemicals (such as lime slurry) or liquid chemicals (acid or caustic soda) must be handled with extreme care. Most of these solutions can cause burns and serious damage to the eyes. Face shields, rubber aprons, rubber boots, and rubber gauntlets should always be worn when working with these solutions. Spills should be cleaned up immediately to prevent accidental contact with other chemicals and damage to equipment. Storage tanks, dilution tanks, pumps, and piping should be inspected routinely to check for leaks.

Operators need to be familiar with the proper first aid procedures to follow for chemical burns from dust or solutions. These procedures should be posted, and first aid training in these procedures should be conducted. The plant should have safety showers and eye-washing facilities available for emergencies.

Record Keeping

Record keeping is especially important for controlling corrosion or scaling because a wide range of information from different sources should be used to monitor the process. For example, records at the treatment plant should be used in conjunction with records from the distribution system to provide an understanding of how the stabilization program is working.

Treatment plant information to be recorded includes the following:

- Current inventory of stabilization chemicals
- Amount of chemical being fed (in milligrams per liter and pounds or kilograms per day)
- Quantity of water being treated
- pH, alkalinity, and other control test results
- Calculation of the LSI (or other index)
- Maintenance of feeders, solution lines, and diffusers

Distribution system records should include the following:

- Water quality analyses
- Results of coupon tests
- Results of examining pipe specimens removed from the system
- Customer complaints relating to corrosion or scaling
- Results of examining pipe specimens taken from household plumbing

Study Questions

1. Which of the following does the saturation point of calcium carbonate primarily depend upon?
   a. pH
   b. Temperature
   c. Concentration of calcium
   d. Total dissolved solids

2. Which of the following will increase pH, hardness, and alkalinity?
   a. Sodium hydroxide
   b. Zinc orthophosphate
   c. Lime
   d. Sodium zinc phosphate

3. Which chemical used for stabilization of potable water is a bacterial nutrient?
   a. Polyphosphate
   b. Lime
   c. Soda ash
   d. Sodium bicarbonate

4. Wastewater water quality discharge requirements could be exceeded if which of the following chemicals are used for corrosion control in the water system?
   a. Silicate
   b. Hexametaphosphate
   c. Calcium hydroxide
   d. Caustic

5. Lime is commonly used for corrosion control because
   a. it can adjust both alkalinity and pH.
   b. it precipitates most particles.
   c. it is more effective than caustic.
   d. most treatment plants already use it for other purposes.

6. Buildup of corrosion products (a process known as ______) or uncontrolled scale deposits can seriously reduce pipeline capacity and increase resistance to flow.
   a. scaling
   b. turbidity
   c. tuberculation
   d. sludge

7. Unslaked lime is also referred to as
   a. sodium carbonate.
   b. calcium dioxide.
   c. quicklime.
   d. calcium hydroxide.

8. Of the metals tin, mild steel, zinc, and aluminum, which is the most likely to corrode?

9. What is the principal source of excessive lead and copper in customers' water?

10. List three reasons that lime is the most common chemical used for pH and alkalinity adjustment.

Chapter 6

Lime Softening

Operation of the Lime–Soda Ash Process

Operation of the lime–soda ash process involves storing and feeding large amounts of chemicals. It also involves processes to perform the actual softening, as well as modifications to the operation of other processes. Finally, careful calculation of chemical dosages is required.

Storing Chemicals

Calcium oxide (unslaked lime, commonly referred to as quicklime) is usually purchased in pebble form to avoid the dust problems associated with the powdered form. However, some dust is created when any form of calcium oxide is used, and operators should always wear protective equipment when handling this chemical to avoid injury.

Calcium oxide is shipped in bulk form by covered railroad hopper cars or by truck. Unloading can be done by screw conveyors, covered belt conveyors, bucket elevators, or pneumatic conveyors. The dry chemical is not corrosive and can be stored in concrete or steel silos or bins.

Calcium hydroxide (hydrated lime) is available in 50- and 100-lb (23- and 45-kg) bags. The bags must be kept intact because the chemical will adsorb carbon dioxide from moisture in the air and change to calcium carbonate.

Soda ash can be purchased in bulk, bags, or barrels. It can be unloaded in bulk by mechanical or pneumatic conveyors, but it must be properly handled because of the dust created by the fine powder. Soda ash is noncorrosive, so it may be stored in ordinary steel or concrete bins.

It is good practice to keep a 30-day supply of lime and soda ash on hand at all times to avoid running out in the event of unexpected shipping problems. Many systems provide a storage capacity of 60 days and take delivery of a 30-day supply every month. Because chemicals lose effectiveness over time, it is important to operate storage facilities in a way that will always use the oldest chemicals first.

Feeding Chemicals

The specific methods of operating chemical feed facilities will depend on the chemicals and feed equipment used. Instructions from chemical suppliers should be followed to ensure proper operation of chemical feed facilities.

As discussed earlier, calcium oxide (unslaked lime) is fed by a gravimetric feeder into a lime slaker. Two types of slakers are used: the detention type and the paste type.

lime–soda ash method
A process used to remove carbonate and noncarbonate hardness from water.

slaker
The part of a quicklime feeder that mixes the quicklime with water to form hydrated lime (calcium hydroxide).

The detention-type lime slaker is the older of the two types and requires a water-to-lime ratio (by weight) of 4:1 to 5:1. It ordinarily takes from 20 to 30 minutes to slake lime. The water ratio is adjustable and is controlled by the operator. Heat generated during slaking is transferred by a heat exchanger to the incoming water, which accelerates the slaking process. A slaking temperature of 160°F (71°C) or higher should be maintained.

The paste-type lime slaker requires a water-to-lime ratio (by weight) of 2:1 to 2.5:1, and slaking is performed at near-boiling temperatures. These higher temperatures accelerate the slaking process. Paste-type slakers take 10–15 minutes to slake lime. The slaked lime is then diluted to approximately 10 percent calcium hydroxide, $Ca(OH)_2$, in a solution chamber, forming a milk-of-lime slurry.

Calcium hydroxide (slaked lime) is fed directly into a solution chamber by a gravimetric feeder. The resulting slurry is similar to the milk-of-lime slurry.

The process of feeding lime slurries presents many problems because the slurries will deposit (or cake) on any surface. Valveless proportioning pumps can provide a satisfactory means of pumping the slurry. The slaker should be located as close to the point of feed as possible, and the slurry is best fed through an open trough that can easily be cleaned when necessary.

slake
The addition of water to quicklime (calcium oxide) to form calcium hydroxide, which can then be used in the softening or stabilization processes.

rapid or flash mixing
The process of quickly mixing a chemical solution uniformly through the water.

flocculation
The water treatment process, following coagulation, that uses gentle stirring to bring suspended particles together so that they will form larger, more settleable clumps called floc.

sedimentation
The water treatment process that involves reducing the velocity of water in basins so that the suspended material can settle out by gravity.

sedimentation basin
A basin or tank in which water is retained to allow settleable matter, such as floc, to settle by gravity.

Operation of Softening Facilities

Rapid Mixing

A high-energy, rapid mixing process called flash mixing is used to dissolve and thoroughly mix lime and soda ash into the water being treated. The optimal mixing speed is that which produces the best chemical reactions; it can be determined only through operating experience. Mixers are operated at varying speeds to produce precipitates that have the best settling characteristics.

A detention time of 5–10 minutes is required when lime and soda ash are mixed with water because lime is slow to dissolve. All mixing equipment and the settling basin should be cleaned frequently to remove precipitates. The concentration of precipitates can be improved by recirculating the sludge from the settling basin. This process allows precipitation to take place on the previously formed solids rather than in the basin.

Flocculation

Flocculation requires slow, gentle mixing to allow the chemical reactions to be completed and to promote the growth and precipitation of the floc. Consequently, flocculation basins usually require a detention time of 40–60 minutes. The speed of the mixing paddles must be adjusted to obtain the best floc. Once the floc has formed, local eddy currents and flow surges must be avoided to prevent breaking up the solid floc particles.

Sedimentation Basins

Sedimentation, which is also called *clarification*, is the removal of settleable solids by gravity. The process takes place in a rectangular, square, or round tank called a sedimentation or settling basin or tank.

To soften water, detention times of 2–4 hours are usually adequate. Surface overflow rates vary, but they are generally higher than the rates in a conventional water treatment plant because the floc settles more readily than alum floc. Because of the large amounts of sludge produced in the softening process, continuous mechanical collection of sludge is essential. The sludge drawoff lines should be backflushed periodically to remove any clogs or caking.

Solids-Contact Basins

In a solids-contact basin (Figure 6-1), the water passes upward through a blanket of flocculated material called the *sludge blanket*, which entraps slowly settling particles that would otherwise escape the basin and end up in the filters. The mixing speed and the level of the sludge blanket are key operational factors for the system. The mixing speed determines how well the chemicals and water are mixed. Increasing the speed shortens the detention time of the initial mix, which can cause inefficient treatment because of incomplete chemical reactions. Inadequate mixing speed, however, can cause poor mixing of the chemicals and water. Settling tests and visual observations must be used to determine the best mixing speed. Many plants run the mixer continuously to keep the sludge blanket in a state of suspension at all times, including periods when the plant is shut down.

In addition to proper dosage and mixing, successful treatment in a solids-contact basin depends on keeping the sludge blanket at desirable levels. If it drops below the "skirt" of the flocculation compartment, water will not pass through a full bed of solids, and some particles could pass through to the filters.

At the same time, the sludge blanket must not be allowed to build up or rise to levels so high that solids will pass over into the main effluent launder and discharge into the filters. The top of the sludge blanket normally should be 4–8 ft (1.2–2.4 m) below the top of the overflow weirs. The height of the sludge blanket is adjusted according to the frequency, duration, and rate of sludge removal (sludge blowdown). Blowdown valves are operated intermittently in response to a timer or a water meter that signals when a given quantity of water has passed through the basin.

Sampling lines should be provided at various levels in the settling zone to permit the operator to determine with accuracy the height of the sludge blanket and the degree of solids present in the water. In the lower level of the blanket, the concentration of solids, by weight, should range from 5 to 15 percent. Maintaining a high concentration of solids, while keeping the sludge blanket at a level low enough to avoid a loss of solids over the weir, minimizes the amount of chemicals needed.

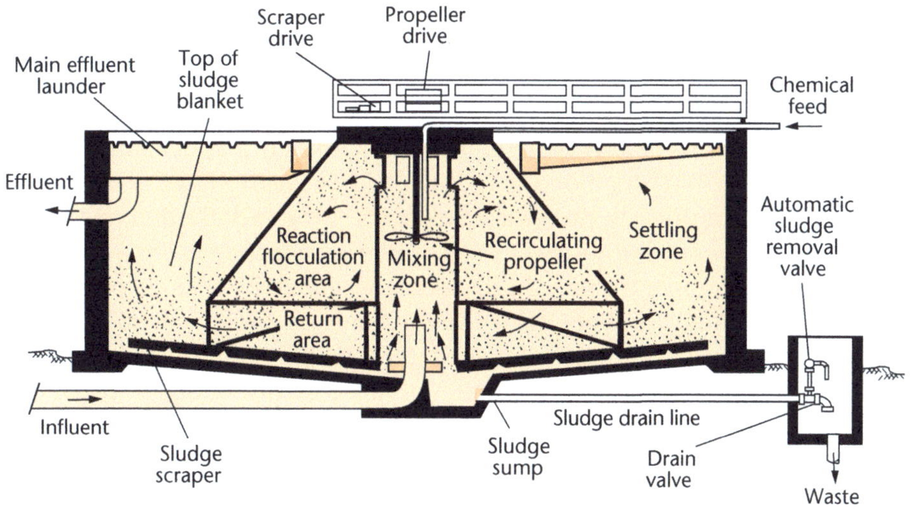

Figure 6-1 Solids-contact basin

Courtesy of General Filter Company, Ames, Iowa.

solids-contact basin
A basin in which the coagulation, flocculation, and sedimentation processes are combined. The water flows upward through the basin. It is used primarily in the lime softening of water.

Solids-contact basins are usually designed for a flow rate of 1–1.75 gpm/ft^2 (0.7– 1.2 mm/sec). Higher flow rates are sometimes possible, depending on the settling characteristics of the floc. Where multiple basins are used, flow must be balanced between the basins so that no individual basin will be overloaded. This balance can be achieved either by flowmeters or by adjusting overflow weirs to provide flow rates that are proportional to the sizes of the basins.

Solids-contact basins must be drained periodically for cleaning. The cleaning must be done once every 3–12 months, depending on the types and amounts of chemicals used. Sludge withdrawal lines and sampling lines should be vigorously flushed and cleaned routinely to remove sludge residues and scale buildup. An alum or chlorine solution will make cleaning of these lines more effective.

Sludge Collection and Disposal Equipment

Sludge collectors normally operate at a slow constant speed of about one revolution every 20–30 minutes. The operator does not usually have to vary the speed. Once the sludge has been scraped into the sludge hopper in the sedimentation basin, the hopper is emptied by pumping or gravity. A portion of the removed sludge is then recirculated, and the remainder is discarded.

Constant-speed centrifugal sludge pumps may have a valve on the discharge for the operator to control the pumping rate. If a variable-speed, positive-displacement pump is used, the operator can control sludge flow by varying the motor speed.

The frequency at which sludge is pumped varies significantly based on the amount of hardness removed and the concentration of sludge solids. As a rule, for each pound of lime used, 2.5 lb (1.1 kg) of sludge will be formed. Sludge concentrations can vary from less than 5 percent to more than 15 percent solids by weight.

The operating procedures for **dewatering** sludge vary depending on the type of process used. If drying beds are used, the sludge should be spread to a depth of 8–12 in. (200–300 mm) over the sand. In good weather, the sludge will dry and crack in about a week. The dried sludge contains about 50–60 percent solids; it can be removed by hand raking or using a front-end loader. The dried sludge is then hauled to a burial or refuse site. The success of the drying-bed process for dewatering sludge depends on clear, dry, warm weather. In cold, wet climates, the process can be used if the beds are covered like greenhouses and have adequate heat and ventilation.

If lagoons are used to dewater sludge, the sludge is pumped directly to the lagoons and allowed to settle. It is best to fill a lagoon at one end and decant the water from the other end. When the depth of sludge reaches 3–5 ft (0.9–1.5 m), use of the lagoons should be discontinued and the water level should be kept as low as possible. After a period of time, the dewatered sludge should contain about 50 percent solids. It is then usually removed by a dragline or clamshell crane.

The sludge may be spread on the ground near the lagoon to dry further before final disposal. In some cases, the sludge solids are left to accumulate in the lagoon. When the lagoon is filled, it is abandoned, and a new lagoon is excavated. In cold climates, freezing temperatures help dewater the sludge by separating water from the sludge solids. After thawing, the solids are small granular particles.

In softening plants that dewater sludge by using a vacuum filter, belt filter, filter press, or centrifuge, sludge is usually pumped into a sludge thickener before proceeding to the mechanical dewatering device. Sludge entering the thickener with 5 percent solids, for example, can be thickened to 15–20 percent solids. This reduced volume of sludge is easier and less costly to treat mechanically. The dried sludge produced by a vacuum filter or centrifuge is a fairly dry, thin, cake-like sludge that has a solids concentration of 40 percent or more.

dewatering
A process to remove a portion of water from sludge.

Final disposal of sludge is usually achieved either by burying it in a sanitary landfill or by spreading it on farmland to serve as a soil conditioner. If the sludge is incinerated, lime can be reclaimed and reused. However, this procedure is economical only for large treatment plants.

Recarbonation Basins

Recarbonation basins are used to stabilize water after softening. To allow adequate time for the chemical reactions to take place, recarbonation basins usually provide 15–30 minutes of detention time. If carbon dioxide is being injected by diffusers, the operator should routinely check the diffusers to see that they are not plugged. The ventilation system in the area should also periodically be checked for proper operation, and the flow of gas should be monitored at least every 8 hours. Fluctuations in the flow rate of gas in the basin can cause excessive scale to form downstream. Control of the recarbonation process should be based on (1) meeting water stabilization goals as determined by the Langelier saturation index (LSI) or other stability index, and (2) using coupons in the distribution system (see Chapter 5).

Filters

The filters used for the softening process are essentially the same as those for conventional filtration, except for one difference that can cause serious operating problems. Because the water being filtered after softening contains calcium carbonate and, in some cases, magnesium hydroxide, precipitates of these chemicals can form on the filter media, causing the grains of media to stick together. If this happens, the effectiveness of the filtration process will be destroyed. To avoid this problem, the filter media should be inspected frequently for scale buildup. If the grains are coated with white (calcium carbonate) or black (magnesium hydroxide) scale, recarbonation is probably inadequate. The presence of scale can be verified by the addition of a small quantity of hydrochloric acid to a small quantity of filter media. If scale is present, the acid will react with it and will boil and foam.

Scale can sometimes be removed from filter media by increasing the carbon dioxide feed rate. Care must be taken during this operation to maintain water quality and to avoid damaging metallic underdrain components by prolonged exposure to corrosive water.

Determination of Chemical Dosages

The amount of lime and soda ash required to soften water depends on (1) the hardness of the water and (2) how much of the hardness is to be removed. Because precipitates are very slightly soluble, some hardness remains in the water, about 50–85 mg/L as $CaCO_3$. The required dosage of lime is based on the combined amount of carbonate hardness, noncarbonate hardness, and carbon dioxide to be removed. The required dosage of soda ash is based only on the amount of noncarbonate hardness to be removed.

As discussed in Chapter 2, there are two methods for calculating dosages: the conventional dosage method and the conversion-factor method. Initial dosages for lime and soda ash can be computed based on these two methods. However, it is important to note that either sludge recirculation or the use of solids-contact basins will allow dosages to be reduced. Adding excessive amounts of chemicals may not increase the amount of hardness removed from the water. An operator should start by using the computed figures as a base and then experiment with various dosages in the same range to find the most effective rates for the water being treated.

The use of a softening curve can be helpful in establishing a base dosage. This curve can be developed by using jar tests, followed by process testing in the plant. Jar tests are used to perform laboratory experimentation with various chemical dosages and mixing rates and to determine the most effective treatment. See Chapter 4 for more information on jar tests.

As shown in Figure 6-2, a softening curve is developed by plotting the lime dosage against the concentrations of carbonate and noncarbonate hardness remaining in the water. At each point on the curve, pH should also be recorded. In the example shown in the figure, a total hardness of 100 mg/L can be achieved with a lime dosage ranging from 7.4 to 10 grains (127–171 mg/L). If the lower dosage is used, a substantial amount of lime and money can be saved.

The dosage of soda ash needed to remove noncarbonate hardness should be based on calculations for either the conventional dosage method or the conversion-factor method. Monitoring the finished water's noncarbonate hardness and alkalinity allows the dosage of soda ash to be adjusted to the most efficient level.

Calculations for the theoretical feed rate for carbon dioxide in the recarbonation process are based on the amount of alkalinity that must be removed in order to stabilize the water. Any adjustments in the carbon dioxide dosage should be based on both the LSI and the results of coupon testing in the distribution system.

WATCH THE VIDEO
Operator Chemistry II—Hardness (www.awwa.org/wsovideoclips)

Operational Control Tests

The lime–soda ash softening process is complex and requires the following control tests:

- Alkalinity tests
- Total hardness tests
- Carbon dioxide tests
- pH tests
- Jar tests
- LSI determination
- Monitoring of coupons in the distribution system

The key to successful control of the softening process is the correct use of these test results. Many of these tests are also discussed in Chapters 4 and 5.

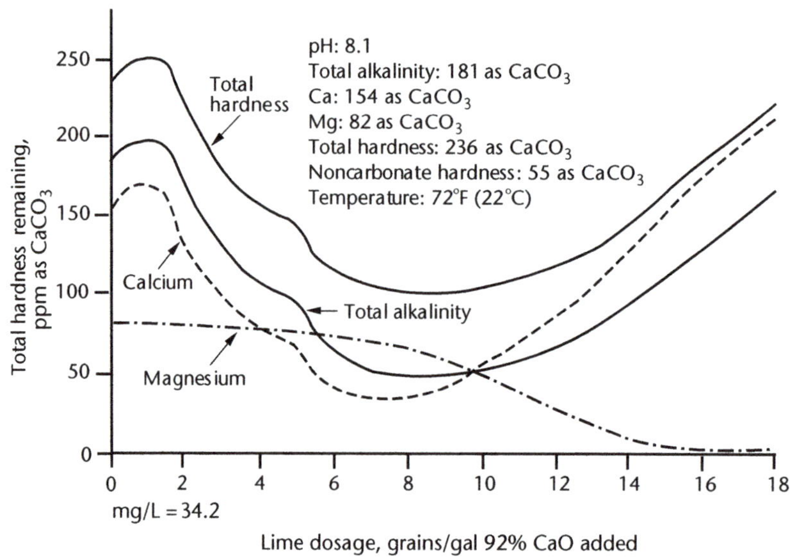

Figure 6-2 Softening curve used to establish most effective lime dosage

Alkalinity Tests

The results of an alkalinity test are used in calculating dosages for both lime and soda ash. There are three specific types of alkalinity:

- Hydroxide alkalinity
- Carbonate alkalinity
- Bicarbonate alkalinity

Total Hardness Tests

The total hardness test indicates the need for softening and provides information vital to the proper operation of the softening process. The test determines the total hardness concentration of a water. This information, together with total alkalinity, is used to calculate carbonate and noncarbonate hardness (Table 6-1).

If noncarbonate hardness is present, the total hardness concentration and the bicarbonate alkalinity are used to calculate the dosage of soda ash needed to remove the noncarbonate hardness. If noncarbonate hardness is absent, no soda ash is needed, and the total hardness concentration is used in place of bicarbonate alkalinity to calculate the correct dosage.

The total hardness test should also be conducted routinely on raw and finished waters to determine how effectively the softening processes are removing hardness from the water.

Magnesium Tests

The magnesium test is used to determine how much magnesium hardness is in the water. This amount, in turn, is used in determining lime dosages. Because the results of magnesium tests indicate the magnesium hardness present, they can be used to calculate calcium hardness when the total hardness is also known.

The magnesium test also provides valuable insight into whether or not the water being treated contains enough magnesium to cause scale in water heaters and boilers. Whenever magnesium hardness exceeds 40 mg/L (as $CaCO_3$), there is a strong possibility that scale will form in hot water heaters and boilers that are operated at temperatures over 140°F (60°C). If magnesium hardness is removed to levels at or below 40 mg/L, scale formation should not be a problem.

Carbon Dioxide Tests

The carbon dioxide test is used to determine the concentration of free carbon dioxide in the water. In turn, this value is used to calculate lime dosages. Although carbon dioxide does not cause hardness, it reacts with lime, precipitating as calcium carbonate. A sufficient amount of lime must be added to the water to allow this chemical reaction to occur and still have enough lime remaining in the water to complete the softening process.

Table 6-1 Carbonate and noncarbonate hardness determination

Laboratory Results	Noncarbonate Hardness (Lime and Soda Ash Used)	Carbonate Hardness (Lime Only Used)
1. TH < TA	0	TH
2. TH = TA	0	TH
3. TH > TA	TH – TA	TA

Note: TH = total hardness, TA = total alkalinity.

magnesium hardness
The portion of total hardness caused by magnesium compounds, such as magnesium carbonate and magnesium sulfate.

pH Tests

Measurement of the water's pH should be made before, during, and following lime–soda ash softening. During the softening process, pH measurements provide valuable information about the success or completeness of the chemical reactions that are occurring. For example, when excess-lime treatment is used, the lime dosage is designed to raise the pH of the water to 10.6, which is the best pH for magnesium hydroxide precipitation. The pH is then reduced by recarbonation to 9.4, which is the best pH for calcium carbonate precipitation.

After soda ash is added and softening reactions have been completed, the final pH is adjusted for recarbonation to whatever pH is required to achieve stabilization (usually about 8.6). This process converts any residual calcium carbonate into soluble calcium bicarbonate, thus preventing after-precipitation either on the filters or in the distribution system mains.

By monitoring pH with an accurate, calibrated pH meter throughout the softening process, an operator can determine whether or not softening and recarbonation dosages are working successfully. In addition, pH and total alkalinity can be used to determine the concentration of carbon dioxide in the water.

Jar Tests

The jar test is a good way to test the performance of lime and soda ash dosages before they are used in the plant. A dosage correctly calculated only from the results of the alkalinity test, total hardness test, and carbon dioxide test is not necessarily the best dosage. Numerous other factors, not easily defined or tested, can greatly influence the selection of the "best" dosage.

By performing the jar test, an operator can check the overall performance of the lime and soda ash dosages, such as the size and condition of floc particles, the ease and speed with which floc particles settle, and the resultant final pH. A softening curve can also be developed to help identify the lime dosage that will soften water to the desired point at the least cost.

The Langelier Saturation Index and Coupons in the Distribution System

The stability of softened water should be monitored to ensure that the water will neither cause corrosion nor form scale. The LSI should be determined daily at the plant, and it should be evaluated in connection with inspection of coupons placed in the distribution system. (The LSI and coupons are described in Chapter 5.)

Operating Problems

The following are several potential problems commonly associated with the operation of the lime–soda ash softening process:

- Excess calcium carbonate
- Magnesium hydroxide scale
- After-precipitation
- Carryover of sludge solids
- Unstable water
- Interference with other treatment processes

Unless recognized and eliminated, these conditions can lead to high maintenance costs, failure of filtration processes, loss of distribution capacity, inefficient operation of water heaters and boilers, and ineffective treatment.

Excess Calcium Carbonate

Extremely small, almost colloidal particles of calcium carbonate can be formed during the coagulation–flocculation process. These particles can pass through the entire softening process to the filters, where they adhere very tightly to the individual grains of filter media. The filter grains can become coated with the calcium carbonate and cement themselves together. In a similar way, calcium carbonate can pass through the media and plug the filter underdrains. Clogged underdrains will usually cause visibly uneven backwash patterns. If these conditions are not corrected, they can cause problems serious enough to require complete replacement of the filter media and underdrains.

In addition, excess calcium carbonate can also precipitate in fine particles on the walls of pipelines. As shown in Figure 6-3, the particles can eventually build up in thickness, reducing the pipeline capacity. The restricted pipe capacity makes pumps work harder, which increases operating costs and shortens equipment life.

Excess calcium carbonate is also the cause of scale deposits on filter walls and wash troughs. Such deposits are both unsightly and costly to remove. These deposits may be caused by inadequate mixing, short detention times, or interference from organic contaminants in the water during chemical precipitation.

The best solution to the problem is to eliminate whatever is causing the incomplete coagulation–flocculation process by improving the mixing, increasing detention times, or pretreating the water to remove organic contaminants. If the fine particles of calcium carbonate cannot be eliminated at the source, they must be either removed by recarbonation or controlled by applying a recommended dosage of 0.25–1.0 mg/L of sodium hexametaphosphate (a sequestering agent). This will usually hold the calcium carbonate in solution and prevent precipitation onto the filter media.

Magnesium Hydroxide Scale

When water softened by the lime–soda ash process contains magnesium hardness in excess of 40 mg/L, magnesium hydroxide scale can form inside boilers and household water heaters. To prevent scale buildup, magnesium hardness should be reduced to 40 mg/L or less through excess-lime treatment.

After-Precipitation

Immediately after the lime–soda ash process is performed, the pH of the treated water is about 9.4. If the pH is left at this level, calcium carbonate will continue to

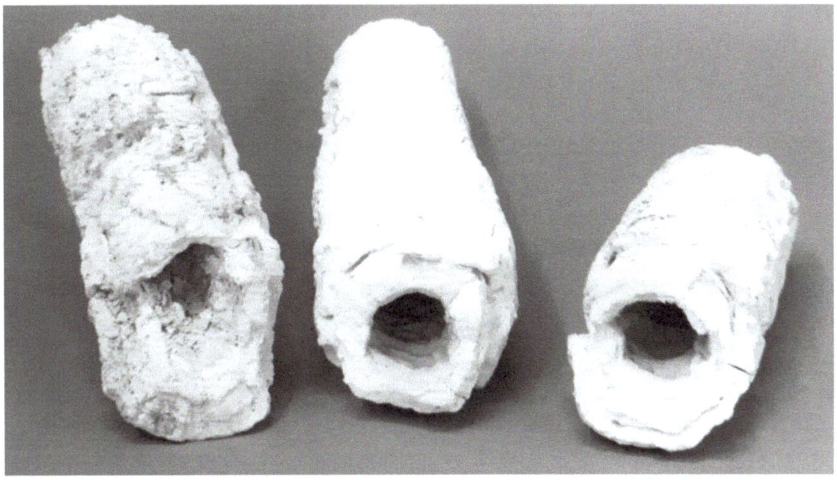

Figure 6-3 Calcium carbonate buildup removed from pipe
Courtesy of Girard Industries.

coagulation–flocculation
The water treatment process that converts small particles of suspended solids into larger, more settleable clumps.

precipitate out of the treated water, even after the water leaves the sedimentation basin or upflow basin. This effect, called *after-precipitation*, can cause the same problems that occur when excess calcium carbonate precipitates out of treated water because of inadequate coagulation–flocculation. To prevent after-precipitation, carbon dioxide should be added to the softened, settled water to decrease the pH to about 8.7, where it will be chemically stable. This procedure converts any unsettled calcium carbonate to soluble calcium bicarbonate. Encrustation can also be prevented by adding sodium hexametaphosphate.

Carryover of Sludge Solids

The primary reasons that sludge solids and properly formed floc particles are carried over the weirs of a sedimentation basin are (1) improper hydraulic conditions and (2) sudden changes in water quality.

Carryover is indicated by the presence of unclear water above the sludge blanket, by sludge solids appearing in the basin effluent, and by short filter runs. Carryover of sludge solids is often most severe when a considerable portion of the settleable solids is composed of magnesium hydroxide floc particles, which are very light. To prevent carryover, it is necessary either to reduce the hydraulic loading on the sedimentation or upflow basin or to improve the settling characteristics of the floc. Recirculation of previously formed calcium carbonate sludge is effective in improving the settling characteristics of the floc.

Unstable Water

If softened water is not properly stabilized, one of two major operating problems will result: scale deposits or corrosiveness. Recarbonation and other stabilization techniques are used to control these operating problems. These techniques are discussed in Chapter 5.

Interference With Other Treatment Processes

Disinfection of water is essential to make water safe for consumption. Although the high pH associated with lime–soda ash softening contributes somewhat to disinfection, it also results in the formation of a chlorine residual, primarily in the form of hypochlorite (OCl^-), when the water is chlorinated. The hypochlorite residual has considerably less disinfecting power than hypochlorous acid ($HOCl$), which exists at lower pH values. Therefore, higher dosages of chlorine or longer contact times may be required to disinfect water softened by the lime–soda ash process.

If the lime–soda ash process is used to treat surface water, the processes for removing tastes and odors (particularly adsorption) may be hindered as a result of the increased solubility of taste- and odor-causing compounds at elevated pH values. If this problem occurs, it can usually be controlled by removing these compounds during pretreatment.

Many surface water treatment plants must treat raw water that has a high color concentration caused by organic compounds. For the most part, color must be removed prior to softening because a pH in the range of 4.0 to 5.5 is often required for alum coagulation to remove color effectively.

The rate of formation of **trihalomethanes (THMs)** increases at pH levels required for the softening process. If THM problems exist, pretreatment processes may need to be improved or disinfection processes may need to be modified. Regulatory agencies or technically qualified consultants should be consulted before a plan to minimize THM formation is adopted.

trihalomethane (THM)
A compound formed when natural organic substances from decaying vegetation and soil (such as humic and fulvic acids) react with chlorine.

Safety Precautions

The lime–soda ash softening process exposes operators to daily contact with large quantities of lime and soda ash. These chemicals pose hazards if not handled properly. As shown in Figure 6-4, operators should be adequately protected when handling chemicals.

Lime Handling Safety

Quicklime is a strong, caustic chemical. Handling the chemical creates dust that can be irritating to the eyes, mucous membranes, and lungs. Prolonged or repeated inhalation can cause lung damage. Operators exposed to the dust should wear goggles and a suitable dust mask.

Skin contact with the dust can cause dermatitis or skin burns, particularly when the dust mixes with perspiration on the skin. Operators should wear heavy cotton clothing with long sleeves, gauntlet gloves, hat, and trousers tied around the shoe tops when handling the chemical. If clothing becomes covered with dust or slurry, it should be laundered immediately.

Operators should always shower immediately after handling quicklime, even if the dust is not visibly present. Dust collection equipment should frequently be checked for proper operation. Chemical spills and dust accumulation should be cleaned up with a dry vacuum, not with a broom.

Calcium oxide is strongly attracted to water. When the two come into contact, the heat generated can be sufficient to start a fire if there are flammable materials close by. The chemical must therefore be stored in a dry place.

Dry calcium oxide and dry alum should never be mixed. The heat that results when they are mixed can generate hydrogen and cause an explosion. To avoid accidental contact, these two chemicals should be loaded, handled, and stored in completely separate locations.

Lime slakers should be equipped with temperature override devices that will stop the lime feed before dangerous temperatures are reached.

Calcium hydroxide (hydrated lime) is less caustic and less irritating to the skin than calcium oxide, but is still a dangerous chemical to handle. It is particularly dangerous to the eyes. In general, the same handling precautions should be exercised for calcium hydroxide as for calcium oxide.

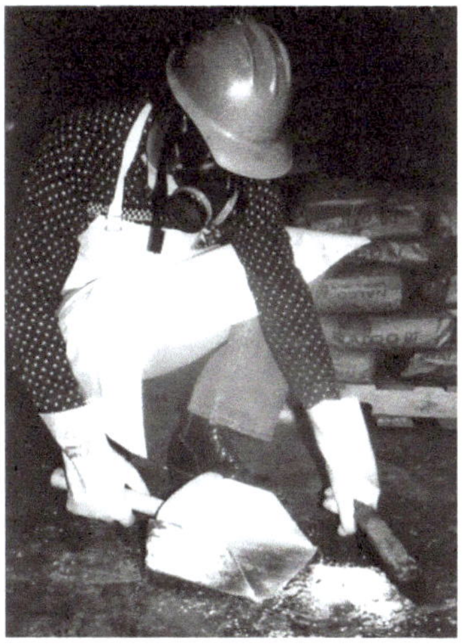

Figure 6-4 Worker properly dressed for handling chemicals

Soda Ash Handling Safety

Moisture will cause soda ash to cake, making it difficult and hazardous to handle. To prevent this problem, soda ash must be stored in a cool, dry place. All soda ash equipment should be fitted with a dust collection system.

Operators working with soda ash should wear protective equipment, including safety goggles for eye protection, a close-fitting dust respirator, and clothing that protects against skin contact. Skin areas that will be exposed unavoidably should be covered with a protective cream or petroleum jelly to minimize the harmful effects of the dust. Some brands of soda ash contain ammonia, and operators who are allergic to ammonia should be particularly careful to protect themselves.

Pumps and equipment that handle soda ash solutions should be equipped with spray or splash guards to protect personnel working in the area. Potential danger areas and storage rooms should all be marked with warning signs.

Carbon Dioxide Handling Safety

Carbon dioxide is a colorless, odorless gas that is heavier than air. If it accumulates, it will displace the air. It can result in oxygen deficiency if inhaled, causing asphyxiation.

Because carbon dioxide is often generated on-site, preventive maintenance of the generating equipment is essential to prevent leaks and to ensure that safety devices and alarms will function properly. The generating equipment should be located in a well-ventilated area. In addition, explosion-proof lighting should be installed, and smoking should be strictly prohibited when carbon dioxide is being used.

Generation of carbon dioxide can also produce carbon monoxide, another highly toxic, odorless, and colorless gas. Good ventilation of the carbonation basin is essential, particularly if submerged-combustion generating units are used. Extreme care should be taken during the cleaning of carbonation basins, particularly if they are enclosed. They should be thoroughly ventilated by blowers before and during the cleaning operation. A carbon monoxide tester should be used to make sure the air is safe before anyone enters the basin.

First Aid for Chemical Exposure

Soda ash dust and the mist from soda ash solutions are very irritating to the respiratory system, mucous membranes, and eyes. Prolonged exposure can damage the nasal passages.

If soda ash dust or solution or lime slurry gets into the eyes, flush the eyes with warm water for at least 15 minutes. Consult a physician immediately. If skin contact with any chemical solutions occurs, flush the area with large quantities of water. In the event that the dust or mist is inhaled, consult a physician immediately. Gargling or spraying the nasal passages and throat with warm water will reduce mild irritation.

Record Keeping

The following information should be recorded daily for the lime–soda ash process:

- Quantity of lime, soda ash, and carbon dioxide fed (in pounds or kilograms)
- Chemical feeder settings
- Amount of water treated (in million gallons per day or megaliters per day)

- Hardness, alkalinity, and magnesium–calcium concentration in raw and treated water
- Free carbon dioxide in raw water
- The pH of
  - raw water,
  - treated water, and
  - stages of the process that are critical to chemical reactions
- Amount of sludge pumped to disposal
- Dosage calculations and results of jar tests (softening curves)
- Depth of the sludge blanket and results of settling tests if a solids-contact basin is used

Study Questions

1. Which chemical is used to stabilize water after it has been lime-softened?
   a. Sulfuric acid
   b. Carbon dioxide
   c. Sodium hydroxide
   d. Zinc orthophosphate

2. Usually, raw water can be softened with lime alone when it contains
   a. mostly calcium hardness.
   b. mostly magnesium hardness.
   c. very little carbonate hardness.
   d. little or no noncarbonate hardness.

3. In the lime–soda ash softening process, the soda ash dosage is based
   a. only on the amount of carbonate and magnesium hardness to be removed.
   b. on the amount of carbonate hardness and carbon dioxide to be removed.
   c. primarily on the amount of noncarbonate and magnesium hardness.
   d. only on the amount of noncarbonate hardness to be removed.

4. Recarbonation after lime–soda ash softening is required to remove calcium carbonate that is in suspension to prevent
   a. deposition of the calcium carbonate in the clearwell and distribution system.
   b. scale formation in distribution piping.
   c. customer complaints of turbidity and color.
   d. cementation of filter media.

5. Which type of hardness usually calls for a two-stage softening process?
   a. Calcium hardness above 100 mg/L
   b. Magnesium hardness
   c. Noncarbonate hardness
   d. Carbonate hardness

6. The concentration of solids in the lower level of a sludge blanket in a softening facility's solids contact basin should be
   a. 2–7%.
   b. 5–15%.
   c. 15–20%.
   d. 25–30%.

7. Softened water has a high pH and a high concentration of calcium carbonate. Therefore, stabilization is essential in order to prevent the calcium carbonate from precipitating out
   a. in household plumbing.
   b. in the clearwell.
   c. in the distribution system.
   d. on the filters.

8. It is impossible to produce waters with a hardness of less than ______ when using the lime–soda ash process.
   a. 9 mg/L
   b. 17 mg/L
   c. 25 mg/L
   d. 50 mg/L

9. Soda ash is added to water in the lime softening process for what purpose?
   a. To disinfect the water and kill the vast majority of protozoans, viruses, bacteria, and other multicellular organisms
   b. To raise the pH of the water
   c. To decrease the CO_2 in the water
   d. To remove noncarbonated hardness

10. Flocculation requires ______ mixing to allow the chemical reactions to be completed and to promote the growth and precipitation of the floc.
    a. rapid, continuous
    b. steady, forceful
    c. slow, gentle
    d. sporadic

11. If soda ash dust or solution or lime slurry gets into the eyes, flush the eyes with warm water for at least 15 minutes, and consult a physician
    a. immediately.
    b. if redness of the eyes occurs.
    c. if policy requires.
    d. within 24 hours.

12. List three specific types of alkalinity.

13. Carryover of sludge solids is often most severe when a considerable portion of the settleable solids is composed of what type of floc particles?

14. To prevent scale buildup, magnesium hardness should be reduced to what level through excess-lime treatment?

Chapter 7
Ion Exchange

Ion Exchange Softening Process

Calcium and magnesium can be removed from water by exchanging calcium and magnesium ions for others that do not impart hardness to the water. The ion exchange process involves the use of a resin that has been saturated with nonhardness-producing ions that readily participate in this exchange process. The process used in water treatment plants is reversible and therefore renewable for many cycles. The method of resin renewal is called *regeneration*. A bed of resin is said to be "exhausted" when it can no longer remove hardness. At that time, regeneration using salt water (brine) is initiated.

Exchange resins vary in their capacity to soften and be regenerated; it is common to describe them based on the number of bed volumes that can be exchanged before the resin must be regenerated. Also, the amount of calcium and magnesium in the source water dictates the amount of water that can be softened. Harder water exhausts a resin bed sooner than a softer water. If an operator knows the exchange capacity of the resin bed, the hardness of the water, and the flow rate of water through the bed, he or she can predict the length of a softening cycle. This prediction is an important part of the operational scheme in ion exchange softening plants because it allows operators to schedule time for other tasks and even allows for ion exchange water softening units to operate unattended. More importantly perhaps, because most ion exchange plants have more than one unit, it allows for staggered operation (Figure 7-1). In this way, all units are not regenerated at the same time.

Design of Resin Beds and Exchange Units

The hardware involved in the ion exchange process (vessels, piping, valves, etc.) is important; however, it will not be covered in this discussion. Rather, an explanation of the materials used (resin, regenerant) is presented.

Ion Exchange Media

The earliest known materials used to soften water by ion exchange were naturally occurring zeolites found in certain soils. Eventually, synthetic resins were developed that had faster exchange rates and higher capacities to soften. These natural and synthetic materials are rated according to their ability to exchange calcium and magnesium for sodium, an ion that does not impart hardness to the water. The

Ion exchange water softener
A treatment unit that removes calcium and magnesium from water using ion exchange resins.

Figure 7-1 Ion exchange units
Source: US Filter.

rating, or exchange capacity, can be expressed in terms of kilograins (kgr) of hardness per cubic foot of resin. Typically, the modern synthetic resins have exchange capacities of 20–40 kgr/ft^3; natural zeolites were rated at 3–5 kgr/ft^3. Natural zeolites could process about 200,000 bed volumes before replacement was needed. Modern synthetic resins can tolerate about 10 times that number of bed volumes before replacement is indicated.

Multiple terms are used to express hardness, and operators need to know the relationship between these. Hardness expressed as milligrams per liter $CaCO_3$ can be converted to grains per gallon hardness using the following formulas:

1 grain/gal hardness = 17.1 mg/L hardness, and

1,000 grains = 1 kilograin

An ion exchange unit in a water treatment plant that has a resin rated at 30 kgr/ft^3 can theoretically remove 513,000 mg of hardness (30 × 1,000 × 17.1) for each cubic foot of resin before it is exhausted. In practice, resin beds normally remove a high percentage of the theoretical exchange capacity because of several factors, including flow rates, bed fouling, ionic interference, and efficiency of regenerant distribution. Dead space in the vessel is also a factor because it is nearly impossible to evenly distribute flow through it. However, it is important to realize that until the bed is exhausted, it is considered able to produce water with zero hardness. Beds are usually operated at flow rates of 1–5 gpm/ft^2 (8–40 bed volumes per hour) and will remove 50–75 percent of the rated hardness-exchange capacity.

Flow Rate Through Bed

Each gallon of water that flows through the bed carries an amount of hardness with it. It is necessary to know the hardness and flow rate values because they will dictate the time of exhaustion. If a bed is loaded at 120 gpm (454 L/min) and the source water has a hardness of 14 gr/gal (240 mg/L), 1,680 gr (120 gpm × 14 gr/gal) of exchange capacity will be used each minute of operation. The operator can see that the bed will exhaust eventually and that the exhaustion time will be reached sooner if the flow rate is increased. Doubling the flow rate will theoretically exhaust the bed in half the time.

As stated, water leaving the ion exchange vessel should have zero hardness. However, because water with zero hardness tends to corrode the system and the customers' plumbing, ion exchange water plants must produce water that is not of zero hardness. The simplest way to do this is to bypass some of the source water around the softeners and then blend it with the zero-hardness water at a predetermined rate to achieve the desired level of finished water hardness. This predetermined rate is called the **percent bypass** and is calculated by dividing the desired finished water hardness by the source water hardness and converting that value to a percentage. This number represents the approximate amount of water that should bypass the softeners. For example, if the source water hardness is 20 gpg (342 mg/L) and the desired finished water hardness is 8 gpg (137 mg/L), the percent bypass is calculated at 8/20, or 40 percent. If the plant flow is 1,000 gpm (2.2 ft^3/sec), the operator knows that 40 percent of 1,000, or 400 gpm (0.9 ft^3/sec), should bypass the softeners to get the desired results.

This bypass water will have the effect of lengthening the life of the softener. More finished water is produced during the cycle because not all of it had to be softened. The greater the desired hardness of finished water, the more efficient the plant becomes with respect to softening expense.

Regeneration

When the bed is exhausted, it must be regenerated. Because the resin has been stripped of sodium ions, these ions must be replaced. This is done by rinsing the bed with a salt solution (brine) at sufficient volume and concentration for a set time. Brine is concentrated sodium chloride (NaCl) solution, which is approximately a 25 percent concentration at 60°F (15.5°C). The specific gravity of this solution is about 1.192, which yields a working solution of about 2.479 lb of salt per gallon. The brine is usually diluted further before being applied to the softener in the regeneration cycle. As an example, if a softening unit contains 1,000 ft^3 of ion exchange material and it has a hardness-removal capacity of 3,000 gr/ft^3, it will have a softening capacity of 3,000,000 gr. If it is assumed that 0.5 lb of salt is required per 1,000 gr of hardness removed, 3,000 × 0.5, or 1,500 lb of salt will be required to regenerate the unit. Each gallon of saturated brine contains 2.479 lb of salt; therefore, approximately 605 gal of brine solution will be required to regenerate the unit. To produce a 5 percent solution (1.03624 specific gravity), the 605 gallons of brine must be diluted to 3,480 gal.

An exhausted resin bed is first backwashed to remove any foreign particles and to restratify the bed. The diluted brine is then introduced through the bed slowly, usually for 20–30 minutes. The flow rate may be about 1 gpm/ft^2 (0.7 mm/sec). It is very important to provide adequate time for the regeneration effect to take place. The bed is then rinsed to remove unused brine before the unit is placed back into service.

percent bypass
The predetermined amount of source water that is routed past water softeners then blended water to achieve the desired level of finished water hardness.

Operating an Ion Exchange Plant

Figure 7-2 is a schematic of an ion exchange plant. Note that this plant treats well water that contains iron. Therefore, the plant is equipped with an aerator and iron-removal filters, which is a typical arrangement for ion exchange plants

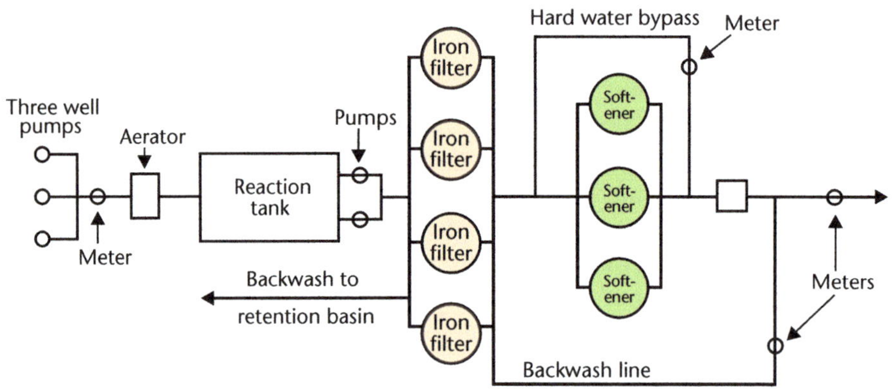

Figure 7-2 Schematic of the ion exchange process used in the example problems

treating water with iron problems. The iron must be removed before the water is passed through the exchangers; otherwise, the resin will be fouled and rendered useless.

Operators running this type of process must know the design and operating data of the plant. Designers of these plants use a much more complex set of variables, but the following examples represent a basic operational view of ion exchange softening.

There are three ion exchange softeners; each measures 6 ft (1.8 m) in diameter and has 7 ft (2.1 m) of media depth in it. The exchange capacity of the media is 15,000 gr/ft^3, and it takes 0.3 lb of salt per 1,000 gr of hardness removed to regenerate. The source water hardness is measured at 325 mg/L and the finished water hardness is 143 mg/L. One 500-gpm (1.1-ft^3/sec) high-service pump is operating. With this information, solve the following sample problems.

What is the total hardness-removal capacity of the plant?

The volume of media in the three exchangers is calculated as $\pi R^2 \times$ depth $\times$ number of units, or $3.14 \times (3^2)$ ft $\times$ 7 ft $\times$ 3 units = 593.46 ft^3 of media. At 15,000 gr/ft^3, the hardness removal capacity would be 593.46 ft^3 $\times$ 15,000 = 8,901,900 gr removal capacity. This would also be stated as 8,901.9 kgr removal capacity. There are three units, so each unit has one-third of the total capacity, or 2,967,300 gr.

What is the percent bypass?

Percent bypass is calculated by dividing the finished hardness by the source hardness. In this case, it would be 143 mg/L ÷ 325 mg/L, or 0.44. Expressed as percent, this is 44 percent.

With one high-service pump operating, how many gallons per minute are bypassing the softeners?

One pump operates at 500 gpm; 44 percent of 500 gpm is 220 gpm. This is the flow rate of the water bypassing the softeners. (This also means that the rest of the water, or 280 gpm, must be going through the softeners.)

How long will one unit operate before it needs to be regenerated?

If 280 gpm are going through the softeners, it can be assumed that the flow through one softener, if balanced, is 280 gpm ÷ 3 = 93.3 gpm. Of course, meters would be installed on the influent to the softeners to help the operator balance the flow, which is desirable. A hardness of 325 mg/L is converted to grains per gallon by dividing it by 17.1. This yields a hardness of 19 gpg. Therefore, a unit is being loaded with source water at a rate of 93.3 gpm, each gallon of which carries with it 19 gr of hardness. After calculating the removal capacity of one unit to be 2,967,300 gr, divide that capacity by 19 gpg to determine that about 156,174 gal of water can be softened before exhaustion. At 93.3 gpm, it will take about 28 hours to exhaust the unit (156,174 gal ÷ 93.3 gpm/60 min/hour).

Approximately how much finished water is produced during this time?

This is a tricky, but simple, question. Regardless of the hardness being removed, if all is going well, the plant is operating at 500 gpm (ignoring backwash, plant use, etc.). Therefore, 500 gpm × 60 min/hour × 28 hours = 840,000 gal.

Assume the plant is producing 720,000 gal of finished water per day before plant usage. How many pounds of salt will be used in a 30-day month?

Only 56 percent of that water had to be softened; so in a month, it is calculated (0.56 × 720,000 gal/day × 30 days) that approximately 12.1 mil gal flowed through the exchange units. First calculate the amount of hardness, in grains, that was removed: 12,100,000 gal × 19 gpg = 229,900,000 gr, or 229,900 kgr. Use 0.3 lb of salt per kilograins of hardness removed, so 0.3 lb/kgr × 229,900 kgr = 68,970 lb needed for the month.

How many gallons of saturated brine is this?

At about 2.5 lb/gal, 27,588 gal (68,970 lb × gal/2.5 lb) would need to be used.

Ion Exchange Residuals

Spent brine used in the regeneration process is typically sent to a sanitary sewer. In coastal regions, water utilities may obtain permits to discharge to the ocean. Other methods that have been used to dispose of brine are discharged to evaporation ponds in arid regions or to brine disposal wells if they exist. Disposal of regenerant brine solutions used in cation radionuclide removal must be approved by regulatory agencies.

Other Softening Processes

There are several additional processes that are sometimes used to soften drinking water, including electrodialysis, membrane softening, and split stream treatment (and blending strategies). Electrodialysis is used mainly for desalination rather than exclusively for softening. Membrane softening is usually accomplished using specially designed nanofiltration membranes (reverse osmosis membranes also remove hardness). Membrane systems and reverse osmosis are discussed in Chapter 10. The operation and design of membrane softening systems are analogous to other membrane systems.

Split stream and blending are not separate treatment processes but rather are strategies to adjust water hardness where the opportunity exists. Some treatment

electrodialysis
A water treatment method used for desalination rather than exclusively for softening.

processes like reverse osmosis and electrodialysis can remove calcium and magnesium almost completely for the water stream. The resulting water may be "too soft" for the intended purpose. Some systems blend this demineralized water with unsoftened water to provide a water that meets the hardness goals. This process, is termed *split treatment*. A similar approach is to blend hard water with a softer water source before it is distributed. One example where this can be accomplished is in a system that has both surface water (soft) and groundwater (hard). The two are blended and medium hardness water is distributed to customers. Where this is possible, it can be a cost-effective approach.

Ion Exchange

Groundwater is known to commonly contain high amounts of minerals and other natural substances that are present in the subterranean rocks and soil. Frequently high levels of hardness (calcium and magnesium) and dissolved radioactive ions are encountered and must be removed. Although precipitative softening (lime softening) can be used to effectively remove hardness, ion exchange is often used instead for smaller flows from many groundwater systems. Ion exchange can often be used directly on groundwater if the turbidity and iron level is not too high.

Radioactive ions, usually encountered in groundwater, are generally highly charged and, therefore, are ideal candidates for ion exchange removal processes. Regulated radionuclides susceptible to ion exchange removal include radium 226 and 228—beta and photon emitters that are usually present as cations (positively charged). The ion exchange process used to achieve their removal is similar to the ion exchange softening process.

Anion Ion Exchange

Some radionuclides are present in water as anions. Special test methods must be used to identify the radionuclide and determine the predominant species that is present. The most common radionuclide anion is uranium. In water, uranium is often present as the ion $UO_2(CO_3)_3^{4-}$. (Depending on pH, the $UO_2(CO_3)_2^{2-}$ ion may also be present.) Therefore, an anion exchange process can be very effective for its removal. The main differences between cation and anion exchange processes are that an anion exchange resin is used and although sodium chloride is often the regenerant, it is the chloride (Cl^-) that is being exchanged. The disposal of concentrated regenerant for uranium is also a concern as it is with other radionuclides.

A strong anion exchange resin is used for uranium removal. In many cases, the resin beds are able to effectively remove uranium for more than 200,000 bed volumes. After the bed is exhausted, the bed resin can be disposed (care must be exercised not to exceed the radioactive limits for disposal). If the bed is regenerated instead, the run length is limited (often to 10,000–50,000 bed volumes) so that the used regenerant meets disposal limits.

The concentrations of other anions (like nitrate, nitrite, or organic tannins) may also be removed by anion exchange. Depending on the conditions and amounts, these may compete with uranium and reduce the effective BV throughput. It may, however, be advantageous to plan on removing other anions in conjunction with uranium ion exchange treatment.

Study Questions

1. Natural zeolites used for softening that have become exhausted with use are regenerated by a strong solution of which chemical?
   a. NaCl
   b. NaOH
   c. HCl
   d. H_2SO_4

2. Softening resin exchanges what for the calcium and magnesium that it removes from the water?
   a. Sodium
   b. Hydroxide
   c. Potassium
   d. Hydrogen ions (H^+)

3. Ion exchange processes can typically be used for direct groundwater treatment as long as turbidity and _______________ levels are not excessive.
   a. calcium carbonate
   b. iron
   c. carbon dioxide
   d. sodium sulfate

4. What are the exchange capacities of natural zeolites?
   a. 1–2 kilograins/ft^3
   b. 3–5 kilograins/ft^3
   c. 5–8 kilograins/ft^3
   d. 9–12 kilograins/ft^3

5. The method of resin renewal is called
   a. regeneration.
   b. recycling.
   c. renewal.
   d. replenishment.

6. In converting hardness, 1 grain/gal =
   a. 14 mg/L.
   b. 100 mg/L.
   c. 17.1 mg/L.
   d. 27.8 mg/L.

7. What should be the hardness level of water leaving the ion exchange vessel?

8. Electrodialysis is used mainly for what process rather than exclusively for softening?

9. If an operator knows the exchange capacity of the resin bed, the hardness of the water, and the flow rate of water through the bed, he or she can predict what?

Chapter 8
Activated Carbon Adsorption

Operating Problems

Powdered Activated Carbon

The greatest operating challenge with powdered activated carbon (PAC) is handling it. It is a fine powder, so dust can be a major problem. If it is used continuously or if large quantities are used periodically, a slurry system should be considered.

PAC passing through the filters and entering the distribution system can result in complaints of "black water." Carbon passing through the filters is usually caused by inadequate coagulation–sedimentation, resulting in carbon carryover to the filters. It can also be caused by adding heavy doses of PAC just before the filters.

An occasional problem is that taste-and-odor problems persist, or even worsen, regardless of the PAC dosage. This problem is usually caused by adding the PAC and chlorine in stages too close to one another. The chlorine reacts with the organic compounds to produce additional compounds that are more difficult to adsorb. One solution to this problem is to add the PAC so that it has at least 15 minutes of contact time before any chlorine is added.

Granular Activated Carbon

When granular activated carbon (GAC) is used as a filter medium, it is important that the coagulation, flocculation, and sedimentation processes be operated continuously to maximize the removal of suspended matter going to the filters. The adsorption capacity of the carbon can quickly be lost if the carbon becomes coated with floc, resulting in shorter bed life and higher operating costs.

In addition, filter rates should be carefully controlled so that they do not fluctuate rapidly. Fluctuations can drive previously deposited material through the filter and can form channels through the bed. The channels will reduce the filtering capacity of the bed and greatly reduce the contact time of water passing through the filter medium.

Proper backwashing is essential for effective filtration and adsorption. The rate must be sufficient to achieve a bed expansion of about 50 percent for proper cleaning. If the bed is not adequately cleaned, both filtration and adsorption capacity will be lost, and mudballs will begin to form. This is particularly a problem where polymers are used as coagulants and filter aids.

GAC is lighter than other filter media, so it can easily be washed away during backwashing. If an excessive amount of GAC is being lost, backwash rates may have to be reduced. The water pressure to surface washers may also have to be decreased if it is found that they are contributing to excessive GAC loss. If carbon loss is over 2 in. (50 mm) per year, the backwashing procedures should be reviewed to determine how the loss can be reduced. If the height of the wash-water

powdered activated carbon (PAC)
Activated carbon in a fine powder form. It is added to water in a slurry form primarily for removing those organic compounds causing tastes and odors.

granular activated carbon (GAC)
Activated carbon in a granular form, which is used in a bed, much like a conventional filter, to adsorb organic substances from water.

bed life
The time it takes for a bed of adsorbent to lose its adsorptive capacity. When this occurs, the bed must be replaced with fresh adsorbent.

mudball
An accumulation of media grains and suspended material that creates clogging problems in filters.

troughs above the filter beds is not adequate, the troughs may have to be raised to prevent further carbon loss.

A special problem with GAC filters is the rapid growth of bacteria within the bed. The carbon removes and holds organic compounds, and bacteria can feed and thrive on this material. The carbon also adsorbs chlorine, so any chlorine added ahead of the filters to control bacterial growth is ineffective. As a result, the bacterial concentration in the filter effluent may be thousands of times higher than in the filter influent.

Research indicates that the high bacterial populations in the bed actually enhance organics removal because the bacteria break down complex compounds to simpler products that are more easily adsorbed by the GAC. Some treatment plants backwash with heavily chlorinated water to help control the bacterial levels. In general, it is not good practice to allow high concentrations of bacteria to pass through the filters. The final chlorination must be closely controlled to ensure that the bacteria are destroyed before they enter the distribution system.

If GAC contactors are used after conventional filters, it is essential that only a minimum of suspended material comes through the filters. If the length of time between backwashing a contactor begins to decrease, or if the bed life is much shorter than it should be, floc carryover from the filters may be a problem. The turbidity of filter effluent should be monitored continuously to prevent reducing the life of GAC in the contactors. Bacterial growth in contactors can also be a problem, but it can be handled in the same manner as that for GAC in conventional filters.

Control Tests

Testing of the effluent from carbon adsorption systems to determine if there is a breakthrough of organic chemicals requires sophisticated water quality analyses using a gas chromatograph. In general, only very large water systems have the equipment and expertise to run these analyses themselves. Other systems should contract with a convenient private laboratory that will perform the analyses.

Determining PAC Dosage Rates

The required dosage of PAC can be approximated by modifying the jar test. The stirring apparatus and all glassware should be cleaned with an unscented detergent and rinsed thoroughly with odor-free water. (The standard jar test and taste-and-odor testing are described in Chapter 4.)

One-liter samples of the raw water are then dosed with varying amounts of a well-shaken stock PAC solution—for example, 5, 10, 20, and 40 mL. The stock solution is prepared by adding 1 g of PAC to 1 L of odor-free water. Each milliliter of this solution, when added to a 1-L sample of raw water, is equivalent to a dosage of 1 mg/L.

The four dosed samples, and a fifth sample to which no PAC is added, are stirred for a period that approximates the contact time the PAC will have with the water as it passes through the plant. At the end of that time, each sample is filtered through glass wool or filter paper to remove the PAC. The first 200 mL of each sample through the filter is discarded, and the remainder is subjected to the threshold odor test to arrive at a threshold odor number (TON) for each sample.

The results can then be plotted, as illustrated in Figure 8-1, to obtain an optimal PAC dosage. In the example, the dosage needed to reduce the TON to 3 (which is recommended by USEPA's secondary drinking water regulations) would be 29 mg/L. Experience has shown that plant-scale PAC application is more

breakthrough
Treatment cycle at which contaminants, such as organics or turbidity, begin to pass through the treatment process.

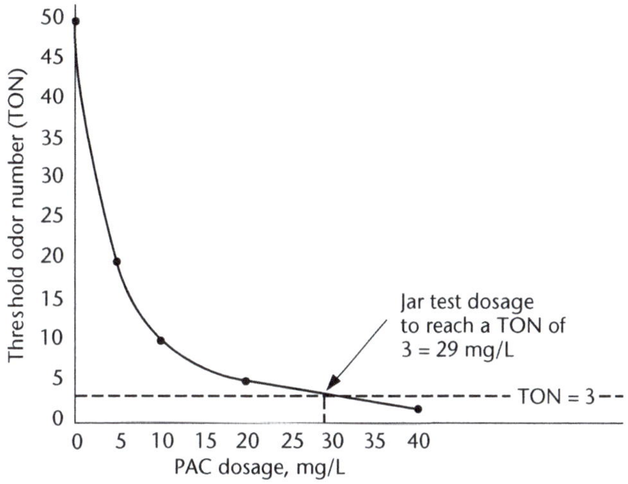

Figure 8-1 Example determination of optimal PAC dosage

efficient than indicated by jar tests. However, a plant should begin with the dosage indicated by the jar test result and then gradually reduce it while sampling the plant effluent. To maintain the optimum PAC dosage, the threshold odor test should be conducted by each shift, or at least daily, on the raw and treated water while PAC is being fed.

If PAC is used to remove organic compounds other than those causing tastes and odors, different procedures are needed to determine the proper dosage.

Tests of GAC Used in Filters

GAC is used in a conventional gravity filter for both filtration and adsorption, so most of the control tests are the same as those used for filtration. The condition of the filter bed should be checked frequently, particularly during and after backwashing, to determine if there are problems with cracks, shrinkage, channeling, or mudballs.

The distance between the top of the carbon and the top of the wash-water troughs (or other reference mark) should be measured at least every 3 months to determine the rate of carbon loss. Samples of backwash water can also be collected and tested to see if they contain excessive amounts of carbon. Excessive loss may indicate a need for changes in the backwashing procedures.

Core samples of the carbon bed should be taken at the time of installation and at least every 6 months to determine both the amount of bed life remaining and the condition of the medium. The sample should represent the carbon from top to bottom of the bed. GAC manufacturers can provide the necessary information on sampling, testing, and interpreting results.

If the filter is designed to remove taste- and odor-causing compounds, the threshold odor test should be conducted routinely on the raw and finished water as a check on the effectiveness of the filter. If the filter is intended to remove other organic compounds, special analyses of samples will periodically have to be performed for these compounds. Taste-and-odor testing cannot be used as an indicator of the removal of other organic compounds. In general, the holding capacity of GAC for other organic compounds is much less than it is for taste- and odor-causing compounds.

Because bacteria can thrive in GAC filters, standard plate count analyses should be performed each day on the filtered water and the water after final chlorination. These checks will help determine the required final chlorination dosage and will indicate whether chlorination of the backwash is needed. The plate count should be kept well below 500 organisms/mL.

Tests of GAC Used in Contactors

Because GAC contactors are typically used for removing organic compounds other than those causing tastes and odors, more sophisticated control tests are usually needed. There is normally a specific compound or group of compounds that are to be removed, so periodic analyses of the contactor effluent must be made to check on the effectiveness of the treatment and to determine when the carbon is exhausted.

Head loss through the contactors should also be monitored and recorded continuously so that backwashing can be performed at the proper time. Core samples of the carbon bed should be taken at least every 3 months so that the approximate remaining bed life can be determined. Turbidity of the effluent should also be monitored and recorded continuously to determine if carbon fines or other suspended matter are passing through the GAC bed. Standard plate count analyses of the contactor effluent before and after chlorination will also help determine what chlorine dosages are needed to control bacterial growth.

Safety Precautions

Bags of PAC and GAC should be stored on pallets in a clean, dry place so that air can circulate underneath. The bags should be stacked in single or double rows with access aisles around every stack to allow easy handling and fire inspection. They should never be stored in stacks over 6 ft (2 m) high.

Although activated carbon is not considered explosive, it will burn like charcoal without producing smoke or flame, and it glows with the release of intense heat. A carbon fire is difficult to detect and to extinguish. Avoidance of this problem is another advantage of using slurry storage rather than dry feed.

The storage area for dry carbon should be fireproof and have self-closing fire doors separating it from other storage areas. Storage bins for dry bulk carbon storage should be fireproof and equipped for fire control. Smoking must be prohibited during the handling and unloading of carbon and in the storage and feeding areas.

Burning carbon should not be doused with a large stream of water because doing so will only cause carbon particles to fly in all directions and spread the fire. A fine mist of spray from a hose or a chemical foam extinguisher is most effective in controlling a fire of this type.

Activated carbon must never be stored near gasoline, mineral oils, or vegetable oils. When mixed with carbon, these substances will slowly oxidize until the ignition temperature of 600–800°F (316–427°C) is reached.

Stored carbon should also be kept well away from chlorine compounds and potassium permanganate because spontaneous combustion can occur when they are mixed. Carbon is an electrical conductor, so explosion-proof light fixtures and electrical wiring should be used in all carbon storage and feeding areas. As an added precaution, the electrical equipment should be cleaned frequently.

All tanks receiving dry carbon should be vented and provided with dust-control equipment, such as bag-type dust collectors. Even slurry storage tanks should be so equipped. If a PAC dry feed system is used, the hoppers and feeders should be enclosed in a separate room so that dust will be confined.

Oxygen is removed from air in the presence of wet activated carbon. As a result, slurry tanks or other enclosed spaces containing carbon may have seriously reduced oxygen levels. Personnel must be careful when entering these spaces to ensure that adequate oxygen is available. Devices to indicate the amount of

Figure 8-2 Safety clothing to be worn in the handling of carbon

oxygen present should be used before anyone enters this type of closed space. Personnel entering a tank or other enclosure should also have attached safety belts and another worker standing by to pull them from danger if necessary.

Personnel who unload or handle carbon should be provided with dust masks, face shields, gauntlets, and aprons (Figure 8-2). In addition, shower facilities should be provided for workers to use if they get unduly dirty from handling carbon.

Record Keeping

Good record keeping can help prevent taste- and odor-causing compounds from reaching consumers, while preventing waste of PAC. Suggested records include the following:

- The type of odor in the water (such as musty, septic, or rotten egg) and the taste (such as sweet or bitter)
- Dates on which PAC was added
- The calculated dosage based on jar tests
- The actual plant dosage that was effective
- The TON values for the raw and treated water

This information may give some clue to the source of the problems, how long they may last, how they may be prevented, and when they may occur again.

Records of GAC in filters should include the following:

- Hours of filter operation for computing the quantity of water processed
- Monthly measurements of the carbon level to monitor GAC loss
- Dates when GAC is replaced so that bed life can be monitored
- Results of periodic threshold odor tests

Records of GAC contactor operations could include the following:

- Quantity of water treated
- Times when the contactor is backwashed and the amount of backwash water used
- Dates of GAC replacement for computing the bed life
- Results of analyses of raw-water and finished-water organics concentrations

Study Questions

1. When should core samples be collected from filters using granular activated carbon?
   a. Every 6 months
   b. At the time of installation and every 6 months thereafter
   c. Every year after installation
   d. At the time of installation and every year thereafter

2. Which bed expansion during backwashing must be achieved for filters using granular activated carbon?
   a. 20%
   b. 25%
   c. 30%
   d. 50%

3. What is the best approach for feeding powdered activated carbon?
   a. Along with prechlorination
   b. After filtration
   c. After at least 15 minutes following chlorine addition
   d. In the chlorine contact chamber

4. Backwashing rate procedures should be reassessed to determine the cause of granular activated carbon loss if the loss per year exceeds
   a. 2 in. (50 mm).
   b. 4 in. (100 mm).
   c. 6 in. (150 mm).
   d. 8 in. (200 mm).

5. In regard to safety, wet activated carbon will remove which of the following from the air?
   a. Organic gases and hydrogen sulfide
   b. Oxygen
   c. Carbon dioxide
   d. Carbon monoxide

6. Which of the following is the most probable cause of tastes and odors persisting or worsening when powdered activated carbon (PAC) is used?
   a. PAC is added at the inlet
   b. PAC is added during flocculation
   c. PAC and chlorine are added too close to each other
   d. PAC is added to water with too high of an alkalinity

7. Because powdered activated carbon is a fine powder,
   a. it can dissolve too quickly in solution.
   b. dust can be a major problem.
   c. it is sold only in small supplies.
   d. its cost is prohibitively expensive.
8. If the bed is not adequately cleaned, both filtration and adsorption capacity will be lost, and ______ will begin to form.
   a. disinfectant by-products
   b. floc
   c. chlorine residuals
   d. mudballs
9. The distance between the top of the carbon and the top of the wash-water troughs (or other reference mark) should be measured at least how often to determine the rate of carbon loss?
10. Because bacteria can thrive in granular activated carbon filters, standard plate count analyses should be performed how often on the filtered water and the water after final chlorination?
11. What is the greatest operating challenge with powdered activated carbon?

Chapter 9
Aeration

The three basic control tests involved in the aeration process are

1. dissolved oxygen (DO),
2. pH, and
3. temperature.

A description of each test can be found in Chapter 4.

The DO test is used to monitor the amount of oxygen being dissolved in the water. Knowing the needed DO concentration and monitoring the amount of DO present will prevent over- or underaeration.

The pH of the water can be used as an indicator of carbon dioxide removal; pH increases as carbon dioxide is removed. It can also be used to monitor the effective pH range for hydrogen sulfide, iron, and manganese removal. The best pH range for hydrogen sulfide scrubbing is 6 or less. Iron and manganese are best treated in a pH range of 8 to 9.

The amount of oxygen that will dissolve in water (the saturation concentration) varies depending on water temperature; the amount increases as the water temperature drops. As water temperature drops, the operator must adjust the aeration process to maintain the correct level of DO.

Additional control tests for the following are used to measure how effective the aeration process is in removing troublesome constituents:

- Iron
- Carbon dioxide
- Manganese
- Tastes and odors

Test frequency depends on both the variability of raw-water quality and the characteristic measured. For example, well water tends to be very constant in quality and very slow to change, so daily testing is not usually necessary.

Surface water quality is subject to more frequent change. Temperature, DO, pH, carbon dioxide, and tastes and odors can vary daily. Therefore, sampling and testing frequencies should be set up so that all significant changes can be detected and monitored.

In addition to these specific control tests, it is advisable to make frequent visual inspections of the process. Clogged diffusers, for example, are a common problem that is most easily identified visually.

If aeration equipment is being used to remove volatile organic chemicals or radon, periodic analyses of the finished water will be necessary to ensure that the process is achieving the required degree of removal. These tests are quite

aeration
The process of bringing water and air into close contact to remove or modify constituents in the water.

complicated, are relatively expensive, and must be performed by a qualified laboratory. State authorities will require a minimum frequency of sampling to ensure that state requirements are being met, but more frequent sampling may be necessary for operational control.

Samples must be collected very carefully. Special containers must be used, and the laboratory will specify procedures to follow. The samples must be analyzed soon after collection, so they must be shipped by a method that will ensure delivery to the laboratory within the required holding time. If a certified laboratory is within a reasonable distance, it is often best to deliver the samples in person.

Operating Problems

The following are some common operating problems associated with aeration:

- Corrosion
- Floating floc in clarifiers
- False clogging of filters (air binding)
- Hydrogen sulfide removal
- Algae
- Clogged diffusers
- Energy consumption

Corrosion

Aeration easily increases the DO content of water, and excessive levels in a water distribution system can cause corrosion. Corrosion can occur whenever water and oxygen come into contact with metallic surfaces. In general, the higher the DO concentration, the more rapid the corrosion.

The solution to the problem of excessive DO concentration is simply not to overaerate. Unfortunately, there is no definite rule as to what constitutes overaeration. The amount required will vary from plant to plant, and may vary seasonally as water quality changes. The correct amount must be determined from experience. (See Table 9-1 for the saturation levels for DO in water at various temperatures.)

Two things can be done to prevent or retard corrosion. First, adequate protective coatings can be applied and maintained on all exposed metallic surfaces. The application and reapplication of paint and other protective coatings are a vital part of any preventive maintenance program. Second, the aeration process can be operated so that it provides an adequate but not excessive level of DO. As a general rule, DO concentrations of 2–4 mg/L should be acceptable.

Floating Floc in Clarifiers

If clarification and filtration are performed after aeration, too much aeration may cause sedimentation problems. Small bubbles of excess air can come out of solution and attach to the particles of floc in the clarifier, causing the particles to float rather than settle. This destroys the efficiency of the sedimentation process and creates an added load on the filters.

False Clogging of Filters

In the same way that bubbles of excess air can attach to floc particles, they can also attach to the media in a filter. This will occur if the water warms as it passes through the filter, releasing the air that was in solution. It can continue until the

Table 9-1 Oxygen saturation (or equilibrium) levels in water

Temperature, °C	Saturation Concentration, mg/L	Temperature, °C	Saturation Concentration, mg/L
0	14.621	26	8.113
1	14.216	27	7.968
2	13.829	28	7.827
3	13.460	29	7.691
4	13.107	30	7.559
5	12.770	31	7.430
6	12.447	32	7.305
7	12.139	33	7.183
8	11.843	34	7.065
9	11.559	35	6.950
10	11.288	36	6.837
11	11.027	37	6.727
12	10.777	38	6.620
13	10.537	39	6.515
14	10.306	40	6.412
15	10.084	41	6.312
16	9.870	42	6.213
17	9.665	43	6.116
18	9.467	44	6.021
19	9.276	45	5.927
20	9.092	46	5.835
21	8.915	47	5.744
22	8.743	48	5.654
23	8.578	49	5.565
24	8.418	50	5.477
25	8.263		

Note: Values given for atmospheric pressure (101.3 kPa).

Adapted from *Standard Methods for the Examination of Water and Wastewater.* 15th ed. 1981.

spaces between the media particles begin to fill with air bubbles, a process called **air binding**. Air binding causes the filter to behave as if it were clogged and in need of backwashing. Serious disruption of the filter media can also occur as the air bubbles burst within the bed.

air binding
A condition that occurs in filters when air comes out of solution as a result of pressure decreases and temperature increases. The air clogs the voids between the media grains, which causes increased head loss through the filter and shorter filter runs.

Hydrogen Sulfide Removal

Hydrogen sulfide is most efficiently removed by the physical scrubbing action of aeration, not by oxidation. Which of the two removal processes actually takes place depends on the pH of the water. At a pH of 6 or less, hydrogen sulfide exists primarily as the gas hydrogen sulfide (H_2S). In this form, it is easily scrubbed away by aeration. However, at pH values of 8 or above, the hydrogen sulfide gas ionizes and the ionized form (HS^- and S^{-2}) cannot be removed by aeration. In fact, the oxygen provided by aeration reacts with the ionized hydrogen sulfide to release the element sulfur, which occurs as a fine colloidal particle and gives water a milky-blue turbidity.

One of the better procedures for eliminating hydrogen sulfide is to lower the pH to 6 or less before beginning aeration. This will favor hydrogen sulfide removal by scrubbing. The pH can be lowered by bubbling carbon dioxide gas into the water. After the hydrogen sulfide is removed, additional aeration will remove any remaining carbon dioxide.

Algae

In warmer climates, the slat-and-coke-tray and positive-draft aerators are often located in the open air and exposed to direct sunlight. The wetted surfaces of these aerators create excellent environments for the growth of algae and slime. Although it is not practical to eliminate these problems completely, growth rates can be greatly reduced if a roof or canopy is provided to shade the aerator from direct sunlight.

Clogged Diffusers

Air diffusers can become partly clogged, either from dust in the air or from oil, debris, or chemical deposits that can collect around the diffuser opening. Operators can minimize clogging by doing the following:

- Maintaining clean air filters
- Not overlubricating blowers
- Preventing backflow of water into diffusers

Techniques for cleaning diffusers vary with the type of diffuser. Most ceramic types can be heated to burn away trapped particulates. Both fine- and large-bubble types can be cleaned with a brush and detergent. Manufacturers of ceramic diffusers should be consulted to determine the best cleaning procedure to follow.

Energy Consumption

Regardless of the method used, the aeration process will require some supply of energy. The water usually must be pumped both before and after cascade aeration; diffused aeration requires power to drive the blowers, and mechanical aeration requires power to drive the aerator and the blowers. Operators can minimize power costs by not overaerating, keeping motors and blowers properly maintained, and keeping air diffusers clean and free of debris.

Safety Precautions

Both hydrogen sulfide and methane can be produced by aeration processes, and operators must be made aware of how dangerous these gases can be.

Hydrogen sulfide is a poisonous gas and is colorless, flammable, and explosive. It kills by paralyzing a person's respiratory system; 30 minutes of exposure to a concentration of hydrogen sulfide as low as 0.1 percent by volume in air is fatal. Higher concentrations, above 4.3 percent, are explosive. A very dangerous characteristic of hydrogen sulfide is that its foul, rotten-egg odor is not evident at high concentrations, because these concentrations paralyze one's sense of smell (the olfactory nerves).

Methane, or natural gas, is a colorless, odorless, and tasteless gas; it is flammable and highly explosive. If the scrubbed gases containing methane are allowed to accumulate in a confined space, an explosion or fire can result. Methane can also kill by simple asphyxiation (suffocation). Because the gas has no color, taste, or odor, the victim may be completely unaware of its presence.

The safest way to deal with hydrogen sulfide and methane gases is to ensure that the aeration process is well ventilated and that there are no locations within the process where gases can accumulate. Methane is lighter than air and will accumulate at the top of an enclosure. Hydrogen sulfide is heavier than air and will collect at the bottom of confined areas. Ideally, aerators should be located in an open area with good ventilation. Where aerators must be enclosed because of climatic conditions, the enclosure must be well ventilated at all times.

Record Keeping

Following are some important aeration process records that should be maintained:

- Results of periodic analyses of raw water to monitor for changes in constituents the aeration process is designed to alter
- Daily quantity of water treated by the aeration process
- Results of periodic analyses of finished water to determine if the aeration process is successfully altering or removing constituents from the raw water as designed
- Inspections and monitoring for safety hazards, such as the release of poisonous or explosive gas from the aerators
- Details of the maintenance of aeration equipment
- Changes made in the equipment or process operation
- Observed changes in other treatment processes that might be caused by failure or improvement in the aeration process

Study Questions

1. Aeration can cause what problem in filters?
   a. Precipitation of odor compounds
   b. Increased color
   c. Mudball formation
   d. Air binding

2. Which of the following is *not* one of the three basic control tests involved in the aeration process?
   a. Dissolved oxygen
   b. pH
   c. Temperature
   d. Iron

3. As water temperature ______, the operator must adjust the aeration process to maintain the correct level of DO.
   a. drops
   b. increases
   c. stabilizes
   d. approaches freezing
4. The process by which spaces between the media particles begin to fill with air bubbles is known as
   a. air stripping.
   b. oxidation.
   c. air binding.
   d. oxygen saturation.
5. One of the better procedures for eliminating hydrogen sulfide is to lower the pH to ______ or less before beginning aeration.
   a. 8
   b. 6
   c. 7
   d. 4
6. What are the basic control tests for aerators?
7. What removal process is most effective for hydrogen sulfide?
8. The quantity of water treated by the aeration process should be measured how often?

Chapter 10

Membrane Treatment

Reverse Osmosis Facilities

This section describes the equipment and chemicals associated with reverse osmosis (RO). Figure 10-1 shows a schematic for a simple RO installation. Figure 10-2 shows a typical package RO unit.

Membranes

The two membranes most commonly used for water treatment are cellulose acetate and polyamide–composite. Each type has specific operating characteristics, such as efficiency of salt rejection, best pH operating range, resistance to degradation if exposed to chlorine or other oxidants, susceptibility to biological attack, and resistance to hydrolysis.

Over time, the performance of all membranes will change, mainly as a result of compaction and fouling. Compaction, which is similar to plastic or metal "creep" under compression, gradually closes the pores in the membrane. The compaction rate increases when membranes are operated at higher pressures and higher temperatures. Membrane fouling can be traced to substances in the feedwater, such

reverse osmosis (RO)
A pressure-driven process in which almost-pure water is passed through a semipermeable membrane. Water is forced through the membrane and most ions (salts) are left behind. The process is principally used for desalination of sea water.

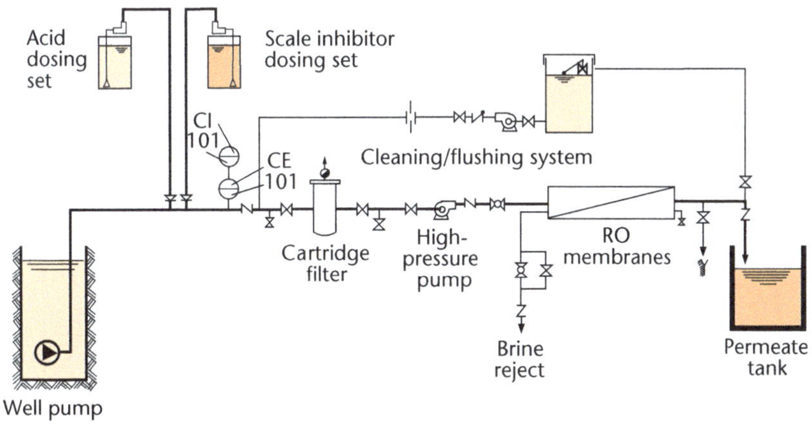

Figure 10-1 Typical reverse osmosis flow schematic
Courtesy of American Engineering Services, Inc., Tampa, Florida.

Figure 10-2 Typical package reverse osmosis treatment unit
Courtesy of Ionics, Inc.

WATCH THE VIDEO
WSO Filtration—Reverse Osmosis (www.awwa.org/wsovideoclips)

as calcium carbonate or calcium sulfate scales, fine colloids, iron or other metal oxides, and silica. Pretreatment of water before RO is usually necessary to prevent premature fouling of the RO membrane.

In some cases, fouled membranes can be restored by periodic cleaning with acid. Otherwise, they must be replaced.

Four membrane configurations are currently available: spiral wound, hollow fiber, tubular, and plate and frame. Only the first two configurations are used in commercial drinking water treatment plants.

Spiral-Wound Membranes

As illustrated in Figure 10-3, spiral-wound membranes consist of two flat sheets of membrane material separated by porous sheets. These layers are sealed on three sides to form an envelope. Feedwater enters at one end, and the open side of the envelope is attached to a plastic tube that collects the product water.

Hollow-Fiber Membranes

As illustrated in Figure 10-4, hollow-fiber membranes consist of a compact bundle of thousands of fibers that surround the feedwater distribution core. Each fiber is laid in a U-shape in the bundle, and both ends are encapsulated in an end sheet.

Feedwater Concerns

The term feedwater describes raw water that has undergone pretreatment (acidification or scale inhibitor addition) prior to entering the membrane arrays. Feedwater quality and pressure are important issues for the RO process. Feedwater treatment is almost always necessary.

feedwater
Raw water that has undergone pretreatment (acidification or scale inhibitor addition) prior to entering the membrane arrays.

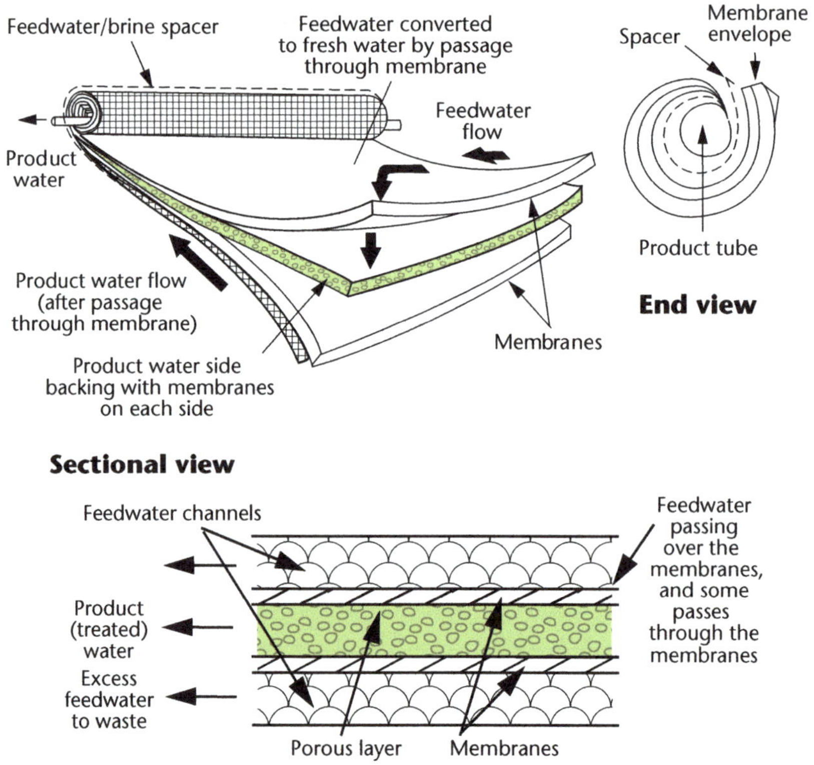

Figure 10-3 Details of a spiral-wound membrane
Courtesy of the Dow Chemical Company.

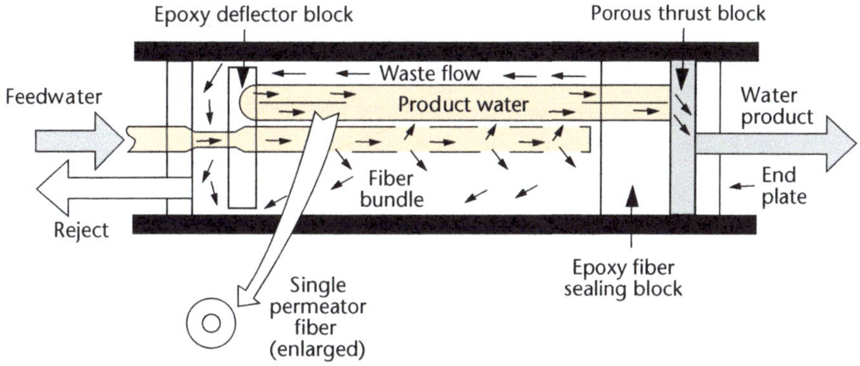

Figure 10-4 Details of a hollow-fiber module
Courtesy of US Filter/Permutit.

Feedwater Quality

The quality of feedwater passing through RO units is critical for prolonging the life of membranes. Improperly treated feedwater can rapidly cause irreversible damage to the membrane. Surface water usually requires more pretreatment and closer monitoring because its quality is not as stable as groundwaters. To maximize membrane life, pretreatment of RO feedwater may be needed for any one or more of the following purposes:

- Turbidity reduction
- Iron or manganese removal
- Stabilization to prevent scale formation
- Microbial control
- Chlorine removal
- Hardness reduction

The types of materials used in feedwater piping and the pressure pump can adversely affect membrane life or performance by releasing trace amounts of metals; therefore only noncorrosive materials, such as stainless steel, PVC, and fiberglass, are preferred. To prevent air ingress, either vertical-turbine types that have mechanical seals or submersible pumps are used.

Feedwater Treatment

In addition to any special pretreatment for contaminant control, sulfuric acid is usually added to the feedwater to prevent calcium and magnesium carbonate scaling of the membrane. Cartridge filters are also commonly installed immediately ahead of RO units for final removal of suspended matter down to the 5- or 10-mm level. This is necessary regardless of whether the supply is surface water or groundwater. Cartridge filter containers are furnished with inlet and outlet pressure gauges to monitor head loss. Cartridges should be replaced when the difference in pressure between the gauges reaches 15 psi (100 kPa).

Feedwater Pressure

The pressure differential applied to RO membranes depends on the desired recovery. Although the process can be designed to operate with a pressure as low as 50 psi (340 kPa), commercial installations are commonly operated at 300 psi (2,100 kPa) or higher, because the osmotic pressure increases as the recovery is pushed to higher levels.

head loss
The amount of energy used by water in moving from one point to another.

Posttreatment

Additional treatment following RO treatment typically includes the following:

- Degasification for removal of carbon dioxide, hydrogen sulfide, or other undesirable gases if they are present
- pH adjustment to minimize corrosion in the distribution system
- Disinfection

Reverse Osmosis Membrane Cleaning

As membranes foul, water production rates and removal efficiencies decrease and higher pressure differentials are required. The following are some indications that a membrane needs to be cleaned:

- The passage of salt through the membrane increases by 15 percent.
- The pressure drop through the unit increases by 20 percent.
- Feed pressure requirements increase by 20 percent.
- Product water flow drops or increases by 5 percent.
- Fouling or scaling is evident.

Each membrane manufacturer has recommended procedures and chemicals that should be used to rejuvenate its membranes. These guidelines should be carefully followed; otherwise the membranes could be damaged. Cleaning chemicals are usually corrosive, so all cleaning system components must be made of stainless steel or other noncorroding materials. Obviously, the discharge of these cleaning chemicals is another disposal issue that takes careful planning.

Reject Water

The amount of reject water resulting from the operation of a commercial RO plant can range from 20 to 50 percent of the feedwater to an RO unit. The amount of reject water depends on the number of stages in which the membranes are configured and the feed pressure.

Reject water presents two potentially significant problems. The first is that the water source must be capable of supplying up to twice the amount of water needed by the system. In areas with limited groundwater availability, other treatment processes that do not waste as much water may need to be considered, even if those processes are more expensive.

The other problem is waste disposal. One acceptable method is discharge to the local waste treatment system. However, the relatively large quantity of wastewater produced and the high mineral content of the reject water may overtax the local treatment system. If the source water is subject to agricultural runoff high in pesticides and fertilizer concentrations, or it is well water with high arsenic levels or fluoride, or even if the source water is salty seawater, the concentrated reject water may limit the disposal options. If disposal to the local wastewater system is out of the question, other common disposal methods are deep-well injection (Class V injection wells) and evaporative ponds. With ponds, care must be taken to protect any aquifers, and the utility needs to know that they have only delayed the eventual disposal. Discharge to surface waters is possible if a National Pollutant Discharge Elimination System (NPDES) permit can be obtained. The amount of reject water from an RO installation can be reduced to a limited extent by an increase in the feedwater pressure. However, this usually results in a shorter membrane life and increased operating costs.

reject water
The water that does not pass through a membrane, carries away the rejected matter, and must be disposed of.

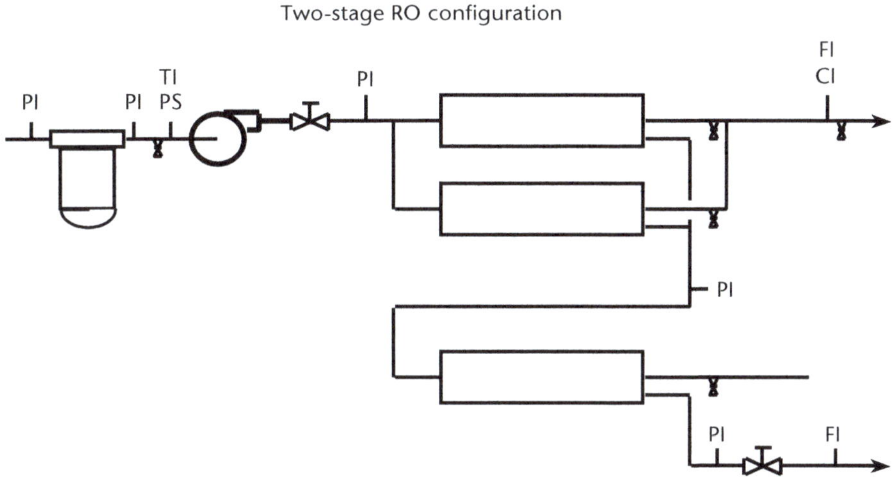

Figure 10-5 Schematic of a two-stage reverse osmosis treatment system
Courtesy of US Filter.

A two-stage system is shown in Figure 10-5. The two first-stage modules purify 50 percent of the water fed to the system. The reject water from the first stage is then processed by the second-stage unit, which purifies it another 50 percent. The final flow ends up being 75 percent purified water and 25 percent reject water.

Operation of the Reverse Osmosis Process

Most RO systems are controlled by programmable logic controllers (PLCs). However, operators are required to monitor PLCs and system operation to ensure maximum efficiency. Some of the principal requirements in the operation of an RO treatment system are as follows:

- If surface water is being treated, the raw-water quality must be routinely monitored for changes in chemical and biological quality.
- The operation of the pretreatment acid feed system must be regularly inspected and the pH level monitored to ensure that the proper proportion of acid is being fed.
- Cartridge filter drains must be flushed daily and the loss of head across the filters recorded.
- Continuous monitoring of feedwater quality and membrane product water must be done to optimize critical membrane operating parameters, including product water recovery rate (proportion of raw water recovered as potable water) and feed pressure.
- Posttreatment systems—which may include degasification, disinfection, and corrosion control treatment—must be monitored and adjusted for proper operation.

A system should have an automatic shutdown capability in the event of a problem that could damage the membranes. Problems that require shutting down the system include the following:

- Higher than normal pump pressure
- Low feedwater pump suction pressure
- Abnormally high or low feedwater pH
- Improper feedwater flow rate

- High feedwater turbidity
- Higher than normal feedwater conductivity
- Abnormally high or low product flow rate
- High product water conductivity
- Low antiscalant level or flow

Each of these controls should be maintained periodically and checked for proper operation.

Operating Problems

Some of the major operating problems experienced with RO installation include the following:

- Premature fouling
- Flux decline
- Unit shutdown
- Bacteria and algae fouling
- Oil and grease fouling
- Leakage between modules

Premature fouling of membranes as a result of any of the factors previously mentioned typically occurs because the installation either was not designed properly or has not been operated in accordance with the design. The unit's manufacturer or membrane supplier should be consulted at the first sign of fouling to see if the problem can be corrected.

Membrane performance will gradually change as a result of compaction; this is often called *flux decline*. The manufacturer should be able to predict what changes can reasonably be expected and what the approximate life of the membrane should be. If the changes appear to be more drastic than predicted, the manufacturer should be consulted for advice on any changes that can be made in the operation to help prolong the membrane life.

If the plant is shut down for any period of time, membranes should be flushed with acidified feedwater or unchlorinated product water. Concentrated brine must be removed from prolonged contact with the membrane because precipitation will eventually occur, resulting in membrane scaling. If shutdown is to be for more than a few days, accumulation of biological growth must be prevented. This can be accomplished by either operating the units for a short time each day or sterilizing the membranes according to the manufacturer's recommendations.

Bacteria, bacterial slimes, and algae can cause fouling of some membranes. Some membrane materials have a limited tolerance for chlorine, and others have essentially none at all. It is safest to obtain the manufacturer's recommendation on whether controlled amounts of chlorine can be used to inhibit bacterial and algae growth or whether biocides should be used.

Oil and grease in very low concentrations can reduce membrane performance by forming a film on the membrane surface. Only lubricants recommended by the membrane manufacturer should be used on pumps, parts, and O-rings that are installed ahead of the membrane units.

WATCH THE VIDEO
Brazos RO Case Study (www.awwa.org/wsovideoclips)

Record Keeping

Records that should be kept on the operation of an RO treatment system include the following:

- Records of the operation of any pretreatment being provided
- Records of feed rates and residual monitoring of acid and antiscalant chemicals
- Daily records of the pressure drop on each cartridge filter
- Daily records of the pressure drop across each membrane unit
- Detailed records of each membrane-cleaning operation
- Daily records of the chlorine feed rate and amount of chlorine used
- Daily records of the metered flow of treated water and reject water
- Records of the operation of posttreatment systems
- Periodic records of the salinity of treated water as an indicator of membrane fouling

In many cases, the membrane system computer or PLC systems can automate this data collection and provide trends in the operating data.

Glossary of Membrane Filtration Terms

Cartridge or filter cartridge. A filtration or separation device (usually in the shape of a cylinder or disk) that is designed for easy installation and removal. The cartridge–cylinder design contains a porous material or media through which a liquid or gas is passed in order to separate the fluid from suspended particulate matter. The media may be a membrane or another material that contains pores of set or varying size.

Concentrate. The retained sample after filtration is complete.

Concentration. Crossflow filtration in which the desired substance (product) remains on the upstream side of the membrane (does not pass through). The opposite of *diafiltration*.

Concentration polarization. The accumulation of retained solute (see also *gel layer*) on the surface of the membrane. This is caused by a combination of factors, including transmembrane pressure, viscosity, solute concentration, and crossflow velocity.

Crossflow. Flow of solution adjacent or parallel to the surface of the membrane (see also *tangential flow filtration*). This contrasts with *direct flow* seen in traditional filters, in which the liquid flows perpendicular to the surface of the filter. In crossflow, only a fraction of the crossflow solution passes through the membrane.

Diafiltration. A crossflow filtration process allowing for the transfer of low-molecular-weight species, water, or solvents through a membrane without changing the solution volume. This process is used for purifying retained large-molecular-weight species, increasing the recovery of low-molecular-weight species, buffer exchange, and simply changing the properties of a given solution.

Differential pressure (ΔP). The difference between the pressure on the influent side of a filter (upstream pressure) and the pressure on the effluent side (downstream pressure). The differential pressure increases as the filter removes contaminants from the process fluid.

Direct flow filtration. Filtration in which liquid flow is directly through the filter medium.

Effective filtration area (device size). Filter media and devices are available in a wide range of sizes with different effective filtration areas (EFAs). EFA is the filter area that is available for filtration; the larger the filter area, the higher the flow rate at a given *differential pressure*.

Filter efficiency. The amount of removal of particles, at or above a certain diameter, from the fluid by the filter. Efficiency is generally measured in terms of the percentage removal of the particles by weight and therefore does not reveal the number of particles that may pass through the filter.

For high-efficiency filters, filter efficiency is often replaced by a beta (β) rating. Efficiency can be calculated from a beta value as follows:

$$\text{Efficiency \%} = \frac{(\beta - 1)}{\beta} \times 100$$

Submicron filters are often rated using titer reduction values.

For filters that remove microbes from a fluid, efficiency is often stated as titer reduction. Titer reduction is calculated by the following equation:

$$T_R = \frac{\text{Total number of cfu influent to the filter}}{\text{Total number of cfu effluent to the filter}}$$

Where:

T_R = titer reduction

cfu = colony-forming unit

For compactness, these ratings are often stated as powers of 10 or as logs. If a filter removes 999,999 out of 1,000,000 *Brevundimonas diminuta*, that filter has a *TR* value of 106, or a 6-log reduction.

Filter life. A measure of how long a filter will last before it needs to be replaced or cleaned. It can be stated either in terms of time (e.g., 30 days between changes) or volume of fluid filtered (e.g., 10,000 L processed between filter changes). A filter's actual life will depend on what particulates and conditions it is exposed to in actual usage; therefore, filter life ratings from lab testing with standard contaminants can be used for comparison but do not necessarily predict actual service life. To predict actual life, testing with the actual application fluids under actual operating conditions is required.

Filtered water/permeate/filtrate. The treated or clean water produced by membrane filtration.

Flux. The amount of solution that passes through a unit of membrane area in a given amount of time. For example, a filter might have a flux of 1 L/min/cm^2.

Fouling. Contamination (plugging) of the membrane, either decreasing *flux* or increasing *differential pressure* or *TMP*. Membrane fouling may necessitate a flux maintenance procedure or a membrane cleaning.

Gel layer. Thin layer of retained molecules that forms on a membrane surface during filtration (see also *concentration polarization*).

Membrane filtration. A pressure-driven process that uses a semipermeable (porous) membrane to separate particulate matter from soluble components in the carrier fluid, such as water.

Microfiltration (MF). A size-exclusion, pressure-driven membrane process that operates at ambient temperature. It is usually considered an intermediate between *ultrafiltration* (UF) and multimedia granular filtration. It is an effective barrier for particles, bacteria, and protozoan cysts. The systems operate at pressures between 5 and 45 psi (34–241 kPa).

Microporous membrane. A thin film that has pores (holes) passing through it. The size of these holes is related to the pore size rating of the membrane (0.01–10 mm). Membranes can be used in the separation or filtration of suspended matter from liquids and gases.

Molecular weight cutoff (MWCO). Referred to as the molecular weight above which a certain percentage (e.g., 90 percent) of the solute in the feed solution is retained by the membrane. It is typically expressed in units called *daltons* and used as a measure of the nominal pore size of *ultrafiltration* and *nanofiltration* membranes.

Nanofiltration. Filtration that removes both particles and small dissolved molecules and ions. Finer than *ultrafiltration*; not as fine as *RO*.

Pore rating/nominal/absolute. Expressed as microns or daltons, membrane filters are given pore ratings to provide the end user with an indication of their filtering capabilities. Nominal rating is arbitrary, describing particulate size ranges for which a filter manufacturer claims some percentage removal. Absolute rating is the diameter of the largest hard spherical particle that will pass through a filter under specified test conditions. Comparisons of filters should not be based on nominal ratings, which vary from manufacturer to manufacturer, but on absolute ratings, which are based on the largest opening in the filter medium.

Reclamation/recycle/reuse. Reclamation is the process of altering the condition or characteristics of water to make the water more useful. Recycle is the use of water or wastewater internal to a facility prior to exit from the facility. Reused water is water used more than once or recycled.

Reverse osmosis (RO). The process of forcing a liquid through a nonporous membrane, removing particles and dissolved molecules and ions. RO is the finest form of membrane separation and is used to desalinate water for drinking and to prepare ultrapure water for various industries.

Tangential flow filtration (TFF). Filtration in which liquid flows tangential to (along) the surface of the membrane. Synonymous with the term *crossflow*. The sweeping action of fluid acts to minimize *gel layer* formation and surface fouling. Contrast with *direct flow filtration*.

Transmembrane pressure (TMP). The force that drives liquid flow through a crossflow membrane. In tangential flow devices, the TMP is calculated as an average related to the pressures of the inlet, outlet, and permeate ports.

The TMP can be expressed as

$$TMP_{avg} = \frac{P_{in} + P_{out}}{2} - P_{permeate}$$

where:

TMP_{avg} = transmembrane pressure
P_{in} = pressure at the inlet port
P_{out} = pressure at the outlet port
$P_{permeate}$ = pressure at the permeate port

Ultrafiltration (UF). A membrane process that retains particles, bacteria, protozoa, viruses, and organic molecules greater than their rated *molecular weight cutoff*. These systems operate at pressures between 10 and 50 psi (69–345 kPa).

Study Questions

1. Reverse osmosis membranes will compact faster with
   a. higher iron content.
   b. higher chlorine contact.
   c. higher pressure.
   d. higher pH.

2. Reverse osmosis membrane cleaning solution feed systems are constructed of
   a. stainless steel
   b. copper piping
   c. black iron pipe
   d. teflon

3. Reverse osmosis reject water cannot be disposed of by:
   a. discharge to wastewater treatment plant.
   b. deep well injection.
   c. using as drinking water.
   d. evaporation ponds.

4. The amount of reject water from a reverse osmosis unit is dependent on the number of stages in which the membranes are configured and the
   a. feed pressure.
   b. amount of cations.
   c. amount of cations and anions.
   d. pH of the water.

5. Over time, the performance of all membranes will change, mainly as a result of compaction and
   a. swelling.
   b. festering.
   c. fouling.
   d. disintegrating.
6. Flow of solution adjacent or parallel to the surface of the membrane is referred to as
   a. crossflow.
   b. direct flow.
   c. flux.
   d. differential flow.
7. What is the term for raw water that has undergone pretreatment (acidification or scale inhibitor addition) prior to entering the membrane arrays?
8. What term refers to the process of forcing a liquid through a nonporous membrane, thereby removing particles and dissolved molecules and ions?
9. What is flux decline?

Chapter 11

Ozone, Ultraviolet, and Other Oxidant Facilities

Ozone Equipment and Facilities

Because ozone cannot be stored, it must be generated on-site as it is needed. It is generated by passing an electrical current through air or pure oxygen. An ozonation system consists of the following components:

- Air compression and drying unit or oxygen source
- Ozonator or ozone generator
- Contactor to introduce the ozone into the water
- Residual ozone destruction unit

Air Preparation

When air is used as the feed gas for an ozone generator, it must be extremely dry. Special compressors and drying systems must be installed to treat the air properly before it is used in a generator.

When oxygen is used as the feed gas, it is simply piped to the generator from a storage tank of compressed oxygen.

Ozone Generators

Most ozone generators for water treatment use one of two designs. The most common type used in large plants is a bank of glass-tube generators, as illustrated in Figure 11-1. Smaller plants may use either this type or a plate-type generator in which the ozone is generated between ceramic plates.

Ozone generators can use pure oxygen, oxygen-enriched air, or air as the feed gas. If air is used, the product is about 2 percent ozone, whereas a generator using oxygen can produce 5–7 percent ozone. Figure 11-2 is a typical flow diagram for a treatment plant using oxygen-enriched air for ozone generation.

Ozone Contactors

Ozone is only sparingly soluble in water. It has a very short life, generally from a few seconds to a few minutes, depending on a number of variables. It must therefore be efficiently introduced into the water to be treated. This is critical to the success of the system.

ozone generator
A device that produces ozone by passing an electrical current through air or oxygen.

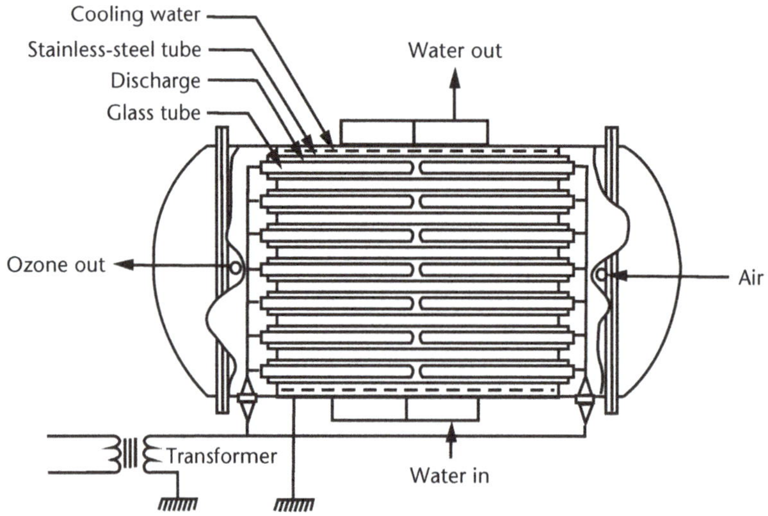

Figure 11-1 Large-scale, tube-type ozone generator

Source: *Water Quality and Treatment*. 5th ed. (1999).

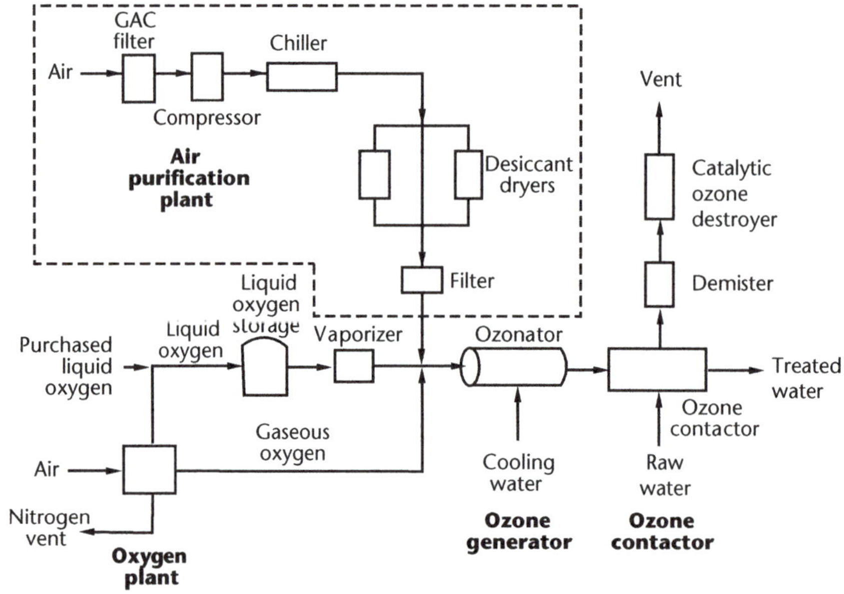

Figure 11-2 Flow diagram for air and oxygen purification for ozone production

Source: *Water Quality and Treatment*. 5th ed. (1999).

A number of techniques and technologies are available to introduce ozone into the water to be treated, including the following:

- Conventional fine-bubble diffusion
- Turbine mixers
- Injectors
- Packed columns
- Spray chambers

ozone contactor
A tank used to transfer ozone to water. A common type applies ozone under pressure through a porous stone at the bottom of the tank.

The most commonly used **ozone contactor** is the fine-bubble diffuser. Its principle is similar to the porous stone used in aquariums. The gas containing ozone is applied under pressure to a porous stone at the bottom of the unit, where it forms small bubbles. As the bubbles rise in the tank, ozone is transferred into the water.

Residual Ozone Destruction

Ozone contactors must have a system to collect ozone off-gas. Ozone is toxic and must be kept within allowable limits—set by the Occupational Safety and Health Administration (OSHA)—in the treatment plant and in the area surrounding the plant. In some areas of the United States, ozone discharge from treatment plants may be further regulated.

As a result, ozone-generating installations must include a thermal or catalytic ozone destroyer. In some installations, the off-gas is also reused by piping it to the air compressor.

Ozone is so highly corrosive that only certain materials can be used in constructing treatment plant equipment. Suppliers of ozone generators should be consulted about appropriate materials.

Potassium Permanganate Equipment and Facilities

Permanganate is shipped as crystals in pails, drums, returnable bins, and bulk truckloads. In water plants using less than 25 lb/d (11.3 kg/d) of permanganate, the dry chemical is usually dissolved in batches in a solution tank. The facility normally includes two solution tanks and at least two chemical feed pumps. Each pump should be capable of meeting peak-day needs, and each tank should have a capacity for a 3-day supply. When making up solutions in day tanks, operators need to be aware that the solubility of permanganate at low water temperatures is only a fraction of the solubility found at 70°F (21°C). For this reason, it may be necessary to make solutions at 0.25 lb/gal (30 grams/L) to avoid accumulation of crystals at the bottom of the tank, which will clog the feeders and interrupt flow.

For larger plants, dry volumetric feeders with hoppers and dust collectors are normally used. A float-controlled solution tank with a mixer is usually located under the feeders, and injectors are used for powering the flow to the feed point. Permanganate dosages of 0.5–2.5 mg/L are usually sufficient to control most oxidizable taste- and odor-causing materials in raw water.

As the purple color of permanganate in water dissipates, a residual of manganese dioxide is left in the water that can give it a yellow-brown color. For this reason, permanganate must be fed only as a pretreatment step so that the manganese dioxide can be removed by clarification. If the water is still purple when it reaches clarification treatment, too much permanganate has been applied. The ideal application point is the raw-water intake, which allows the maximum reaction time for oxidation and complete coagulation of the manganese dioxide.

If possible, permanganate should be fed ahead of chlorine. The oxidation of organics before chlorination will reduce both the amount of chlorine needed and the formation of total trihalomethanes. If powdered activated carbon is used in addition to permanganate, it should, if possible, be fed after the permanganate has completed its oxidizing action. If both are fed together, the permanganate consumption increases substantially.

Chlorine Dioxide Equipment and Facilities

Chlorine dioxide is usually generated on-site by the reaction of chlorine and sodium chlorite. One method of preparation is to introduce the chlorine to the water and then add the water to a solution containing hydrochloric acid and sodium chlorate. This method is likely to leave substantial amounts of chlorine in the solution, which may lead to the formation of chlorinated by-products and undesirable amounts of chlorate and chlorite. Another method is to inject the chlorine as a gas under vacuum into a stream of chlorite solution. Chlorine dioxide can also be prepared by adding acid, such as hydrochloric acid, to a chlorite solution. This method can be employed where the use of chlorine is not convenient, but for large-scale applications, the second method is usually preferable.

Ultraviolet Light

Unlike ozone, the inactivation of microorganisms by ultraviolet (UV) light is well understood. The reaction is purely physical and therefore not significantly affected by pH. UV light is generated by mercury or antimony vapor lamps and is commonly used to treat wastewater effluent. UV disinfection is often considered for disinfection due to its effectiveness in inactivation of *Cryptosporidium* and because it does not produce disinfection by-products (DBP)s. However, because it does not produce a lasting disinfectant residual, other technologies for residual maintenance (like chlorination or chloramination) are necessary.

UV disinfection is ideally suited to groundwater treatment. The process requires a stable, low-turbidity water. UV does not provide a protective residual so it is useful in groundwater systems that have a variance that allows them to forego disinfection. UV may then be used as an extra level of protection where residual is not required. Small water systems are also a good fit for UV because most small systems require only a basic level of operator skill to operate.

Process Description

UV light effectively destroys bacteria and viruses. However, how well the UV system works depends on the energy dose that the organism absorbs. An effective dose is measured as a product of the lamp's intensity, including radiation concentration, proper wavelength, exposure time, water quality, flow rate, and the microorganism's type and source, as well as its distance from the light source.

Performance

The Long-Term 2 Enhanced Surface Water Treatment Rule (LT2ESWTR) lists log credits for UV disinfection of *Cryptosporidium*, *Giardia lamblia*, and virus. These values are listed in Table 11-1. Additionally, UV light disinfection does not form any significant DBPs.

UV effectiveness is relatively insensitive to temperature and pH. UV application does not convert nitrates to nitrites, or bromide to bromines or bromates. Studies show that UV-treated drinking water inhibits bacterial growth and replication in the distribution system; however, conditions within distribution systems, such as leaks, still require additional residual disinfection (e.g., free chlorine).

Table 11-1 UV dose (mJ/cm²) table with inactivation credit

Log Credit	*Cryptosporidium* Dose	*Giardia lamblia* Dose	Virus Dose
0.5	1.6	1.5	39
1.0	2.5	2.1	58
1.5	3.9	3.0	79
2.0	5.8	5.2	100
2.5	8.5	7.7	121
3.0	12	11	143
3.5	15	15	163
4.0	22	22	186

Design and Operation

At a minimum, drinking water systems should install two UV units, both of which should be capable of processing the amount of water the system was designed to handle. Having two units in place ensures continuous disinfection when one unit is being serviced. Two units also can ensure operation during periods of low flow demand. Modular units designed for small drinking water systems are easy to install and operate (two plumbing connections per unit and one electrical hookup). They should be equipped with automatic cleaners and remote alarm systems. For systems in isolated areas, operators should maintain and store a set of spare parts on-site and should consider a telemetry system for monitoring treatment.

WATCH THE VIDEO
Ozone & Ultraviolet Light (www.awwa.org/wsovideoclips)

Study Questions

1. Ozone is not introduced into water by
   a. air stripping towers.
   b. fine-bubble diffusers.
   c. spray chambers.
   d. injectors.

2. Potassium permanganate should be fed
   a. after filtration.
   b. after chlorination.
   c. as pretreatment.
   d. together with chlorine.

3. Ozone generators
   a. are capable of producing extra electricity
   b. create ozone by exposing oxygen to an electric discharge
   c. can use water as an oxygen source
   d. discharge any excess ozone to the air

4. Which oxidant has the potential of producing chlorate (ClO_3) by-products?
   a. Chlorine dioxide
   b. Chlorine
   c. Chloramines
   d. Calcium hypochlorite

5. UV disinfection
   a. converts nitrate to nitrite.
   b. changes bromide to brominated organic compounds.
   c. is insensitive to temperature and pH.
   d. leaves a lasting residual in treated water.

6. Which method for chlorine dioxide generation is most likely to have substantial amounts of chlorine in solution?
   a. Add chlorine to water, then this water to sodium citrate and sodium chlorate.
   b. Add chlorine to water, then this water to hydrochloric acid and sodium chlorate.
   c. Inject chlorine gas under vacuum into a stream of chlorite solution.
   d. Add hydrochloric acid to a chlorite solution.

7. Because ozone cannot be stored,
   a. it simplifies the treatment process for facilities.
   b. it must be generated on-site as it is needed.
   c. it must be used immediately upon delivery from the manufacturer.
   d. it cannot be used in water treatment.

8. When air is used as the feed gas for an ozone generator, it must be
   a. warmed first.
   b. moisturized first.
   c. extremely dry.
   d. converted to pure oxygen.

9. List three devices available to introduce ozone into the water to be treated.

10. Which authority sets the limits for the amount of ozone allowed in the treatment plant and in the area surrounding the plant?

11. Permanganate dosages of what level are usually sufficient to control most oxidizable taste- and odor-causing materials in raw water?

Chapter 12

Plant Waste Treatment and Disposal

Residuals

All basins, whether conventional or high-rate, require sludge disposal. This is usually the most troublesome operation in a water treatment plant. In the past, operators commonly discharged residuals to a stream or lake. However, discharge regulations, both for the environment and for wastewater plants, now restrict an operator's ability to waste residuals with impunity. The process of sludge disposal is now as tightly controlled as any other process in the plant. Residuals are removed manually or with an automatically controlled vacuum or raking systems (Figure 12-1).

Residuals are comprised of all the unwanted material that has been removed from the water and all of the compounds that are created by addition of treatment chemicals (floc, lime solids, etc.) (Figure 12-2).

Alum sludge is the most common residual produced in water plant sedimentation basins, but ferric sludge also is common. Both wastes come from the

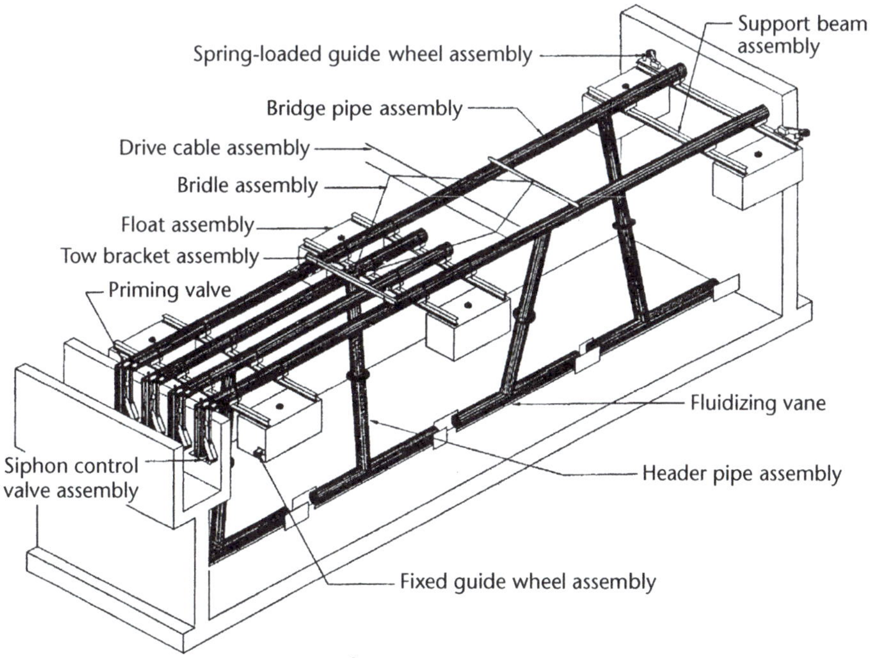

Figure 12-1 Floating sludge collector
Source: F. B. Leopold Inc., Zelienople, PA, 16063.

sludge
The accumulated solids separated from water during treatment.

alum
The most common chemical used for coagulation. It is also called aluminum sulfate.

Figure 12-2 Sludge collectors at Philadelphia Water Treatment Plant. Rakes help to concentrate the solids.

Source: Philadelphia Water Department.

sedimentation basins at low concentrations by weight: normally less than 2 percent. It is inefficient to dispose of this low concentration of sludge; thus, water plant operators dewater and concentrate the residual.

Dewatering is the process of removing water from the thin sludge, allowing the solid material to remain (thickening) for ultimate disposal. It is difficult to dewater alum and ferric sludge because the water is chemically bound to the aluminum and ferric hydroxide floc. These flocs are gelatinous and thus difficult to handle. As a rule of thumb, the thicker the sludge, the less space it takes up, and so less storage space will be needed. Additionally, the thicker the sludge, the less expensive its disposal. Wastewater plants and landfills usually charge by volume of material they receive (there may be other surcharges for strength of solids or for special considerations due to the nature of the sludge). When disposing of thin sludge, therefore, the water plant operator is also paying to dispose of a larger proportion of water than may be necessary.

Sludge drawn from sedimentation basins varies with the coagulant used, of course, but also with the nature of the material found in the raw water. Sludge produced in plants that draw water from lakes and other sources with relatively low turbidity is very gelatinous, whereas sludge produced from a source high in silt and organic content is not as gelatinous and, thus, is more easily handled. The percent solids concentration of the sludge also affects its characteristics: low-solids sludge is more liquid than high-solids sludge, which tends to handle like clay.

In many water treatment plants, sludge from sedimentation basins is pumped to a gravity thickener that is designed to dewater the waste to a specific concentration before disposal. This gravity system allows the solids to settle to the bottom and provides for removal of the clear supernatant at the top of the unit. Supernatant is either returned to the process or sent to waste. The thickener also serves as an equalization tank, allowing for continuous or intermittent wasting of solids that might otherwise burden the receiving sewers or treatment facility.

Further treatment of sludge after thickening can take place in the plant or at the disposal site (Figure 12-3). Water treatment plants often treat thickened sludge with polymers or lime and may subject the sludge to flotation or belt-press thickening to obtain the desired concentration. When sand drying beds, lagoons, or freeze–thawing beds are employed, the process is referred to as *nonmechanical dewatering*. Mechanical dewatering is usually accomplished with vacuum filtration, belt presses, or centrifuges. All of these processes are effective but costly. Each water treatment plant needs to determine the most cost-effective procedure

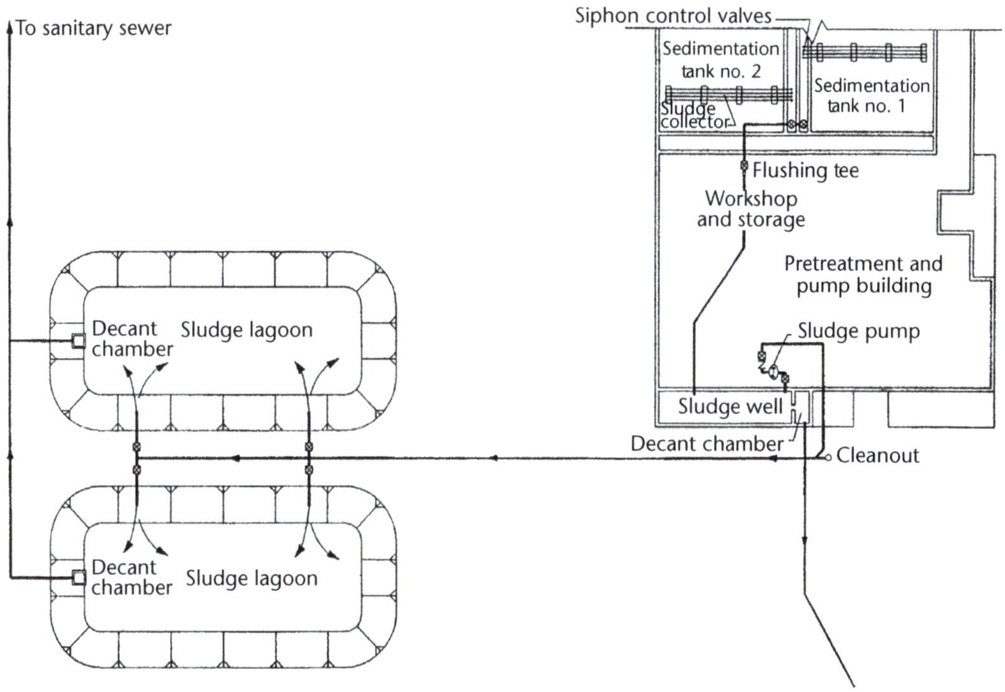

Figure 12-3 Typical sludge lagoon system. Lagoons can be operated independent of each other, allowing for maintenance.

for its situation, taking into account all economic factors at its disposal. Local regulations will also be a factor.

For simplicity and operator convenience, the sand-drying bed method of solids handling is difficult to beat. Costs of the process vary by site and size, but beds with greenhouses built around them to trap heat, and with fans to hasten drying, show great promise for freeing plant operations from the burgeoning costs imposed on them for residuals disposal when discharging to wastewater systems.

Some utilities with warm and sunny climates have been using simple sand beds for years. Ongoing research is leading engineers to design these systems for utilities in varying climates. One process refinement is simple in its application: thickened sludge is pumped to the bed as polymer is injected into it. Sludge is loaded onto the beds at rates of about 3 dry lb of solids per square foot of bed. Polymer dosages are typically found to be 3 lb of polymer per dry ton of sludge. The cakes that are formed can quickly dry to 30 percent or more in a few days and are easily handled and disposed of.

WATCH THE VIDEO
Wastewater Treatment Secondary—Sludge Collection and Disposal (www.awwa.org/wsovideoclips)

Sludge Calculations

Thickening Process

Thickening involves using methods that remove the water from the sludge and leave behind a residual that is more concentrated and smaller in volume. As stated, the thickening process can be a gravity basin, a sand bed, or any of the several methodologies in use. If the concentration of solids coming from the beginning process is known, an approximation of the amount of solids that the thickener will produce can be calculated using the formula

$$V_2/V_1 = P_1/P_2 \tag{12-1}$$

where V_1 and V_2 are the beginning and ending volumes of the sludge, and P_1 and P_2 are the beginning and ending percentages of the sludge. This calculation assumes that any amount of polymers or chemicals added as thickening agents is negligible when compared to the volume of sludge.

Amounts Produced by Alum Coagulation

There are other formulas that operators use to predict the amount of sludge that will be produced by the chemical coagulation process. These formulas are for alum, ferric salts, and softening chemicals, such as lime and soda ash.

For alum,

$$S = (8.34Q)(0.44Al + SS + A) \tag{12-2}$$

where:

S = sludge produced, lb/d

Q = plant flow, MGD

Al = dry alum dose, mg/L as 17.1% Al_2O_3

SS = raw water suspended solids, mg/L

A = solids from other treatment chemicals, such as powdered activated carbon, etc., mg/L

Example Calculation

The "SS" portion of the preceding formulas is either measured in the laboratory or approximated by the use of a "B value." The B value is a ratio of the suspended solids of the raw water to the turbidity of the raw water. It is usually found to be about 0.7 to 2.2, but can be much higher if large amounts of total organic carbon are present. The ability of the operator to obtain a quick analysis for turbidity makes the B value a great help in the prediction of solids production. Every water plant needs to develop a history of solids measurement in order to rely on a usable B value.

Using Equation 12-2, calculate the pounds of solids produced per day if a treatment plant processes 25 mil gal with 15-mg/L dry basis alum. The raw water turbidity is 44 ntu, and the historic B value for this plant is 1.5. The plant is adding 5-mg/L activated carbon for taste and odor control.

Using the B value, multiply it by the SS: 1.5 × 44 = 66. The calculation becomes

$$(8.34 \times 25)[(0.44 \times 15) + 66 + 5] = (208.5)(77.6) = 16{,}180 \text{ lb}$$

Amounts Produced by Ferric Coagulation

For iron salts,

$$S = (8.334)(2.9Fe + SS + A) \tag{12-3}$$

where Fe = the iron dose in mg/L as Fe.

Lime–Soda Softening Sludge Calculation

Amounts of sludge produced by the lime–soda softening process can be calculated using this formula:

$$S = 8.34(Q)(2.0Ca + 2.6\ Mg) \tag{12-4}$$

where Ca and Mg are the calcium and magnesium hardness removed as mg/L calcium carbonate.

Study Questions

1. Which national law regulates underground disposal of wastes in deep wells?
   a. The Toxic Substances Control Act (TSCA)
   b. The Comprehensive Environmental Response Compensation, and Liability Act (Superfund)
   c. The Resource Conservation and Recovery Act (RCRA)
   d. The Safe Drinking Water Act
2. What is typically the percentage of solid material of alum sludge?
   a. 0.1–2.0%
   b. 2.0–5.0%
   c. 4.0–7.0%
   d. 5.0–15.0%
3. Which of the following is the most troublesome operating problem for sedimentation basins?
   a. Algae and slime growths
   b. Short-circuiting
   c. Density currents
   d. Sludge collection and removal
4. A plant produces 100,000 gal of 2% sludge from its sedimentation basin and then thickens it to 30%. What is the volume of 30% sludge?
   a. 50,000 gallons
   b. 3,333 gallons
   c. 1,500,000 gallons
   d. 6,667 gallons
5. Sludge produced by water drawn from sources with relatively low turbidity
   a. is high in silt and organic content.
   b. is high in liquid content.
   c. can be very gelatinous and difficult to handle.
   d. handles like clay.

6. Which of the following is *not* a type of equipment used in mechanical dewatering?
   a. Vacuum filters
   b. Belt presses
   c. Centrifuges
   d. Hydrologic siphon
7. The thicker the sludge,
   a. the less expensive its disposal.
   b. the less efficient its disposal.
   c. the more expensive its disposal.
   d. the greater the volume.
8. In the past, to what locations were treatment residuals typically discharged?
9. What type of sludge is the most common residual produced in water plant sedimentation basins?
10. What term refers to the process of removing water from the thin sludge, allowing the solid material to remain (thickening) for ultimate disposal?

Chapter 13
Water Supply Management

Water Impoundment Management

Some utilities construct impoundment reservoirs or use water from naturally occurring reservoirs that are fed from surface streams or that catch precipitation. Constructed reservoirs serve several useful purposes. They can store amounts of water in excess of daily demand and therefore act as a hedge against drought conditions. They can serve as pretreatment reservoirs when aeration systems or chemical treatments are added, or they may be used simply as large settling ponds prior to treatment. Reservoirs also allow the operator to draw water from nearby streams for storage. In this way, water can be pumped from the stream on days when water quality is excellent. When quality is poor, such as after a rainstorm or when there is an upstream spill, pumping can be discontinued if the reservoir is large enough to store sufficient water for treatment. When conditions return to normal, pumping from the stream can resume.

Reservoirs can also be a source of water quality problems because they are exposed to the elements. Sunlight penetration into the depths of the reservoir can promote algae growth, which causes taste-and-odor problems. Runoff from the ground can bring with it chemical and biological contaminants. Extreme wet and dry conditions affect water quality. For example, heavy precipitation can resuspend bottom sediments and increase turbidity. Such conditions can load the treatment plant with huge amounts of solids, making treatment expensive and management of residuals difficult. Drought conditions can cause stagnation and minimize the otherwise healthy dilution of contaminants because of low water levels.

The depth of a lake or reservoir is subject to two conditions that can impact source water quality. The location of the intake with respect to these depths is important. The first condition, known as the trophic cycle, involves three stages where nutrients and biological activity influence water quality. The *oligotrophic* stage is associated with low nutrients and limited algae production and biological activity. The *mesotrophic* stage involves a moderate amount of these constituents, and the *eutrophic* stage is rich in nutrients and biological activity. This eutrophic stage is associated with low oxygen levels, high turbidity and color, and natural organic matter. Because oxygen is depleted, usually by algae, an anoxic condition can develop. This creates a reducing environment where insoluble iron and manganese bottom deposits can become soluble ions. If this water is brought into the plant, water quality problems can occur that can require treatment.

The second condition that can impact source water quality is thermal stratification. Layers of the lake or reservoir, called *limnion*, develop due to temperature differences. Oxygen depletion can take place within these layers and cause problems similar to those described previously. In colder climates, water at the top

trophic cycle
A three-stage process that describes how nutrients and biological activity influence water quality. The *oligotrophic* stage is associated with low nutrients and limited algae production and biological activity. The *mesotrophic* stage involves a moderate amount of these constituents. The *eutrophic* stage is rich in nutrients and biological activity.

thermal stratification
The layering of water as a result of temperature differences.

of a lake or reservoir can cool to around 39.4°F (4°C), a temperature at which water reaches its greatest density. This cooler, denser water sinks and forces the warmer water at the bottom to the top. This mixing can be so violent that the bottom of the lake or reservoir is riled, causing poor-quality, high-turbidity water to be brought into the plant. Operators refer to this phenomenon as *lake turnover.* It can occur each fall and spring. Some operators have witnessed this phenomenon in uncovered sedimentation basins.

River bank filtration is considered groundwater under the direct influence of surface water. Treatment plants with this source of supply receive water of changing quality similar to the river, but often the quality is more stable. The filtration of the soil either in the bottom or adjacent to the river provides beneficial pretreatment. In some situations, the degree of pretreatment has been assessed and the regulatory agency has granted some treatment credit to this process. This decision is site specific and at the discretion of the regulatory agency.

Direct river intakes often challenge water supply sources. Although there are advantages to drawing water directly from a river, such as contamination incidents that may flow downstream quickly beyond the intake, water quality can change rapidly compared to supplies from impoundments. Many possible contaminant situations may impact water quality, including chemical spills from industrial discharges, river shipping accidents, floods, droughts, and surface runoff. Operators who work at treatment plants with this source of supply need to be alert to these possibilities. Online, continuous monitoring is often used to provide warning of changes that may require adjustments to the water treatment system. Many treatment plants with direct river intakes have procedures in place to shut down production if certain conditions develop. Protecting the public water supply is of primary importance.

Source Water Protection Program

A utility's **watershed** management program is the first barrier of protection against pathogen intrusion. Such a program can also protect drinking water quality by minimizing disinfection by-product precursors and protecting against algal growths that cause taste-and-odor problems. Deterioration of a watershed not only results in more costly treatment requirements at the plant but also can erode public confidence. If a consumer perceives that the watershed providing drinking water is not managed well, the consumer may seek alternative sources.

In an ideal setting, water utilities would own all of the land that comprises the watershed for their source. In fact, land ownership was the principal method of protecting source water in early protection programs. Currently, some utilities can still claim virtual ownership of their source's watershed; New York, Boston, Portland, Seattle, and San Francisco are among those. These utilities have avoided filtration requirements due in part to the high quality of their source water, which is a direct result of extensive land ownership and watershed management programs.

However, there is concern about microbial contaminants that are resistant to chlorination. By understanding the watershed and its sources of contamination, both active and potential, drinking water utilities can become advocates for improving their watershed and source water. Using tools such as those provided under the Clean Water Act (CWA) to achieve water quality goals, operators can develop a watershed protection program. In effect, Storm Water Pollution Prevention (SWPP) plans are simply an old idea that has received new attention to help

river bank filtration
Extraction of water through the bank of a river or through the bottom using wells or buried collection piping.

watershed
An area from which surface runoff is carried away by a single drainage system.

guard against "new" threats. In general, a good watershed management program contains the following components:

- Identification and characterization of the source
- Characterization of potential impacts to the source
- Determination of the intake's vulnerability to contaminants
- Established protection goals
- Development of protection strategies
- Program implementation
- Program monitoring and evaluation

Many utilities, including the Philadelphia Water Department, have initiated SWPPs. Philadelphia's experiences are described as an example SWPP. In colonial times, Philadelphia drew its water from wells, but by the mid-18th century, the expanding city of nearly 30,000 had polluted this first source. Tanneries, which used unlined pits as part of the industrial process, and the large number of privies in the city had rendered the groundwater undrinkable. In 1799, after a series of yellow fever epidemics, Philadelphia decided to draw from the then "uncommonly pure waters of the Schuylkill River" by building the city's first waterworks at Centre Square (where City Hall stands today). In 1815, the city opened the Fairmount Waterworks to distribute water from the Schuylkill River throughout the city. As Philadelphia grew during the first half of the 19th century, its water supply was expanded to draw more water from the Schuylkill River to meet demand. The city then made efforts to ensure the purity of its source water. In addition to taking legal action to close upstream industries, Philadelphia began purchasing land along the banks of the Schuylkill River and Wissahickon Creek, a major tributary. This land was eventually set aside as a public park.

This watershed management experiment was successful for four decades, but upstream development eventually overwhelmed the city's program. With increased upstream industrial growth, particularly the coal mining industry, the Civil War, and population growth spurred by successive waves of immigrants, water quality continued to decline. By the 1880s and 1890s, Philadelphia experienced the worst typhoid and cholera epidemics of any American city as a result of deteriorated water quality. In 1900, the city decided to filter its water supply; by 1913, all of Philadelphia's drinking water was both filtered and chlorinated at five plants on the Schuylkill and Delaware rivers. An unfortunate by-product of this decision to invest in treatment plants, however, was a de-emphasis on source water protection. Not coincidentally, the period following construction of the water treatment plants saw considerable deterioration of the Schuylkill and Delaware Rivers. This deterioration occurred in most major rivers across the country.

Eventually, Congress passed the CWA in 1972. In the 40 years since the passing of the CWA, most of the country's rivers and streams have been rehabilitated due to an emphasis on installing controls at municipal and industrial point source discharges (e.g., wastewater treatment technologies).

However, because of the threat from emerging pathogens such as *Cryptosporidium* and the realization that nonpoint sources such as urban and agricultural runoff now merit serious attention, Philadelphia and other cities around the country now realize that watershed management programs are critical to the long-term protection of public health. Watershed management programs represent the best opportunity for reaching beyond a city's limits and focusing on point sources to further enhance the source water. Consequently, in 1999, Philadelphia established a new SWPP to formalize and coordinate many ongoing research efforts, such as *Cryptosporidium*, geosmin, MIB (2-methylisoborneol), and bromide occurrence studies. The program also enabled the water department to have a single, cohesive voice when dealing with upstream stakeholders.

The Philadelphia Source Water Program began by collecting and analyzing data relating to existing water quality trends within the watershed. Additionally, the program began to actively complete total maximum daily loads (TMDLs) in the watershed by supplying sampling data and analyses, funding flow monitoring gauging stations, and providing technical assistance to the agencies conducting the TMDLs.

Another major milestone for the program was partnering with the Pennsylvania Department of Environmental Protection (DEP) to implement Source Water Assessment Programs (SWAPs), mandated by the 1996 Safe Drinking Water Act (SDWA), for Philadelphia's 3 treatment plants and 47 others within their watershed. DEP contracted with the Philadelphia Source Water Protection Program to perform SWAPs for 42 surface water intakes on the Schuylkill River. Additionally, DEP contracted with Philadelphia to conduct eight assessments on the tidal portion of the Delaware River, including Philadelphia's Baxter Water Treatment Plant.

The general approach (see Figure 13-1) for the assessments was to first identify all potential sources, if possible; develop and apply a screening procedure to narrow down the list to important sources; prioritize the important sources; and then follow up with data collection and reality checking of the top-priority sources.

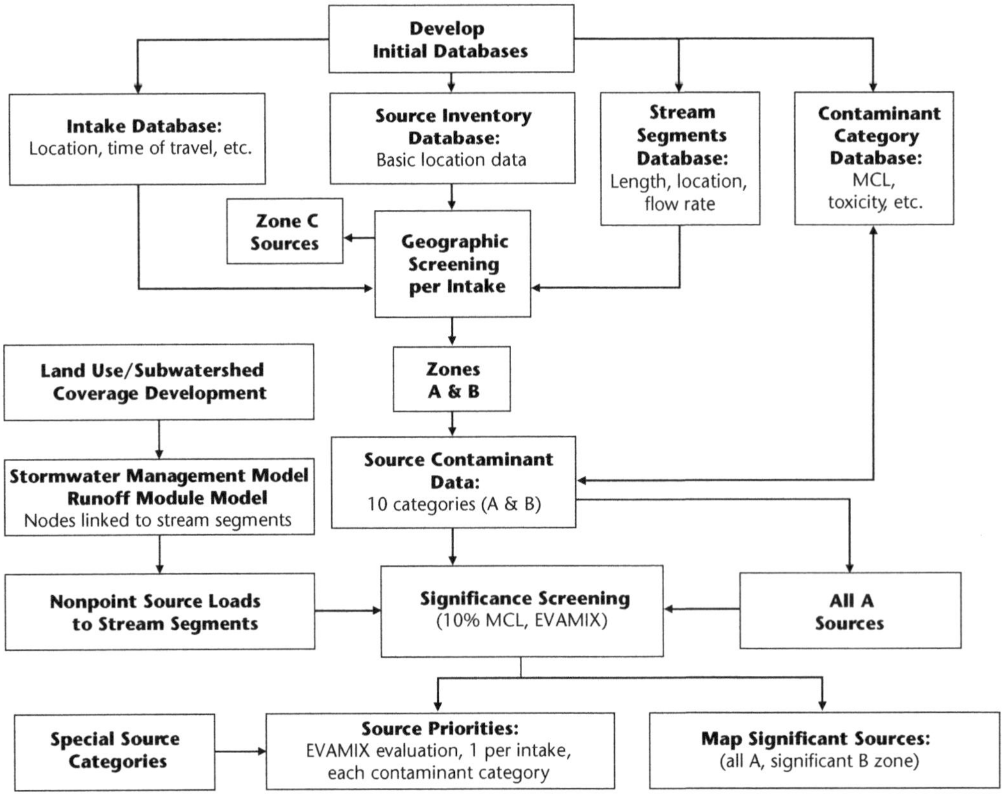

Figure 13-1 Source prioritization flow diagram used by Philadelphia Water Department

Source: Philadelphia Water Department.

The process for the Schuylkill River, for example, started with development of a database of more than 3,500 potential sources. The data were collected from various national databases such as the Permit Compliance System; Resource Conservation and Recovery Act Information System; Comprehensive Environmental Response, Compensation, and Liability Act Information System; Toxic Release Inventory; Envirofacts; Right to Know Network; and Business Pro. Based on contaminant information for each source, the sources were assigned 1 or more of 10 pollutant categories.

Additionally, nonpoint sources are entered into the evaluation using a variation of the US Environmental Protection Agency's (USEPA) Storm Water Management Model, which utilizes land use coverage data for 300 subwatersheds to estimate annual loadings of contaminants as the result of stormwater runoff. Geographic screening is accomplished using delineated time-of-travel zones (A, B, or C) for each water plant intake. Finally, a computerized decision-support model (EVAMIX) is used to prioritize sources based on stakeholder-defined weighting criteria to determine top-priority source rankings. All high-ranking sources are flagged for follow-up data collection in later phases to verify results.

Source Water Quality Considerations

Many possible water quality issues confront water treatment facilities. Recognizing these challenges and being prepared to effectively adjust to them is the primary task for water plant operations personnel. It is the reason for extensive education and training. When the water quality changes and the treatment plant systems must be adjusted, experience and training are called on to meet the threat.

A treatment challenge is presented when the source water quality *changes*. A steady predictable water quality, even if it is not the greatest, can often receive effective treatment. The treatment system can be refined over time and excellent drinking water quality can be produced from even highly impaired waters.

Both surface and groundwater can change and potentially create a situation where water quality is at risk. The following are some of the water quality issues that may result from changes in source water quality:

- Bacteriological—e.g., fecal contamination
- Biological—e.g., zebra mussels, algae
- Chemical—e.g., radioactive substances, spills
- Physical—e.g., temperature, pH
- Agricultural—e.g., pesticides, fertilizer
- Groundwater conditions—e.g., wellhead contamination
- Lake stratification—e.g., taste and odor, iron or manganese

Direct river intakes can be very challenging water supply sources. Conditions can change quickly, so operators must be alert to these conditions. Treatment for known changes should be documented and plant adjustments rehearsed to ensure that acceptable drinking water quality is maintained.

Lake or other impoundments often stabilize water quality for lengthy periods. However, changes can occur, such as lake stratification and turnover. Another common cause of water quality changes in lakes is the growth of algae and other aquatic organisms. Some of these organisms can create disagreeable taste and odors. Others may contribute chemical by-products that can be toxic. Treatment of these conditions is best performed in the lake and controlled, if possible, to avoid these problems. Control of nutrients such as nitrogen and phosphorus compounds have been used to reduce algae and aquatic weed growth. In most instances, however, chemical control of algae is necessary. Copper sulfate is often the chemical used at a generic application rate of 1.1 pound per acre-foot (lb/ac-ft). Also, strategies have been employed to reduce stratification by mixing or oxygen (air) application. These practices are often successful in reducing potential water quality problems but not eliminating them completely.

Groundwater sources can be contaminated by surface runoff entering an unprotected wellhead. This sort of incident can be serious because the well may not be monitored for this unforeseen occurrence. Disinfection strategies that include continuous monitoring and pacing chemical addition can be an effective defense when demand increases expectantly.

Understanding the characteristics of the water supply source and preparing treatment practices to combat quality changes are the best ways to maintain high-quality water production. Treatment plant systems must be adjusted to meet changing water source conditions. Water plant operators should have a proven plan in place to deal effectively with water source quality changes.

Contingency Planning

Emergency and Alternative Water Sources

It is very important that the operation of a public water system be uninterrupted. One of the most compelling reasons to maintain continuity of service is the protection of public health. Any loss of system pressure or interruption in service is an opportunity for microbiological or chemical contamination to enter the water system. And because water is used for sanitary purposes, a lack of water for more than a few days can contribute to the spread of disease. The possibility of water contamination and the loss of the raw-water source require that every water utility develop an emergency response plan.

Table 13-1 illustrates a disaster effects matrix for a typical water utility. It lists the disasters that can affect a water system and the potential magnitude of the hazards associated with each of those disasters. More information on emergency planning is contained in AWWA Manual M19, *Emergency Planning for Water Utilities*.

An emergency response plan should be developed by each utility in cooperation with neighboring utilities and with the appropriate local and state agencies. Agencies that have a stake in the emergency response plan include local police, fire, and emergency or disaster preparedness departments, as well as the corresponding groups at the county and state levels. The discussion of emergency response plans in this chapter is not intended to address the entire emergency planning program but only to examine possible emergencies involving source water. Alternative water sources for nonpotable use are also discussed.

Causes of Source Disruption

In general, catastrophes that can disrupt the operation of a water source fall into one of two categories: natural disasters, including earthquakes (Figures 13-2 and 13-3), floods, hurricanes, tornadoes, forest fires, landslides, snow and ice storms,

Table 13-1 Hazard summary for a hypothetical water system

Type of Hazard	Estimated Probability	Estimated Magnitude	Comments
Earthquake	1 in 60 years	7.0 (Richter scale)	
Fault rupture	Medium	2 ft (0.6 m)	Meridian fault
Ground shaking	High		
Liquefaction	Medium–low	Vertical and horizontal accelerations	Fill areas
Densification	Medium		Fill areas
Landslide	Medium–high		In slopes of 30 percent
Tsunami and seiche	None		
Hurricane	None		
Wind			
Storm surge			
Flooding			
Tornado	Low		
Flood	Low–medium	100-year flood to elevation = 1,020 ft (311 m)	At treatment plant
Forest or brush fire	High		Dry Creek Watershed
Volcanic eruption	1 in 300 years	150 miles away (241 km)	Mount Nueces
Other severe weather			
Snow or ice	None		
Extreme heat	High	100-year drought	Reservoirs depleted
Wind	Medium	60–80 mph (97–129 km/hr)	Usually in winter
Lightning	Low		
Other			
Waterborne disease	Low		*Cryptosporidiosis*
Hazardous-material release			
Chlorine	Medium–high	1-ton containers	Earthquake damage
Other spill	Medium	Tanker car	Dry Creek Reservoir
Structural fire	Low		
Construction accident	Medium	Line damage	In older area of system
Road	Low		
Rail	Medium		Rail yard near warehouse
Water	Low		
Air	Low		
Nuclear power plant accident	Low	Contamination	Lake West Reservoir
Nuclear bombs explosion	Low		
Vandalism, terrorism	Medium		Storage tanks
Riot	Low		
Strike	Low		

Figure 13-2 A 400,000-gal (1.5-million-L) steel tank uplifted during the 1992 Landers, California, earthquake
Source: M. J. O'Rourke, RPI, NCEER.

Figure 13-3 Flocculator/clarifier center mechanism damaged during 1989 Loma Prieta earthquake
Source: D. B. Ballantyne.

and tsunami; and human activities, including vandalism, explosions, strikes, riots, terrorism, warfare, and water contamination caused by leaks, spills, or dumping of hazardous materials. Risk of operator error persists under any circumstances, and machines sometimes break down on their own. Disruptions can affect the delivery of water either by directly damaging components of the treatment and distribution systems or by contaminating the water itself.

Sources of Contamination

The contamination of source water as a result of a disruption may be caused by microbiological, chemical, or radiological agents. Both groundwater and surface water sources can be affected. Some relatively common examples of contamination incidents include the following:

- A pipeline that has broken upstream of a surface water intake, allowing several hundred gallons (liters) of fuel oil to be spilled into the river
- A tank truck that has overturned on a bridge and has spilled a quantity of toxic chemical into the river
- A spill of chemical solvent on the ground near a well that has penetrated to the aquifer and is now showing up in the well at increasing concentrations
- Excessive rainfall resulting in excessive runoff entering a combined sewer; the result is a combined sewer overflow that sends untreated sewage into a nearby body of water

When contamination of a water source occurs, the first thing that must be determined is the length of time of the contamination episode. In the previous examples, the tank-truck spill will probably pass the river intake in a matter of hours. The fuel-oil spill could last for several days.

Contamination of an aquifer can create a very long-term, if not permanent, problem. Whether the contamination plume passes by naturally or is deliberately cleaned up, the process may take years. In some cases, special wells can be operated around the contaminated portion of an aquifer so that the pollutant is pulled toward those wells and does not spread further.

contamination
Any introduction into water of microorganisms, chemicals, wastes, or wastewater in a concentration that makes the water unfit for its intended use.

When source water becomes contaminated, a utility can respond in several ways to ensure a continuous and wholesome water supply: it can stop drawing water and operate on stored water until the episode is over, change to an alternative water supply until the episode is over, or treat the contaminated water.

The concentration of most chemical contaminants in water can be reduced to acceptable levels with the proper type of treatment, such as aeration or carbon adsorption. Microbiological contaminants can generally be controlled by proper coagulation, sedimentation, and filtration followed by adequate doses of disinfectant. However, the type and extent of equipment required for special treatment during a contamination episode may not be required for normal source treatment operations. Therefore, water system operators should anticipate the contamination problems that might occur and have plans for bringing any special equipment online in the case of an emergency.

Notification of Possible Contamination

Utilities should coordinate with their local emergency planning department to establish specific procedures for contacting local, state, and federal officials should an incident occur. If the contamination is the result of an intentional act, utilities should also notify local law enforcement, as well as state and federal authorities (including homeland security agencies). USEPA provides further information in its document titled "Guidance for Water Utility Response, Recovery, and Remediation Actions for Man-made and/or Technological Emergencies."

Loss of Water Source

A complete loss of a water source can occur under several scenarios: when an earthquake has damaged all of a water system's wells, when a flash flood on a river has destroyed a system's intake facilities, or when systems fail because of a pipeline break or tunnel collapse.

If a water system's usual water source is no longer available, the utility will need to determine quickly whether the damaged component is repairable and whether a temporary solution can be put in place. Damaged wells, for example, may not be readily repairable, in which case new wells may have to be drilled, and that could take months. But if a surface water intake is destroyed, it may be possible to install a temporary pipeline or bring in portable pumps and have them operational within a few days.

Short-Term Options The following short-term options should be considered when a water source is damaged or destroyed:

- If a limited supply of water is still available, commence conservation or rationing.
- Supply customers with water from tank trucks.
- Supply customers with bottled water.
- Draw water from adjoining water systems.

Long-Term Options The following long-term options should be considered when a water source is damaged or destroyed:

- Drill new wells.
- Construct a new surface water source.

- Clean up the source of contamination.
- Install a connection to draw water from another water system.
- Impose permanent conservation requirements.
- Develop additional sources for nonpotable uses.
- Reuse wastewater for nonpotable requirements.
- Install a dual potable–nonpotable water system.
- Construct new raw-water storage, such as a new reservoir.
- Perform aquifer recharge.

Evaluating the Options Emergency response options are best evaluated by carefully considering and comparing various factors:

- Technical and logistical feasibility
  - What procedures are required to implement the option?
  - Is the required technology available and properly developed?
  - How much water can the option provide?
  - Will the option meet only the system's priority water needs?
  - Can the option meet the system's current total water needs?
  - Can the option be expanded to meet future community water needs?
  - How quickly can the option be made operational?
  - What equipment and supplies are needed?
- Reliability
  - How reliable is the option?
  - Does the option require special operation and maintenance skills?
- Political and legal considerations
  - What administrative procedures are required to implement the option?
  - Is property ownership an issue?
  - Are water rights an issue?
  - Will the option be acceptable to the public?
- Cost considerations
  - What initial investment is required?
  - What will the operating costs be?
  - Who will bear the cost of the design, construction, and operation of the option?

Alternative Water Sources

If at all possible, alternative water sources should be identified before an emergency that requires an alternative source actually happens. Some typical alternative sources that should be considered are discussed below.

- Surface water systems
  - Provide two or more intakes at different locations in the lake or stream.
  - Provide intakes in more than one water source, if possible.
  - Construct wells for backup of the surface source.
- Groundwater systems
  - Provide enough wells to meet demand when one or more wells are out of service.

    - Locate wells in different aquifers—or far enough apart in the same aquifer—so that contamination affecting one well will not affect other wells.
    - Provide a surface water emergency source.
- All systems
    - Establish interconnection with other water systems that have sufficient capacity to meet at least minimum needs.

Some emergency situations can be avoided by good practices in facility design and operation. For instance, good security at water facilities can reduce vandalism; providing standby power eliminates complete dependence on the commercial electric power grid; and locating all facilities far enough away from and well above floodplains (of both rivers and coasts) can prevent inundations and many contamination problems. In addition to averting problems that can possibly be prevented, water system operators should be prepared to act swiftly and efficiently in the event of an emergency.

Intersystem Connection

In urban and suburban areas, it is not unusual for water mains belonging to different water utilities to lie close to each other in some locations. Even without prior planning, emergency connections can be made between the distribution systems using fire hoses or temporary piping. Unfortunately, this tactic consumes precious time during an emergency and usually provides only a fraction of the water needed.

Consequently, it is often advantageous to interconnect adjoining distribution systems or raw-water transmission systems. If water mains are large enough at adjacent sites and water pressure is about the same, the interconnection can be relatively simple. A valve can be kept closed until an emergency need arises to share water between the systems.

The managers of adjoining water systems should carefully plan for either system to furnish water to the other in an emergency. Before establishing a connection, system managers should determine whether the separate water supply systems have a compatible water chemistry and quality to allow intermingling. A satisfactory connection will require piping of sufficient size to provide the needed amount of water in either direction. If system pressures are not compatible, it may be necessary to install pressure-reducing valves and a booster station. It may also be desirable to provide meters for flow in either direction, for purposes of cost reimbursement and water accountability.

Emergency Provision of Water

When the distribution system is providing only a reduced amount of water, emergency supplies can be brought in by tank trucks and as bottled water. Rationing can also be used while the distribution system is operating at reduced capacity.

Tank Trucks

If a water distribution system suffers massive failure and no alternative piped source is available, nonpiped water must be used. Tank trucks (Figure 13-4) can be used for this purpose, and every utility should maintain a record of where suitable units are available. The utility must be able to confirm the chemical and

Figure 13-4 Emergency potable water provided by tanker truck in Kauai, Hawaii

Source: Ray Sato.

biological safety of the water supplied by the trucks and must ensure that adequate disinfection is maintained.

Drivers of tank trucks must take the essential sanitary precautions and, most importantly, must bring water from a wholesome water source. If water is to be taken from a neighboring utility, communication and monitoring are necessary to ensure that the water-filling point is satisfactory to both utilities.

Bottled Water

Bottled water is also frequently made available to customers during an emergency. Because demand can quickly exceed the normal supply of bottled water in an area, it may be necessary to tap into supplemental supplies of bottled water. In some cases, local dairies and beverage bottlers can provide water in their own milk cartons or bottles during an emergency. Beverage companies may also have trucks available that can be used for delivery. In any case, the cost of bottled water will be very high compared to the cost of water from the public distribution system.

Water Rationing

Some disasters reduce the supply of water, even when the distribution system is still functioning. Typical situations include the following:

- The system can reestablish the water source soon and has enough water in storage to supply basic customer needs if use is drastically reduced.
- An interconnection has been established with another utility, but it provides only enough water to supply basic needs.
- The water system must operate at very low pressure because of pump failure or other problem. Care should be taken if the system operates at a low water pressure because the risk of contamination is higher due to backfill or groundwater infiltration.

In these cases, a drastic reduction in demand can help alleviate the crisis. The utility's conservation plan or emergency response plan should have all categories of water use rated by priority. This situation is comparable to that faced by communities with severe or critical drought conditions. Table 13-2 is an example of a plan prepared for use in such situations.

Table 13-2 Example of a drought or emergency conservation plan

Stage	Degree of Drought or Emergency	Consumption Reduction Goal, %	Public Information Action	Public-Sector Action	User Restrictions	Penalties for Noncompliance
I	Minor	10	Explain drought or emergency conditions. Disseminate technical information. Explain other stages and possible actions. Distribute retrofit kits at central depots. Request voluntary reduction.	Increase enforcement of hydrant-opening regulations. Increase meter-reading efficiency and meter maintenance. Implement intensive leak detection and repair program.	Voluntary installation of retrofit kits. Restriction of outside water use for landscape irrigation, washing cars, and other uses.	Warning
II	Moderate	15–18	Use media intensively to explain emergency. Explain restrictions and penalties. Explain actions in potential later stages. Request voluntary reduction.	Reduce water usage for main flushing, street flushing, public fountains, and park irrigation.	Mandatory restriction on all outside uses by residential users, except landscape irrigation. Prohibition of unnecessary outside uses by any commercial users.	1. Warning 2. House call 3. Installation of flow restrictor 4. Shutoff and reconnection fee
III	Severe	25–30	Public officials appeal for water use reduction. Explain actions and consequences of emergency.	Prohibit all public water uses not required for health or safety.	Severely restrict all outside water use. Prohibit serving water in restaurants. Prohibit use of water-cooled air conditioners without recirculation.	Same as Stage II
IV	Critical	50 or more	Same as Stage III	Reduce system pressure to minimum permissible levels. Close public water-using activities not required for health or safety.	Prohibit all outside water use and selected commercial and industrial uses. Terminate service to selected portions of system as last extreme measure.	Same as Stage II

Use and Conservation of Water

WATCH THE VIDEO
Water Conservation and Efficient Use (www.awwa.org/wsovideoclips)

While water is one of our most vital natural resources, the availability of clean water at the tap has often been taken for granted. But recent shortages, quality problems, and cost increases in various parts of the country have resulted in a new appreciation for the challenge of providing a continuous and wholesome water supply.

A substantial part of the cost of water is due to the expense of treatment and distribution. Additional costs are incurred by utilities to comply with the SDWA and other environmental regulations. Drinking water is being treated and tested more than ever. Increased health-related concerns and ongoing advances in commercially available technology continue to drive improvements in the effectiveness of treatment and monitoring performed by water utilities, resulting in increased capital and operating costs.

It is worth emphasizing, however, that the value of water as a resource cannot be measured simply by the costs associated with water treatment and distribution or the price paid by customers. Wise management of water recognizes not only the cost of obtaining it in a useful form, but also the value of the benefits it delivers—not only in society, but in the natural world.

Water Use

Public water supply uses can be grouped into three categories: domestic, industrial, and public.

Domestic Use

The category of domestic water use includes water that is supplied to residential areas, commercial districts, and institutional facilities.

- *Residential uses.* Water provided to residential households serves both interior uses (toilets, showers, clothes washers, faucets, etc.) and exterior uses (car washing, lawn watering, etc.). The amount of water used by households varies according to such factors as the number and ages of occupants in a household, income level, geographic location (which influences climate), and the efficiency of water fixtures and appliances in the home. Typical data for rates of interior water use are presented in Figure 13-5. Outdoor water use is strongly influenced by the annual weather patterns; water use in hot, dry climates is significantly higher than in colder, wetter climates. Table 13-3 shows a typical breakdown of water use by a typical family of four people living in a single-family house.
- *Commercial uses.* Water use can vary widely among commercial facilities depending on the type of commercial activity. A small office building would be expected to use less water than a restaurant of the same size, for example. In addition to the amount of water required by employees and customers for drinking and sanitary purposes, commercial users may need water for such diverse uses as cooling, humidifying, washing, ice making, vegetable and produce watering, ornamental fountains, and landscape irrigation.
- *Institutional uses.* Estimates of water demand for facilities such as hospitals and schools are usually based on the size of the facility and the amount of water needed for each user in the facility. Water use by schools is highly dependent on whether the students are housed on campus or are day students.

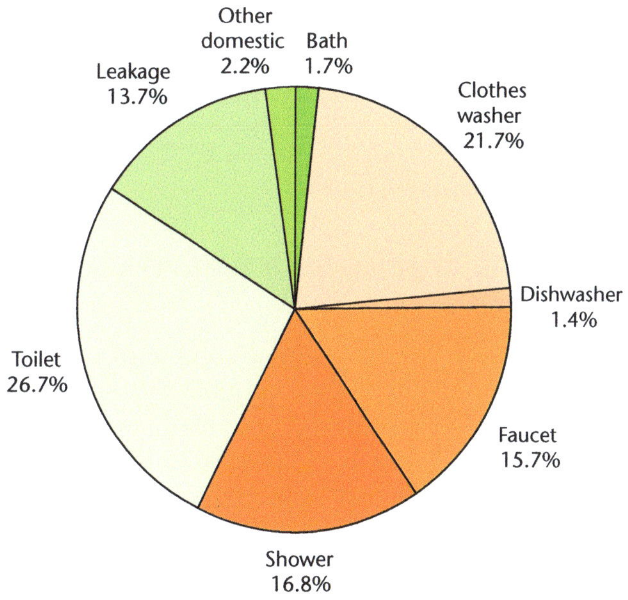

Figure 13-5 Indoor per capita use percentage by fixture

Table 13-3 Typical residential end uses of water by a family of four*

Fixture/End Use	Average Gallons per Capita per Day	Average Liters per Capita per Day	Indoor Use, %	Total Use, %
Toilet	18.5	70	30.9	10.8
Clothes washer	15	56.8	25.1	8.7
Shower	11.6	43.9	19.4	6.8
Faucet	10.9	41.3	18.2	6.3
Other domestic	1.6	6.1	2.7	0.9
Bathtub	1.2	4.5	2.0	0.7
Dishwasher	1	3.8	1.7	0.6
Indoor Total	59.8	226.4	100.0	34.8
Leakage	9.5	36	NA	5.5
Unknown	1.7	6.4	NA	1.0
Outdoor	100.8	381.5	NA	58.7
TOTAL	**171.8**	**650.3**	**NA**	**100.0**

*Data collected from the following cities: Boulder, CO; Denver, CO; Eugene, OR; Seattle, WA; San Diego, CA; Tampa, FL; Phoenix, AZ; Tempe, AZ; Scottsdale, AZ; Cambridge, Ontario, Canada; Waterloo, Ontario, Canada; Las Virgenes, CA; Walnut Valley, CA; and Lompoc, CA.

Adapted from AwwaRF (1999).

Industrial Use

In the industrial sector, rates of water use depend on the type of industry, cost of water, wastewater disposal practices, types of processes and equipment, and water conservation and reuse practices. In general, industries that use large quantities of water are located in communities where water quality is good and the cost is reasonable.

Historically, major industrial water users (such as the steel, petroleum products, pulp and paper, and power industries) have provided their own water supply. Smaller industries and industries with low water usage are more inclined to purchase their water from public systems.

Public Use

Municipalities and other public entities provide public services that require varying amounts of water. Water is used for the following:

- Public parks, golf courses, swimming pools, and other recreational areas
- Municipal buildings
- Firefighting
- Public-works uses such as street cleaning, sewer flushing, and water system flushing

Although the total annual volume of water required for firefighting is typically small, the rate of flow required during a fire can be very large. The flow rate needed to fight a large fire in a small community can put a significant strain on the water system and may threaten to reduce pressure in the system substantially. For example, a community that uses an average of 1 mgd (4 ML/d) or 694 gpm (2,600 L/min.) might reasonably expect a fire flow demand of 3,000 gpm (11,400 L/min.) for up to 3 hours. The rate of flow for firefighting in this case is more than four times the average daily flow, and it will likely be required from a few water mains near the fire.

Variations in Water Use

Water use for a municipality or community can vary due to several factors, including the following:

- Time of day and day of the week
- Climate and season of the year
- Type of community (residential or industrial) and the economy of the area
- Presence or absence of customer meters
- Dependability of supply and dependably high quality of the water
- Condition of the water system (leakage and losses)
- Water conservation/demand management

Time of Day and Day of Week

Water use rates vary considerably according to the time of day and the day of the week. On a typical day in most communities, water use is lowest at night (11 p.m. to 5 a.m.) when most people are asleep. Water use rises rapidly in the morning (5 a.m. to 11 a.m.), and usage is moderate through midday (11 a.m. to 6 p.m.). Usage then increases in the evening (6 p.m. to 10 p.m.) and drops rather quickly around 10:00 p.m. A diurnal curve showing the hourly rate of water use for a community on a typical day is shown in Figure 13-6.

Total water use rates for a community typically vary by the day of the week as well, depending on the habits of the community. Some water systems have significant increases in water use on Monday because in many households Monday is washday. Water systems supplying industries that do not operate on weekends observe significantly reduced usage rates on Saturday and Sunday.

Climate and Season

Water use is typically highest during the summer months when the weather is hot and dry and the need to provide water for exterior activities and landscape irrigation is at its peak. Particularly significant seasonal variations are commonly

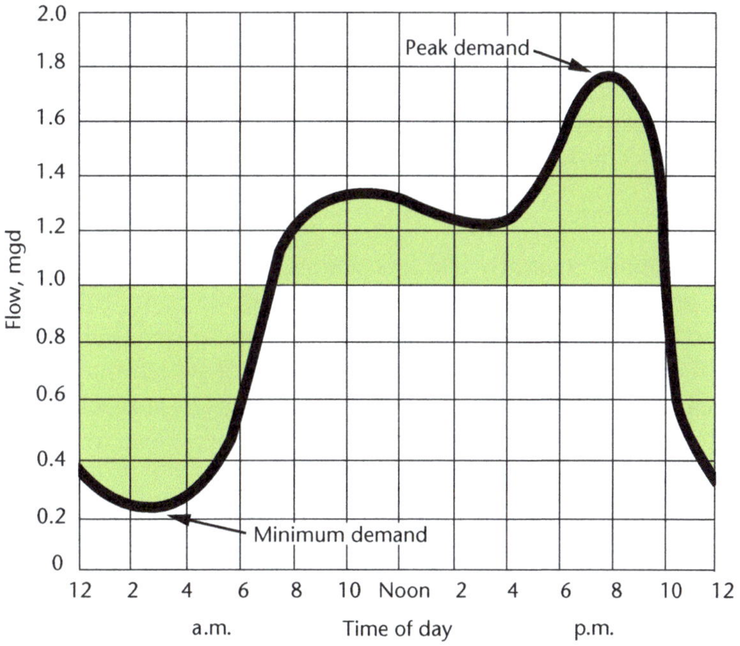

Figure 13-6 Typical daily flow chart showing peak and minimum demands

observed in resort areas, small communities with college campuses, and communities with seasonal commercial or industrial activities.

Most systems have relatively low water use in winter when there is no need for lawn or garden sprinkling and there is less outside activity requiring water. However, in extremely cold weather, consumers may occasionally run water faucets continuously to prevent water pipes from freezing.

Air-conditioning units and cooling towers operate by evaporating water and are used extensively by homes and large commercial and industrial facilities located in hot, dry climates. The demand for water to support air conditioners and cooling towers generally coincides with periods of highest water demand for lawn and garden sprinkling, pushing the annual water use to its peak for the year.

Type and Size of Community

The type of housing in the service area will affect rates of water use. Areas that have individual homes on large lots with gardens and lawns typically have a much higher water use per person than areas with multiple-family dwellings such as townhomes, condominiums, and apartment complexes. Households with automatic washing machines, dishwashers, and garbage disposals tend to use more water than households where these appliances are not installed. Economically depressed areas generally have lower water use per capita.

Community size typically affects the variability of water use rates over time. The ratio of peak use to average use tends to be greater for systems serving small communities than for larger water systems.

Metering

Most water utilities install meters that track all water usage. Other systems do not have customer meters and continue to charge customers a flat rate regardless of the quantity of water they use. The choice of billing scheme significantly influences the amounts of water consumed. Flat-rate customers may use as much as 25 percent more water than is used by customers who have to pay more as they use more. This excess water usage can be caused by unnecessary and unrestricted sprinkling and by personal behavior (e.g., letting a faucet run to get cold water or letting a hose run while washing a car). Another consequence of flat-rate billing is that household leaks are often ignored because wasted water costs the customer nothing.

Dependability and Quality of Water

Customers generally use less water from the public system if the water pressure is poor or if the water has an unpleasant taste, odor, or color. They may also use less if it has a high mineral content. Under such conditions, customers may choose alternative water sources, some of which may be less protective of public health. Bottled water, for example, has been subject to less stringent regulatory standards than tap water. Private wells, likewise, are not as closely regulated as a public utility.

Sewer Connection

The availability of municipal sewer systems also increases water use. Homes with sewer service usually have more water-using appliances and fixtures because there is no concern about overloading an on-site wastewater treatment and disposal system that would be used otherwise. The increase in water use due to sewer availability may be 50–100 percent.

Condition of the Water System

The difference between the average annual water production and the average annual consumption of a system amounts to ongoing unaccounted-for losses. Unaccounted-for losses can result from many factors, including leaks in piping, main breaks, periodic fire hydrant flushing, tank drainage for maintenance purposes, unauthorized use, unmetered services, and inaccurate or nonfunctioning meters. Current industry standard states that unaccounted-for system leakage and losses should be less than 10 percent of production for new distribution systems (those less than 25 years old) and less than 15 percent for older systems. Rates of leakage are affected by the age of the system, the materials used in construction, and the level of system maintenance. Water mains are usually tested for watertightness when they are installed. However, as the water mains get older, leakage at pipe joints, valves, and connections increases.

Unless the system is regularly checked for leaks and unless leaks are promptly repaired, the amount of water lost daily can be significant. Such losses not only place an extra strain on the ability of the utility to furnish water to customers, they also represent an important loss of revenue.

Water Conservation

Definitions of water conservation vary. In this book, water conservation is considered to be any beneficial reduction in water use or in water losses. Over the past four decades, the level of appreciation for water conservation efforts has increased significantly. Water utilities have been among the leaders in conservation efforts. For utilities in areas of prolonged drought, the benefits of reducing water use are obvious, and mandatory restrictions may be put into effect where the drought is severe. The benefits are less evident in an area that has adequate rainfall, a relatively stable population, and a water source with an adequate safe yield. A useful approach to evaluating whether a conservation program is worthwhile is to compare the expected benefits against the anticipated costs.

water conservation
The reduction of water usage, accomplished primarily by preventing waste.

Potential Benefits of Water Conservation

The following are some potential benefits of water conservation:

- *Reduced per capita demand on the supply source.* A reduction in water use will allow a larger population to be served by the water source. This is particularly important where aquifers are being depleted and limits on the amounts of available surface water are being approached by growing communities. Reducing normal rates of water use can also help ensure that adequate supply is available during an emergency.
- *Reduced need for facility capacity.* Reduced flows in the water and wastewater system may allow a water utility to eliminate or downsize new facilities and defer the construction of new ones.
- *Reduced costs for utility and customer.* A reduction in water use can provide energy savings for the water utility. Pumping costs can be reduced, as can the costs for power and chemicals required for all stages of treatment. Cost savings can help defer increases in billing rates for consumers. In addition, when heated water is conserved, energy costs will be reduced in homes, businesses, and industry.
- *Protection of the environment.* Reducing the rate of withdrawals from a stream, lake, or impoundment can help maintain flow rates and water levels needed to sustain a healthy environment for native aquatic plants and animals.

Potential Problems With Water Conservation

The following are some undesirable consequences of implementing a water conservation program:

- *Loss of revenue for the utility.* When a utility implements a water conservation program and does not increase water rates, it will receive less income because of reduced water sales. There will be some offsetting savings to the utility due to lower energy and chemical costs, but the net loss can still be substantial. Often the only way to compensate for this loss in revenue is to raise water rates. Before starting a conservation program, the water system must plan for a decrease in revenue and take steps to ensure that all financial obligations, including bond payments, will still be met. Because it is usually difficult to raise rates on short notice, advance financial planning is particularly important should it ever become necessary to implement an emergency conservation effort during a water shortage.
- *Utility and customer program costs.* The cost for implementing a conservation program will likely be borne by the water utility and the water utility's customers. Costs for programs can vary depending on the implementation method used, the size of the program, and the scope of program evaluation (i.e., surveys, billing data reviews, water audits, etc.). To the extent possible, these short-term and long-term costs should be estimated ahead of time for each component of the water conservation program.
- *Possible delay in developing additional source capacity.* When a water crisis is alleviated through conservation, the public may find it hard to understand why there is still a need for long-range water resource development projects. Lack of public support could then delay (or eliminate from further consideration) development of a new reservoir, well field, or other new water source that is still needed. This delay can, in turn, increase the cost

of the project, as well as worsen the water crisis if there is a drought year before the new facilities can be constructed.

- *Possible stimulation of water service growth.* The water saved by a conservation program will allow more people to be served by the amount of water available. This capacity to take on new customers can act as a stimulant for growth. As long as water still appears to be available for use, additional customers may continue to be connected to the system. However, serving new customers will not be in the best interests of the water system or community, at least until a more ample supply can be tapped, if it increases the risk that shortages will occur.
- *Difficulty in dealing with drought conditions.* When a drought occurs after a utility has a conservation program in effect, the situation can become very difficult. It is sometimes hard to explain why it is necessary to conserve even further. This situation emphasizes the need to update water demand forecasts and continue overall water supply planning beyond any conservation effort. The utility must convince customers that looking to the future water supply needs of a community is as important as any aspect of a conservation program.

Supply Management Techniques

In conjunction with any water conservation behavior expected from the public, a utility should examine its own practices in properly managing the available supply of water. This process is called supply management. Features of supply management include the following:

- *Careful management of all resources.* Maintaining an adequate source of supply should be part of both long-range (25–50 years or more) and short-range (5–10 years) plans. The plans should be in writing and circulated to all local interest groups for comment and review.
- *Analysis of water use data.* A water use database should be available to allow evaluation of records and forecasts concerning average daily production, peak demand, and geographical and seasonal water use patterns. The complexity of this database should grow with utility size so that any unexpected change affecting the utility can be analyzed for its impact on total water use.
- *Complete source and customer metering.* All sources of supply and all customer services should be metered. Complete metering is necessary so that the amount of water used by customers can be compared against the amount pumped to the system.
- *Reduction of unaccounted-for water.* The difference between the amount of water pumped to the system and the amount metered to customers (or otherwise accounted for) is called unaccounted-for water. Unaccounted-for water is generally water that is wasted through leaks or unauthorized use. State utility commissions usually have formulas that can be used to calculate unaccounted-for water. Unaccounted-for system leakage and losses may range from 5 to 10 percent of production for new distribution systems (less than 25 years old) and up to 15 percent for older systems. If the amount of unaccounted-for water exceeds 15 percent, a systemwide leak survey should be conducted. In addition, leak awareness by field personnel can be promoted by a program of listening for leaks at hydrants, services, and meters.

supply management
The implementation of techniques by a utility to save water by managing the available supply.

unaccounted-for water
Water use that does not go through meters and thus is not paid for (including water that is wasted).

Demand Management Techniques

In conjunction with the program to manage the available supply, a utility must develop and implement demand management programs to encourage efficient water use. Demand management can consist of these components:

- Public education
- Distribution of water-saving devices
- Changes in consumer behavior
- Modification of the water rate structure
- Water-saving plumbing fixtures
- Landscaping and agricultural practices that require less water
- Promotion of water conservation by businesses and industries

Public Education

Educating the public is usually a key component of a water conservation program. If water rates are already low or moderate, customers may need some incentive to participate in a water conservation program. Printed materials and suggested programs are available from AWWA, state water organizations, state agencies, and USEPA, as well as environmental and conservation organizations. An effective way of reaching and informing the public about a conservation program is by offering speakers, literature, and videos to local schools, service clubs, and civic groups.

Customers can also be educated directly by bill stuffers and conservation notices. Direct assistance in detecting leaks or developing water use audits can also be offered to customers. Newspapers and radio and television stations can be supplied with public service announcements (PSAs) and press releases explaining the program. A utility's website can be a very effective channel for explaining the program to the public.

Water-Saving Devices

Many water utilities offer water-saving kits to residential customers. Installation of these kits will decrease the flow of water through several household plumbing fixtures. A typical kit includes faucet aerators, a low-flow showerhead or flow restrictor, and a toilet flapper to reduce leakage. Some utilities offer rebates to customers who replace older-model toilets with low-flow models. The cost of these kits can be fully paid by the customer, partially paid by the customer, or fully funded by the utility. Some state regulatory agencies have mandated the use of these kits, stipulating that the cost be shared between the customer and utility.

Changes in Consumer Behavior

Some branches of a water system may experience water shortage as a result of inadequate distribution system capacity. If this occurs, the utility may gain some relief by educating the public about the problem and encouraging customers to use water during off-peak hours. Water use restrictions can be either mandatory or voluntary. Customers may be advised to save water by watering lawns at night, when less water evaporates, for example. Mandating that lawn watering or car washing be done only on alternating days is a common summer water use restriction.

Rate Structure Modification

Water can be priced in a way that encourages efficiency and discourages excessive and wasteful use. Information on various approaches to changing the rate structure for water can be obtained from AWWA and public utility regulatory agencies.

Water-Saving Plumbing Fixtures

The Energy Policy Act of 1992 established water conservation standards for four types of plumbing fixtures: toilets, kitchen and lavatory faucets, showerheads, and urinals. With limited exceptions, the standards apply to all models of the fixtures manufactured after January 1, 1994. Approximately 16 states and six localities had water-efficient standards for at least two of the plumbing fixtures regulated under the Energy Policy Act before the national standards took effect in 1994. Studies have shown that, compared to their less efficient counterparts, low-flow fixtures conserve water, particularly in the case of toilets.

Landscaping and Agricultural Practices That Require Less Water

Landscaping that requires little water is called xeriscaping (Figure 13-7). It is being encouraged for homes and businesses in semi-arid and arid areas. The use of drought-tolerant landscaping also has value in areas where there are seasonal shortages of water.

Although agricultural operations do not require treated water from a municipal water supply system, a municipality and the surrounding agricultural community may both draw raw water from the same source. Therefore, a municipal water system does have a stake in encouraging the most efficient use of water for agricultural purposes. Greater water use efficiencies in agriculture may be obtained by changes in tillage practices, irrigation scheduling, and technology choices such as drip irrigation and by cultivation of crops that require less water.

Figure 13-7 Typical water-efficient landscaping

Source: Doug Hanford, Hanford Company Landscape Design; John Nelson, North Marin Water District; Ali Davidson, Sonoma County Water Agency.

xeriscaping
The selection and use of trees, shrubs, and plants that require small amounts of water for landscaping.

Business and Industry Conservation

Utilities can often achieve significant reductions in water use by providing conservation incentives for business and industry. Increasing water rates is one very strong incentive. Careful evaluation of how water is used, reused, or wasted in a manufacturing plant can also lead to significant reductions. In some instances, installing smaller, more sensitive meters in businesses and industries can generate considerable revenue by more accurately recording low flows and can serve as a stimulus to water conservation. Some industries and businesses can substantially reduce their water use by installing closed-loop processes, which keep filtering and recycling of the water within their facilities.

Droughts

One of the most serious threats to the ability of a water utility to meet the demands of its customers is a drought. To respond to a drought emergency, the utility will have to tap into a supplemental supply, reduce demand, or both.

Every water system should have an emergency response plan to address how water will be provided under all conceivable emergencies. Although a drought can be very serious, an immediate response is not necessarily needed. It can take weeks or months for a drought to develop to the point at which water levels in streams and reservoirs fall too low to make them useful. That time can be used to organize resources to respond before the crisis becomes acute.

Emergency procedures to reduce water use may require limited behavioral change by consumers at first, but several significant water use restrictions may need to be fully enforced if the situation becomes critical. For instance, some communities establish mandatory restrictions and issue warnings and fines for unauthorized uses such as washing a car or watering the lawn during an acute water shortage.

WATCH THE VIDEO
Water Conservation and Drought (www.awwa.org/wsovideoclips)

Water Rights

Conflicts over who owns a source of water and who gets to use it may arise among individuals, among communities, among states, between states and the federal government, and among countries.

Allocation of Surface Water

In the United States, the legal reasoning used for allocating water from surface lakes and streams follows one of two basic lines of thinking:

- Riparian doctrine is commonly used in the water-rich eastern part of the country.
- Prior appropriation doctrine is used in the more arid western states.

Riparian Doctrine Riparian doctrine, which is commonly used in the water-rich eastern part of the country, is sometimes called the "rule of reasonable sharing." All states that follow this doctrine subscribe to the basic theory that all property owners who have land abutting a body of water have an equal right to that water. Among the states that follow this doctrine, however, there are substantial differences in how the law is applied.

Under riparian doctrine, each riparian property owner can use as much water as needed for any reasonable purpose, as long as this use does not interfere with the reasonable use by other riparian property owners.

Consequently, when a water shortage occurs, all riparian property owners must accept a reduction in supply. The amount of water allowed for any one property owned is not affected by how long the land has been owned or by how much water the owner has used in the past.

riparian doctrine
A water right that allows the owners of land abutting a stream or other natural body of water to use that water.

prior appropriation doctrine
A water rights doctrine in which the first user has the right to water before subsequent users.

Riparian rights are a common-law doctrine. This means that they are a part of a large body of civil law that, for the most part, was made by judges in court decisions in individual cases. Statutory law, in contrast, is enacted by a legislative body.

Because the possession of rights arises from ownership of land abutting a water body, all of the owners are treated as equals by the courts. Conflicts are then usually resolved by the court, which determines what reasonable use is. Factors considered in the court's decision include the following:

- Purpose of the use
- Suitability of a use with respect to the water body in question
- The economic and social benefits of the use
- The amount of harm caused by the use
- The amount of harm avoided by changing any one party's use
- The protection of existing values

The growth of water use in riparian-doctrine states has led to more frequent conflicts over water sources. Several of these states have tried to remedy weaknesses in the doctrine by enacting statutes requiring that permits be obtained from a state agency. These permits vary widely in the restrictions they apply and the requirements they place on the riparian owner.

The issuance of permits moves a state away from basic riparian doctrine and closer to prior appropriation doctrine. The permit systems are severely criticized by some and considered simply unnecessary by others. They have not undergone serious challenges in the courts, so their eventual success is not guaranteed. The trend toward permit systems, however, is continuing.

Prior Appropriation Doctrine The appropriation doctrine concept started with the forty-niners in the goldfields of California. When there was too little water for all the mining operations, the principle applied was "first in time, first in right." The principle was given legal recognition by the courts and later made into law by western state legislatures.

The appropriation doctrine rests on two basic principles: **priority in time** and **beneficial use**.

Priority in Time When stream flow is less than what is demanded by all water users along the stream, use is prioritized based on who has been using the water for the longest time. Water withdrawals that began more recently must be discontinued so that withdrawals begun earlier can continue. Unlike the riparian doctrine, the appropriation doctrine does not treat riparian owners equally.

Beneficial Use While appropriation doctrine recognizes no water right based on land ownership alone, a user who has been appropriated water is entitled to it only when it can be used beneficially. Waste is therefore theoretically prohibited, and any available water beyond the amount that can be used by one appropriator is available to others. Furthermore, nonuse of the water for a long period may result in loss of the water right by forfeiture or abandonment.

Defining beneficial use remains difficult, and conflicts arise as a result. Beneficial use is determined based on two broad characteristics: the type of use (such as irrigation, mining, or municipal use) and its efficiency. Most types of water uses have been judged to be beneficial, and the majority of court cases have been concerned with whether a particular water use is inefficient or wasteful.

priority in time
A water rights term in which water use is prioritized based on who has been using the water for the longest time.

beneficial use
A water rights term indicating that the water is being used for good purposes.

Legal Complications Legal solutions to water conflicts can get quite complex because water, unlike property, is a shared resource, causing the exercise of one user's water rights to affect someone else. Boundaries are not obvious and may be constantly changing.

With the appropriation doctrine, the assignment (or not) of a right at any particular time depends on the observed streamflow, the number of prior existing rights, and the amount of water needed by those who have the prior rights. For many western streams, the exercise of rights might have to be adjusted as often as daily in times of heavy use, such as in summer when irrigation demands are high.

Allocation of Groundwater

There are four systems used in determining rights to groundwater:

1. Absolute ownership
2. Reasonable use
3. Correlative rights
4. Appropriation–permit systems

Absolute Ownership

The **absolute ownership** system is based on the principle that the owner of the land owns everything beneath that land, all the way down to the center of the earth. The dynamic nature of groundwater movement makes it impossible to apply this ownership literally. For all practical purposes, the system is a rule of capture. That is, the landowner can use all the water that can be captured from beneath the owner's land.

There are almost no restrictions involved with this rule. The water can be used for the owner's purposes both on and off the land, and it can be sold to others. There is no liability if pumping reduces a neighbor's supply or even dries up the neighbor's well. Some interpretations of absolute ownership allow that water can be wasted as long as it is not done maliciously.

Reasonable Use

As with the rule of absolute ownership, the rule of **reasonable use** holds that a property owner has the right to pump and use groundwater lying beneath the property. However, if this use interferes with a neighbor's use, it may continue only if it is reasonable. Furthermore, an owner is liable if unreasonable use causes harm to others. This rule amounts to a *qualified right* rather than an *absolute right* to use groundwater.

The determination of what is reasonable is the important aspect of making reasonable use legally workable. Reasonable use for groundwater is much easier to determine than reasonable use for surface water under the riparian doctrine. In general, any use of water that is not wasteful for a purpose associated with the land from which the water is drawn (an **overlying use**) is reasonable. Conversely, any use off the property may be considered unreasonable if it interferes with the use of the groundwater sources by others.

The overlying use criterion, however, does not resolve disputes if all parties to the dispute are using the water for overlying purposes. In this case, all uses are reasonable, all can continue, and reasonable use becomes the rule of capture. In other words, the owner with the biggest pump gets the water. On the other hand, because a nonoverlying use is considered unreasonable if it interferes with an overlying use, such uses are forbidden no matter how beneficial.

Correlative Rights

The rule of **correlative rights** holds that the overlying use rule is not absolute but is related to the rights of other overlying users. This rule is used when there is not

absolute ownership
A water rights term referring to water that is completely owned by one person.

reasonable use
A water rights term indicating that the water use is acceptable in general terms.

overlying use
The land use that occurs on top of an aquifer.

correlative rights
A rule that contends that the *overlying use* rule is not absolute but is related to the rights of other overlying users. This rule is used when there is not enough water to satisfy all overlying uses.

enough water to satisfy all overlying uses. In this case, the rule requires sharing. This rule has been applied by allocating rights among property owners in proportion to the size of the overlying land parcel.

Appropriation–Permit System

The appropriation–permit system is sometimes called a groundwater appropriation system. The most important aspect of it is the rule of priority: water rights are based on who has used the water for the longest time. Priority is all-important in the appropriation doctrine as applied to surface water because the variation in streamflow requires frequent changes in allowable withdrawals. However, the flow of groundwater in an aquifer fluctuates very slowly under natural conditions, so that frequent adjustment in allocations is not necessary. As such, priority, as applied to groundwater, mainly involves limiting the number of permits to prevent overuse of the aquifer.

The permit system thus amounts to groundwater management and administrative regulation. By placing limitations on pumping rates, well field placement, and well construction standards and by refusing permits when necessary, for example, the administrating agency manages the groundwater in its jurisdiction. The implementation of permit systems varies widely, however. The only generalization that can be made is that permits can be more relaxed if adequate water is available and more stringent if it is not.

Conclusion

The administration of water rights can vary considerably across the United States. Consequently, a water utility operator should be familiar with the regulations that apply to the specific area that the utility serves.

Public Relations

It is important for a water utility to maintain public confidence. Satisfied customers will pay their bills, are less likely to complain if they are temporarily inconvenienced during system maintenance and construction, and are more likely to be supportive of rate increases and bond issues. Operations personnel are often engaged with customers when they are investigating a complaint. Not only is this an opportunity to solve a problem and gain customer confidence, but also important information can be obtained on the system's condition.

Dealing With the Media

On occasion, distribution system employees may be approached by reporters from the local media asking about the work they are doing. Although one must be courteous, the general rule in talking to reporters is, *Don't unless you absolutely have to!* There are many opportunities for being misquoted, and it is quite embarrassing for the utility to see the wrong information in print.

The best policy is to give a very brief explanation and offer no more information than is requested. Beyond that, the workers should say they are not qualified to go into more detail. Large municipalities and water utilities maintain a public relations department whose job is to deal with the media, so reporters should be referred there for additional details. If such a department does not exist, the reporter should be referred to the utility manager, city manager, or mayor.

appropriation–permit system
A water use system in which permits to use water are regulated so that overdraft cannot occur.

Customer Complaint Response

Responding to customer complaints is an important element in assessing system performance. Information from customers about water quality, reliability of service, and security can be helpful when these topics are of concern. A professional interaction with customers is often the most effective method to convey a positive view of the utility and gain customer satisfaction. Also, customer complaint reporting procedures may be required by some state or local agencies.

The objectives of a customer complaint response program are to (1) address the customer's concerns, (2) gather information that can be used to identify system problems, and (3) catalog data that can be used to assess long-term system performance goals. System operators need specific training to achieve these objectives.

Study Questions

1. The "rule of reasonable sharing" of water is usually called the
   a. riparian doctrine.
   b. legal use doctrine.
   c. priority use doctrine.
   d. beneficial use doctrine.

2. The rule that landowners can "capture" all the water under only their land is called
   a. correlative rights.
   b. absolute ownership.
   c. reasonable use.
   d. appropriation–permit systems.

3. The rule that a landowner has the right to groundwater as long as it does not harm a neighbor is called
   a. correlative rights.
   b. absolute ownership.
   c. reasonable use.
   d. appropriation–permit systems.

4. Which rule applies when the most important issue is priority, where water rights are based on who has used the water for the longest time?
   a. Correlative rights
   b. Absolute ownership
   c. Reasonable use
   d. Appropriation–permit systems

5. Which of the following would be the most probable solution to control algae in the source water if the algae were clogging the filters at the water plant?
   a. Use activated carbon.
   b. Decrease oxygen levels.
   c. Backwash filters more frequently.
   d. Control nutrients.

6. A conventional water treatment plant in a rich farming area has to treat water high in nitrates. Which of the following is its best course of action in the long term?
   a. Manage watershed to prevent nitrates from entering the water system.
   b. Change its treatment to ion exchange.
   c. Change its treatment to reverse osmosis.
   d. Add a strong oxidation chemical before coagulation.

7. When a water source is destroyed or damaged due to some disaster that causes total contamination, which of the following would be a short-term goal?
   a. Install a dual potable–nonpotable water system.
   b. Construct a new raw-water source.
   c. Clean up the source of contamination caused by the disaster.
   d. Draw water from an adjoining system or systems.

8. The main objective of a customer complaint response program is to
   a. gain popularity for rate increases.
   b. address the customer's concerns.
   c. provide company information to the customer.
   d. pacify the customer.

9. Typically, the greatest amount of water in residential use can be attributed to the
   a. toilet.
   b. washing machine.
   c. shower.
   d. faucet.

10. Water use rates for a community are typically highest at what time of day?
    a. 11 p.m. to 5 a.m.
    b. 6 p.m. to 10 p.m.
    c. 5 a.m. to 11 a.m.
    d. 11 a.m. to 6 p.m.

11. What are the two general categories of catastrophes that can disrupt the operation of a water source?

12. What system is based on the principle that the owner of the land owns everything beneath that land, all the way down to the center of the earth?

13. List three undesirable consequences of implementing a water conservation program.

Chapter 14

Treatment Plant Administration

Records of plant operations and maintenance activities provide a basis for predictive decision-making and proof of compliance. Records also help operators to efficiently use chemicals (minimize waste) and equipment.

Process Records

In addition to the records that are required by the local regulatory agency, a water treatment plant should maintain records of activities as they relate to each process or discipline.

The following sections describe the types of records operators should keep when performing their duties.

Coagulation and Flocculation

Maintain records of past performances for coagulants, including dosages, turbidities (sedimentation basin and filtration), particle counts, filterability indices, and zeta potentials as well as the temperatures, pH values, and color values at which the coagulants performed. For softening plants, record which softening chemicals are used. Also record detention times and overflow rates in basins used for softening as well as residual quantities produced. Keep equipment records such as hours run for each mixer and flocculator, basin-cleaning activities, maintenance performed on motors and gears, and electrical use records.

Sedimentation

Calculate and record surface overflow rate, weir overflow rate, turbidity performance of each basin, and quantity of sludge removed from each basin (with a calculation for percent solids).

Maintain records of any operations and maintenance activities for each unit, including blowdown, cleaning, raking, and so on. Safety records for confined space entry may be necessary.

The Partnership for Safe Water suggests that the turbidity performance (and particle count, if available) in each sedimentation basin be recorded periodically (daily, every 4 hours, or hourly). Record the settled water turbidity as a combined flow, but realize that it cannot characterize the performance of an individual basin.

Filtration

For each filter, record the rate of flow, in million gallons per day or gallons per minute; head loss; length of run and unit filter run volume (UFRV); turbidity (in accordance with the Interim Enhanced Surface Water Treatment Rule [IESWTR]); particle counts; and dosage of any filter aid used (Table 14-1). At each backwash, record the amount of backwash water and surface wash water used, the length of the backwash, any observations during backwash, and time of filter-to-waste if used. Record the percentage of water treated that the backwash water represents (make note of amounts greater than 4 percent). All information can be included on one form.

On a quarterly basis, an operator should record bed expansion measurements, condition and depth of media, and an evaluation of the bed surface. On a yearly basis, record the results of solids retention analysis before and after backwash and the results of sieve analysis of the media. Record the addition of any media to the bed, including amounts and types. Keep a record of underdrain inspections and any work that was done on any appurtenances.

Record results of any testing required for filter-exceptions reporting under the IESWTR (this is regulated). Many states have required forms for plant operators to submit for filter performance. Consult the state regulatory agency for specific requirements.

Chlorination or Disinfection

Maintain records of the type of disinfectant used, including ordering information (phone number, address, and shipment amounts and container types), as well as information on costs and current dosage rates. Also results of bacteriological testing should be recorded for each process that uses a disinfectant, the water temperature and pH, and any unusual conditions that may occur. Keep records of safety training.

Precipitative Softening

Record the amount of softening chemicals used and the results of jar testing and lab analyses that justify these amounts. Record the results of analysis for alkalinity, pH, hardness, and related parameters for each shift, preferably every 2 hours for raw water and for all steps in the process. Keep a record of the chemical feeder setting and record the amount of water treated. Put everything on one form and maintain it daily.

Record the amounts of sludge produced. If a solids contact unit is used, record the sludge blanket depth and results of settling tests. Keep records of the amounts and rates of solids returned to the process, if applicable.

Ion Exchange

Record the hardness, alkalinity, magnesium, calcium, and pH of the source and treated waters. Record the total amount of water treated and the amount of water treated by each softening unit, including bypass water. Also, keep a record of the amount of salt used for regeneration and the length of each cycle between regenerations. Record the amount of backwash water used and the amount of waste sent to sewer or other disposal places.

Aeration

Keep process records of raw water and finished water quality to determine if process is accomplishing its goals. Also, record the daily quantity of water treated, the details of safety and maintenance procedures, and changes in other treatment

Table 14-1 A sample filter record

Time	Rate of Flow, mgd or gpm	Head Loss, ft	Hours in Service	Total Volume Filtered (cumulative)	UFRV, gal/ft^2	Turbidity, ntu and particle counts	Backwash Start	Length of Backwash, minutes and amount used (gal)	Surface Wash, amounts and minutes	Filter Aid Applied? Dosage?	Filter to Waste? ntu at Startup
0000											
0200											
0400											
0600											
0800											
1000											
1200											
1400											
1600											
1800											
2000											
2200											

processes that may be affected by aeration. Document the process operating conditions including the air-to-water ratio to calculate performance and cost information.

Adsorption

Records kept regarding activated carbon differ depending on the type of carbon used.

- *Powdered activated carbon (PAC).* Record the types of tastes and odors being experienced, the threshold odor number (TON), and the dates of taste-and-odor occurrences. Keep a record of the coagulant and PAC dosage predicted from jar tests as well as the actual dosage used.
- *Granular activated carbon (GAC).* Record the number of filter hours; UFRV; any losses of GAC from the filter, measured monthly; dates of installation; and periodic TON from each filter. Also record when backwashing was performed and the amount of backwash water used as well as the raw-to-finished organic content removal.

Iron and Manganese Removal

If using ion exchange, the previously referenced records can be used. In addition, record raw and finished iron and manganese levels and results of distribution systems analyses performed. If permanganate is used, record amounts and dosages, including those predicted by jar testing.

If a sequestrant is used to control iron and manganese, record the amount and type used and the data from distribution system flushing efforts.

Fluoridation

Keep a daily record of the amount of fluoride in the finished water and the raw water, if necessary (a weekly raw-water analysis may be sufficient if the regulator allows). Also, keep a daily record of the chemical dosages used and record the chemical feed rate each hour of each shift. Compute and record the daily fluoride chemical feed and compare it to the lab analysis as a double-check. Record safety classes and lectures that were provided to operators.

Corrosion Control

Maintain records of corrosion chemical dosages and any data available as to their performance, such as coupon testing results, direct electrochemical readings, temperatures, flushing activities, and lead and copper compliance results (i.e., 90th percentile levels and maximum levels). If phosphates are used, keep strict records of dosages and amounts of total and dissolved phosphate in the system. Flush mains regularly, especially dead-end mains; sample for phosphates at 5-, 10-, and 15-minute intervals into the flush; and record results. Record the main flushing velocities, in feet per second. Track any customer complaints that may relate to use of corrosion chemicals.

Water Main Flushing

Main flushing records are best kept in a searchable database. Record main location and size; size of flushing pipe or hydrant; flow rate at flush; gallons used for flushing; velocity, in feet per second; chlorine residual; amount of phosphate used, if any; and results of biological testing, such as heterotrophic plate count quantity or quantity of coliform. Query the database by main size and show gallons used and time needed, along with any chemical and biological data, to generate monthly reports. In this way, a profile will emerge that shows the amount of water necessary to flush each main size, the time it takes, and the results obtained.

Reporting

Reporting requirements are generally a function of the compliance reports that local and federal regulators require. In general, compliance testing results must be reported no later than 10 days after the end of the month in which they were accumulated. It is good practice for water treatment plant staff to provide reports to upper management. Even if the administration does not require these reports, they should be generated and kept on file for future reference.

Like a diary, a written report is an excellent reminder of events that took place and the methods used to handle these events. Internal memos of safety issues, personnel and disciplinary actions, purchases, and maintenance events can actually help protect staff from allegations of nonfeasance. Reports should be written as soon as possible after an event takes place so that details can be easily recalled. Do not write reports or internal memos in anger; take time to cool down before recalling an event.

Most compliance reports must conform to a specific format. This enables the regulator to assemble large amounts of data from many utilities in an orderly fashion. Internal or outsourced reports, other than compliance reports to regulators, should include the following:

- Times and dates of the audit and any description of incidents or accomplishments
- Financial or inventory considerations
- Personnel involved
- Conclusions drawn, with supporting data and references
- List of report recipients, with acknowledgment of receipt, if appropriate
- An executive summary if the report is lengthy and involved

The report should be filed in a logical order; for example, file reports under general categories such as research, safety, finance, and so on, so that the report can be easily retrieved in the future. As time permits, reread all reports to gauge progress in reaching goals and to learn from past mistakes.

Reporting Treatment Incidents

There are specific requirements for the time and method of reporting a regulatory violation. These are described in the drinking water regulations for the applicable regulatory agency. However, there are times when it is not clear if there has been a violation or when there is a water quality incident that may affect consumers. The water plant personnel should have a predetermined plan of action to deal with these situations.

The notification and action plan for incident reporting should include plant management, utility management, other utility officials, and regulatory agencies. The utility should have discussions with the regulatory agency representatives prior to any incident to establish an understanding of how they will handle this type of communication. The agency may have technical resources to assist the plant personnel with a response. Also, the agency may be helpful when dealing with mass communication media regarding the incident.

Plant Performance Reports

Periodic (monthly, quarterly, and annual) plant performance reports should be prepared and should be examined to detect trends and identify opportunities for improvement. The operating reports previously described are the basis for these performance reports. Performance reports summarize the operating results and usually include various charts and calculated performance measures. These summary reports do not need to be lengthy and, indeed, can be added to the operating reports if desired. The purpose of the performance reports is to examine the plant results over longer time periods to see if there are trends that require investigation. All plant operating personnel should review these reports to provide feedback and to participate in efforts to meet plant operating goals.

Specifications and Approval of Treatment Chemicals and System Components

Drinking Water Additives

Many chemicals are added to potable water in the various water treatment processes. Chemical additives are commonly used in the following processes:

- Coagulation and flocculation
- Control of corrosion and scale
- Chemical softening
- Fluoridation
- Sequestering of iron and manganese
- Chemical precipitation
- pH adjustment
- Disinfection and oxidation
- Algae and aquatic plant control

It is important that the chemicals added to water during treatment not add any objectionable tastes, odors, or color to the water; otherwise, customers will likely complain. It is of even greater importance that the chemicals themselves, or side effects that they cause, not create any danger to public health.

Coatings and Equipment in Contact With Water

In addition to chemicals that are added to water for treatment, many products come in direct contact with potable water and could, under some circumstances, cause objectionable contamination. The general categories of products that could cause an adverse effect because of their contact with water are as follows:

- Pipes, faucets, and other plumbing materials
- Protective materials, such as paints, coatings, and linings
- Joining and sealing materials
- Lubricants
- Mechanical devices that contact drinking water
- Process media, such as filter and ion exchange media

Objectionable effects that can be caused by construction and maintenance materials include unpleasant tastes and odors, support of microbiological growth, and the liberation of toxic organic and inorganic chemicals.

NSF International Standards and Approval

For many years, the water industry relied on AWWA standards and on approval by USEPA of individual products to ensure that harmful chemicals were not unknowingly added to potable water. However, there was no actual testing of the products, and the water treatment industry had to rely on the manufacturers' word that their products did not contain toxic materials.

In recent years, increasing numbers of products have been offered for public water supply use, many of them made with new manufactured chemicals that have not been thoroughly tested. In addition, recent toxicological research has revealed potential adverse health effects caused by rather low levels of continuous exposure to many chemicals and substances previously considered safe.

In the early 1980s, it became evident that more exacting standards were needed, as well as more definite assurance that products positively met safety standards. In view of the growing complexity of testing and approving water treatment chemicals and components, USEPA awarded a grant in 1985 for the development of private-sector standards and a certification program. The grant was awarded to a consortium of partners. NSF International was designated the responsible lead. Other cooperating organizations were the Association of State Drinking Water Administrators (ASDWA), AWWA, the Water Research Foundation (WRF), and the Conference of State Health and Environmental Managers (COSHEM).

NSF Standards Nomenclature

Today NSF is accredited by the American National Standards Institute (ANSI) to develop American National Standards. These standards, formerly known as "ANSI/NSF Standards," are now called "NSF/ANSI Standards" (e.g., NSF/ANSI Standard 60).

ANSI's accreditation verifies that NSF develops standards in a way that ensures openness and due process.

The NSF Certification

The NSF certification program involves two fundamental documents. One is the standard to which the product is evaluated, including test procedures and pass/fail criteria. All NSF drinking water treatment unit standards are American National Standards, or NSF/ANSI standards.

The second document is the NSF *Certification Policies for Drinking Water Treatment Systems and Components*. This document details the procedures and criteria required to achieve and maintain NSF certification of a company's products. This edition of NSF remarks focuses on selecting the proper standard; a future newsletter will focus on policy requirements.

Two standards were developed by these organizations, along with the help of many volunteers from the water supply and manufacturing industries who served on development committees. These standards have now been adopted by ANSI, so they bear an ANSI designation in addition to the NSF reference.

NSF/ANSI Standard 60 essentially covers treatment chemicals for drinking water. The standard sets up testing procedures for each type of chemical, provides limits on the percentage of the chemical that can safely be added to potable water, and places limitations on any harmful substances that might be present as impurities in the chemical.

NSF/ANSI Standard 61 covers materials that are in contact with water, such as coatings, construction materials, and components used in processing and distributing potable water. The standard sets up testing procedures for each type of product to ensure that they do not unduly contribute to microbiological growth, leach harmful chemicals into the water, or otherwise cause problems or adverse effects on public health.

Manufacturers of chemicals and other products that are sold for the purpose of being added to water, or that will be in contact with potable water, must now submit samples of their product to NSF International or another qualifying laboratory for testing based on Standards 60 and 61. If a product qualifies, it is then "listed." There are also provisions for periodic retesting and inspection of the manufacturer's processes by the testing laboratory.

The listings of certified products provided by NSF International are used in particular by three groups in the water supply industry:

1. In the design of water treatment facilities, engineers must specify that pipes, paints, caulks, liners, and other products that will be used in construction, as well as the chemicals to be used in the treatment process, are listed under one of the standards. This ensures in a very simple manner that only appropriate materials will be used. It also provides contractors with specific information on what materials qualify for use, without limiting competition.
2. Individual states and local agencies have the right to impose more stringent requirements or to allow the use of products based on other criteria, but most states have basically agreed to accept NSF International standards. When state authorities approve plans and specifications for the construction of new water systems or improvements to older systems, they will generally specify that all additives, coatings, and components must be listed as having been tested for compliance with the standards.
3. Water system operators can best protect themselves and their water system from customer complaints, or possibly even lawsuits, by insisting that only listed products be used for everything that is added to, or in contact with, potable water. Whether it concerns purchasing paint for plant maintenance or taking bids for supplies of chemicals, the manufacturer or representative should be asked to provide proof that the exact product has been tested and is listed.

Copies of the current listing of products approved based on Standards 60 and 61 should be available at state drinking water program offices. A copy of the current listing can also be obtained by contacting the nearest NSF International regional office or by calling or writing to the international headquarters:

NSF International
789 N. Dixboro Rd.
Ann Arbor, MI 48113-0140
(734) 769-8010

AWWA Standards

Since 1908, AWWA has developed and maintained a series of voluntary consensus standards for products and procedures used in the water supply community. These standards provide minimum requirements for most aspects of drinking water systems and cover products, such as pipe, valves, meters, filtration media, and water treatment chemicals. They also cover procedures, such as disinfection of storage tanks and design of pipe. As of 2016, AWWA had more than 160 standards. A list of standards currently available may be obtained from the AWWA website, www.awwa.org, or by calling AWWA Customer Service at (800) 926-7337.

AWWA offers and encourages use of its standards by anyone on a voluntary basis. AWWA has no authority to require the use of its standards by any water utility, manufacturer, or other person. Many individuals in the water supply community, however, choose to use AWWA standards. Manufacturers often produce products complying with the provisions of AWWA standards. Water utilities and consulting engineers frequently include the provisions of AWWA standards in their specifications for projects or purchase of products. Regulatory agencies require compliance with AWWA standards as part of their public water supply regulations. All of these uses of AWWA standards can establish a mandatory relationship, for example, between a buyer and a seller, but AWWA is not part of that relationship.

AWWA recognizes that others use its standards extensively in mandatory relationships and takes great care to avoid provisions in the standards that could give one party a disadvantage relative to another. Proprietary products are avoided whenever possible in favor of generic descriptions of functionality or construction. AWWA standards are not intended to describe the highest level of quality available, but rather describe minimum levels of quality and performance expected to provide long and useful service in the water supply community.

While maintaining product and procedural standards, AWWA does not endorse, test, approve, or certify any product. No product is or ever has been AWWA approved. Compliance with AWWA standards is encouraged, and demonstration of such compliance is entirely between the buyer and seller, with no involvement by AWWA.

AWWA standards are developed by balanced committees of persons from the water supply community who serve on a voluntary basis. Product users and producers, as well as those with general interest, are all involved in AWWA committees. Persons from water utilities, manufacturing companies, consulting engineering firms, regulatory agencies, universities, and others gather to provide their expertise in developing the content of the standards. Agreement of such a group is intended to provide standards, and thereby products, that serve the water supply community well.

Study Questions

1. Which of the following details is *not* recorded for each filter run?
   a. Rate of flow, in million gallons per day or gallons per minute
   b. Length of run and unit filter run volume (UFRV)
   c. Expected length of filter life
   d. Dosage of any filter aid used

2. How often should an operator record bed expansion measurements, condition and depth of media, and an evaluation of the bed surface?
   a. Daily
   b. Quarterly
   c. Hourly
   d. Yearly

3. Which authority suggests that the turbidity performance (and particle count, if available) in each sedimentation basin be recorded periodically (daily, every 4 hours, or hourly)?
   a. The Partnership for Safe Water
   b. The American Water Works Association
   c. The US Environmental Protection Agency
   d. The National Institute for Occupational Safety and Health

4. Which standard covers materials that are in contact with water, such as coatings, construction materials, and components used in processing and distributing potable water?

5. How often does AWWA check that its standards are being met by each water system in the United States?

Chapter 15
Additional Study Questions

As you advance through the *WSO Water Treatment* series, it is important that you maintain your competence at lower certification levels. This chapter includes additional study questions for topics discussed in the Grade 1 and Grade 2 books that are relevant to Grade 3 or 4 certification.

Water Sources and Treatment Options

1. Total dissolved minerals in rainwater are typically
   a. low resulting in very soft water.
   b. such that they cause pH to be above 8.
   c. high in calcium.
   d. high in nutrients.

2. A confined aquifer
   a. is the intersection of the water table with the surface.
   b. cannot be a source of groundwater.
   c. is free of pollutants.
   d. is a permeable layer between layers of low permeability.

3. The least important issue when considering the selection of the most suitable water treatment processes is
   a. the water source characteristics.
   b. the product water quality.
   c. legislative initiative.
   d. the capital and operating cost.

4. A treatment plant that incorporates coagulation, sedimentation, filtration, and disinfection is
   a. the most common system for treating groundwater.
   b. an example of the multiple-barrier approach to treatment.
   c. the least costly treatment plant to operate.
   d. mainly used for the removal of iron.

5. Slow surface runoff
   a. can infiltrate into the ground and recharge aquifers.
   b. causes erosion.
   c. dissolves only small amounts of minerals from the soil.
   d. is not susceptible to pollution.

Groundwater Quality and Wells

1. What is the duration of the general test period to confirm design capacity for a public water supply well that is in an unconfined aquifer?
   a. At least 24 hours
   b. At least 48 hours
   c. At least 72 hours
   d. At least 96 hours

2. What contaminants commonly found in groundwater most often affect potability?
   a. Calcium and magnesium
   b. Low amounts of arsenic and radium 228
   c. Low concentrations of iron and manganese
   d. Chloride and sulfate

3. The static water level in a well is
   a. the water surface when the well is being pumped.
   b. the water surface level after 24 hours of use.
   c. the water surface level when no water is being taken from the aquifer.
   d. the water surface level when the pump is off but other wells nearby are operating.

4. What is the specific capacity of a well with a yield of 100 gpm and a drawdown of 20 ft?
   a. 2,000
   b. 0.2
   c. 10
   d. 5

5. A common method of drilling water supply wells is
   a. the cable tool method.
   b. the hydraulic jet method.
   c. the Texas method.
   d. the drive method.

Surface Water Source Treatment

1. Surface water is stored for later use
   a. in a river.
   b. in an impoundment.
   c. in soil.
   d. in a water treatment plant.

2. Water supply intake structures used in lakes and reservoirs do not include
   a. submerged intakes.
   b. multiple port intakes.
   c. riverbed infiltration galleries.
   d. pump intakes.

3. Algae control in a lake or reservoir is most effective if
   a. alkalinity is above 50 mg/L and pH is 7.
   b. alkalinity is below 50 mg/L and pH is 7.
   c. alkalinity is above 50 mg/L and pH is 8–9.
   d. alkalinity is below 50 mg/L and pH is 8–9.

4. Stratification in lakes and reservoirs
   a. occurs when a warm layer of water overlies a colder layer.
   b. separates segments of the water column.
   c. occurs when all of the water is 4°C from top to bottom.
   d. is good because it creates an anaerobic zone near the bottom.

5. Zebra mussels are often controlled by adding
   a. chlorine in the treatment plant finished water.
   b. chlorine at the plant intake.
   c. calcium carbonate in the plant intake.
   d. copper sulfate after filtration in the plant.

Treatment Plant Pretreatment

1. Screens located on an intake structure to remove sticks, logs, and other large debris are usually called
   a. microstrainers.
   b. cyclones.
   c. pleated filters.
   d. trash racks.

2. Screen equipment is constantly wetted and dried, creating
   a. conditions ideal for accelerated corrosion.
   b. opportunities for inspection.
   c. caking and clogging.
   d. a condition where debris can fall off.

3. Systems that have installed ________ have found that the solids loading of the water entering the treatment plant is reduced.
   a. venturi meters
   b. plate or tube settlers
   c. static mixers
   d. chlorination facilities

4. Microstrainers can improve sand filter operation by removing
   a. 50–90 percent of the filter-clogging material.
   b. viruses and bacteria.
   c. dissolved iron and manganese.
   d. *Giardia* and *Cryptosporidium*.

5. Source water treatment with __________is called preoxidation.
   a. alum
   b. lime
   c. ozone
   d. electrolysis

Coagulation and Flocculation

1. Which of the following is a disadvantage of using static mixers?
   a. They do not provide good mixing.
   b. They are not economical.
   c. They increase head loss.
   d. They require too much maintenance.

2. Ion salt coagulants are effective __________ than alum?
   a. over a wider pH range
   b. and less corrosive
   c. but add iron to the water
   d. and are less expensive

3. ______ is not a classification of polyelectrolytes used as polymer coagulation aids.
   a. Cationic
   b. Anionic
   c. Activated silica
   d. Nonionic

4. Which of the following is a type of positive displacement pump used to feed liquid chemicals?
   a. Peristaltic pump
   b. Velocity pump
   c. Centrifugal pump
   d. Turbine pump

5. If the activation process of silica is not carefully controlled,
   a. the silica could splash due to high heat of reactants.
   b. it could inhibit floc formation.
   c. it could corrode and destroy the metal and rubber in the flocculators.
   d. it could deposit silica on the flocculators and the gears, bringing it eventually to a grinding halt.

6. If the flocculation time is too long,
   a. floc is reduced and flows over the weir.
   b. fragile flocs can break up.
   c. floc mixing speed should be increased.
   d. the flow should be decreased.

Sedimentation and Clarifiers

1. At which minimum angle must self-cleaning tube settlers be placed?
   a. 50°
   b. 65°
   c. 70°
   d. 75°

2. At which angle should the parallel inclined plates be installed when using the shallow-depth sedimentation method?
   a. 35°
   b. 45°
   c. 50°
   d. 60°

3. Why do solids-contact basins have much shorter detention times than conventional treatment basins?
   a. Because chemical reactions take place throughout the basin.
   b. Because the settling zone water moves upward, while at the same time the mixing zone moves upward.
   c. Because the gentle upward flow of the water throughout the basin is conducive for producing larger settleable floc.
   d. Because of the recycled materials from the sludge blanket, the chemical reactions occur more quickly and completely in the mixing area.

4. Pulsator clarifiers are used to treat water that is
   a. low in temperature, usually <10°C (<50°F).
   b. high in color and low in turbidity.
   c. low in color and high in turbidity.
   d. high in organic acids.

5. Typically, the time between backwashing of Trident contact adsorption clarifiers averages
   a. 4–8 hours.
   b. 16–24 hours.
   c. 1–2 days.
   d. 3–5 days.

6. Which high-rate process uses microsand?
   a. Tube settlers
   b. Superpulsators
   c. Actiflo process
   d. Dissolved air flotation

7. In dissolved-air flotation units, mechanical rubber scrapers travel over the tank surface and push the float over a ramp called a
   a. trap.
   b. beach.
   c. launder.
   d. loading plate.

Filtration Fundamentals

1. Filtration by granular media depends in part on ____, where suspended particles contact and stick onto the surface of the filter media grains.
   a. adsorption
   b. suspension
   c. flotation
   d. dissolution

2. Typically the amount of backwash water used for dual-media filters should not exceed _____ of the amount of water treated.
   a. 4%
   b. 6%
   c. 8%
   d. 10%

3. Direct filtration is used to treat raw water that has average turbidities
   a. below 10 ntu.
   b. up to 25 ntu.
   c. of 40–50 ntu.
   d. above 50 ntu.

4. Diatomaceous earth filters
   a. have a relatively high installation cost.
   b. have relatively high operating costs.
   c. are used only for water with low turbidity.
   d. produce very little backwash sludge.

5. An advantage of a variable declining-rate filtration system is that
   a. it does not require a loss-of-head indicator.
   b. it does not require a rate-of-flow controller.
   c. individual filters keep a constant filtration rate.
   d. flow rate changes do not impact the system.

Chlorine Disinfection

1. Chlorine is advantageous over chloramines in that chlorine
   a. is a much stronger oxidant.
   b. has a long history of use.
   c. is simple to feed.
   d. has a persistent residual.

2. A single chlorine cylinder is delivering 48 lb/d of chlorine to the water process, causing the cylinder to form a little frost. Which of the following would be the best solution to this problem?
   a. Install a fan to improve air circulation.
   b. Heat the cylinder immediately below the valve with heat tape.
   c. Heat the valve only.
   d. Add another cylinder and feed from both.

3. Which emergency kit is for chlorine tank cars?
   a. Kit A
   b. Kit B
   c. Kit C
   d. Kit D

4. Which chlorine compound will produce the least detectable chloronous taste and odor?
   a. Hypochlorous acid
   b. Monochloramine
   c. Dichloramine
   d. Nitrogen trichloride

5. A disadvantage to using chloramines is
   a. their vulnerability to nitrification.
   b. that they cannot penetrate biofilms in the distribution system.
   c. their propensity to form trihalomethanes.
   d. that their residual persistence in the distribution system is low.

6. A sodium hypochlorite solution at room temperature will lose which percentage of its available chlorine content per month?
   a. 1–2%
   b. 2–4%
   c. 4–6%
   d. 6–8%

7. Where is scale most likely to form in a hypochlorinator?
   a. In the push rod
   b. In the plunger
   c. In the suction and discharge hoses
   d. In the spring

Iron and Manganese Treatment

1. Virgin greensand
   a. does not require regeneration.
   b. requires regeneration with potassium permanganate (1-hour soak with 60 grams $KMnO_4/ft^3$).
   c. requires regeneration with manganese dioxide (2-hour soak with 25% by weight solution of MnO_2).
   d. requires regeneration with manganese hydroxide (4-hour soak with 200 grams $Mn(OH)_2/ft^3$).

2. Bacterial slimes formed due to iron and manganese in the distribution system
   a. can increase pipeline flow capacity.
   b. increase chlorine demand.
   c. lubricate valves so the work is easier for a short period.
   d. decrease discoloration of household fixtures.

3. Granular media filters can be used to remove iron and manganese precipitates without sedimentation if solids are under
   a. 5 mg/L.
   b. 10 mg/L.
   c. 15 mg/L.
   d. 20 mg/L.

4. Another benefit of using potassium permanganate to oxidize both iron and manganese is
   a. it is a strong disinfectant.
   b. pink water indicates that more is needed.
   c. odor-causing compounds can be enhanced.
   d. no trihalomethanes are formed.

5. Manganese greensand filter performance can be improved by adding a filter coal layer on top of the greensand to
   a. increase the filter loading rate.
   b. decrease the filter run time.
   c. reduce solids loading on the greensand bed.
   d. increase the operating head.

Fluoridation

1. At what thickness should the layer of sodium fluoride crystals be maintained in a saturator tank for flows of less than 100 gpm?
   a. 6 in. (15 cm)
   b. 10 in. (25 cm)
   c. 1 ft (30 cm)
   d. 2 ft (60 cm)

2. Which type of fluoride chemical should be used in downflow saturators?
   a. Tablets of sodium fluorosilicate
   b. Granular sodium silicofluoride
   c. Crystalline sodium fluoride
   d. Powdered sodium fluoride

3. Volumetric feeders are
   a. usually more difficult to operate than gravimetric feeders.
   b. more expensive to purchase than gravimetric feeders.
   c. less expensive to maintain than gravimetric feeders.
   d. more accurate than gravimetric feeders.

4. What safety equipment should be worn when handling fluoride chemicals?
   a. No special equipment is required.
   b. A respirator approved by NIOSH
   c. Leather or cotton work gloves
   d. An apron

5. Water used to prepare sodium fluoride solutions should be softened if the hardness exceeds ________as calcium carbonate.
   a. 50 mg/L
   b. 75 mg/L
   c. 100 mg/L
   d. 125 mg/L

Ion Exchange

1. Magnetic ion exchange resin has been developed to remove
   a. total organic carbon.
   b. chlorides.
   c. iron and magnesium.
   d. sulfates and sulfides.

2. Natural zeolites used for softening that have become exhausted with use are regenerated by immersing them in a strong solution of which chemical?
   a. NaCl
   b. NaOH
   c. HCl
   d. H_2SO_4

3. Express a hardness of 100 mg/L as $CaCO_3$ in grains per gallon.
   a. 5.85
   b. 1,710
   c. 140.3
   d. 10.0

4. Regeneration of the ion exchange media in the sodium cycle is called
   a. rinsing.
   b. backwashing.
   c. recalcining.
   d. brining.

5. One common method of removing high levels of fluoride uses
   a. aluminum chlorohydrate.
   b. alum.
   c. activated alumina.
   d. activated carbon.

Electrical Systems

1. Water supply regulations require most treatment plants to record data on
   a. customer relations.
   b. electrical usage.
   c. employee hours.
   d. volume of water treated.

2. An example of a source of direct current is
   a. car battery.
   b. household electrical outlet.
   c. gasoline powered generator.
   d. light fixture.

3. The simplest form of automation is
   a. residual loop control.
   b. start–stop control.
   c. proportional control.
   d. flow combined with proportional control.

4. When both the flow rate and the chlorine demand vary, what is the best type of control system?
   a. Proportional control
   b. Residual control
   c. Floating proportional control
   d. Compound PID loop control

5. Continuous monitoring is required for ______ to meet EPA regulations.
   a. toxic chemicals
   b. turbidity
   c. bacteria
   d. total organic carbon

Pumps

1. A pump loses its prime because the suction line has an air pocket. What is the best solution?
   a. Check the pump's amperage and be sure its strainer is clean.
   b. Clean or repair the priming unit.
   c. Open the suction piping air bleed-off valves.
   d. Check the external water seal unit.

2. When the discharge of a velocity pump is blocked, head builds up, typically greater than the pressure generated during pumping. When this occurs, water recirculates within the pump impeller and casing. This flow condition is called
   a. run-around.
   b. skirting.
   c. slip.
   d. bypass.

3. Vertical turbine pumps
   a. are inexpensive to repair.
   b. cost less than same-sized centrifugal pumps.
   c. are easy to lubricate.
   d. cannot pump water containing sand.

4. The flow restriction between the impeller discharge and suction necessary for all centrifugal pumps is made by
   a. a coupling.
   b. wear rings.
   c. a shaft sleeve.
   d. a packing ring.

5. Centrifugal pump shaft alignment is important because
   a. vibration from misalignment can cause bearing failure.
   b. pump efficiency can be reduced.
   c. the packing can become pinched, creating high running temperatures.
   d. pump stalling can occur.

Treatment Plant Safety and Security Practices

1. SCBA used for emergency purposes require inspection by OSHA regulations
   a. once a year.
   b. quarterly.
   c. monthly.
   d. before each use.

2. Chlorine cylinders should be stored in a temperature-controlled area because
   a. chlorine cannot be withdrawn when the temperature is too cold.
   b. chlorine cylinder valves require a constant temperature.
   c. fusible plugs can melt at high temperature, releasing chlorine.
   d. operators need comfortable working conditions.

3. Treatment plant security requires _____ to prevent intruders from entering and causing harm.
   a. video cameras
   b. physical barriers
   c. motion detectors
   d. police presence

4. Each year approximately what percentage of confined space deaths are would-be rescuers?
   a. 25%
   b. 50%
   c. 67%
   d. 75%

5. Effectively dealing with treatment disruptions due to employee strikes, natural disasters, and on-site emergencies that require quick support
   a. requires careful contingency planning and training.
   b. is the state's responsibility.
   c. is the responsibility of the Office of Homeland Security.
   d. is the responsibility of the Health Department.

Study Question Answers

Chapter 1 Answers

1. **a.** Ozone
2. **c.** indicates that the water is scale forming.
3. **d.** sodium bicarbonate remains in the product water.
4. **b.** the pH level is higher than 10.5.
5. **d.** $MgSO_4$.
6. **b.** Recarbonation
7. **a.** Corrosion
8. Calcium bicarbonate
9. Calcium and magnesium
10. Soda ash

Chapter 2 Answers

1. Effective volume of chamber = 44 ft × 30 ft × 8.3 ft × 7.48 × 0.34 = 27,863.3 gallons

   Effective Contact Time (ECT) = 27,863.3 gal/300 gpm = 92.9 min

   $C \times T$ value = ECT × residual = 92.9 min × 1.6 mg/L = 148.6 mg-min/L

2. **b.** 37 lb/day, F

   First convert gpm to mgd:

   mgd = [(1,750 gpm)(1,440 min/day)] ÷ (1,000,000) = 2.52 mgd

   Next, determine the fluoride (F) required:

   F required = F desired – F raw water: F required = 1.20 mg/L – 0.15 mg/L = 1.05 mg/L F

   $$\text{Fluoride feed rate} = \frac{(\text{Dosage, mg/L})(\text{mgd})(8.34\ \text{lb/gal})}{(\text{Percent F ion})(\text{Chemical purity})}$$

   $$= \frac{(1.05\ \text{mg/L F})(2.52\ \text{mgd})(8.34\ \text{lb/gal})}{(60.6\%/100\%)(98.1\%/100\%)}$$

   = 37.12 lb/day, round to **37 lb/day, F**

3. **d.** 163 mL/min, H_2SiF_6

Solution:

$$\text{F required} = \text{F desired} - \text{F in raw water}$$
$$= 1.20 \text{ mg/L} - 0.20 \text{ mg/L} = 1.00 \text{ mg/L}$$

$$\frac{(\%\ \text{Solution})(10{,}000\ \text{mg/L})}{1\%} = \#\ \text{of mg/L}$$

$$(20.5\%\ \text{solution of } H_2SiF_6)(10{,}000\ \text{mg/L per } 1\%) = 205{,}000\ \text{mg/L}$$

Equation:

$$(\text{F, mg/L})(\text{mgd})(8.34\ \text{lb/gal}) = (\text{Percent, mg/L})(\text{mgd})(\text{Solution, lb/gal})\ (\%\ \text{F ion}/100\%)$$

$$(1.0\ \text{mg/L F})(11.8\ \text{mgd})(8.34\ \text{lb/gal}) = (205{,}000\ \text{mg/L})(x\ \text{mgd Flow})(9.8\ \text{lb/gal})(79.0\%/100\%\ \text{F})$$

$$\text{Solve for } x\text{: } x\ \text{mgd Flow} = \frac{(0.20\ \text{mg/L F})(11.8\ \text{mgd})(8.34\ \text{lb/gal})}{(205{,}000\ \text{mg/L})(9.8\ \text{lb/gal})(79.0\%/100\%\ \text{F})}$$

$$= 0.000062\ \text{mgd}$$

Convert mgd to gallons:

$$(0.000062\ \text{mgd})(1{,}000{,}000) = 62\ \text{gal/day}\ H_2SiF_6$$

Then convert gal/day to mL /min.

$$H_2SiF_6,\ \text{mL/min} = (62\ \text{gal/day})(3{,}785\ \text{mL/gal})(1\ \text{day}/1{,}440\ \text{min})$$
$$= 162.98\ \text{mL/min, round to } 163\ \text{mL/min, } H_2SiF_6$$

4. **a.** gate valve.
5. **a.** 1–2 in.
6. **a.** Ball and socket joint
7. **b.** Push-on joint
8. **d.** 150+.
9. **d.** Pinch valve
10. **c.** cold climates.
11. **b.** Polyethylene pipe
12. **b.** 39.6 hr

Equation: Operating time, hr = Treated water, gal ÷ Flow rate, gph

First, convert flow rate from gpm to gph:

$$(125\ \text{gpm})(60\ \text{min/hr}) = 7{,}500\ \text{gph}$$

$$\text{Operating time, hr} = 297{,}000\ \text{gal} \div 7{,}500\ \text{gph} = 39.6\ \text{hr}$$

13. **b.** 5.26 psi

Equation:

$$\text{psi} = (\text{Depth, ft}) \div 2.31\ \text{ft/psi}$$

$$\text{psi} = 8.95\ \text{ft} \div 2.31\ \text{ft/psi} = 3.874\ \text{psi}$$

Next, find the specific gravity (SG) of the alum:

$$\text{Alum, SG} = 11.32 \text{ lb/gal} \div 8.34 \text{ lb/gal} = 1.357 \text{ SG of Alum}$$

Then, multiply the psi by the alum's SG to determine the psi:

$$\text{psi} = (3.874 \text{ psi})(1.357) = 5.257 \text{ psi, round to } 5.26 \text{ psi}$$

14. **d.** 2,140 gpm

First, find the water production during the 12-hr interval:

$$\text{Gallons of water treated in 12-hr interval} = (2.95 \text{ mgd})(1{,}000{,}000)\ (12 \text{ hr})/24 \text{ hr}$$

$$\text{Gallons of water treated in 12-hr interval} = 1{,}475{,}000 \text{ gal}$$

Next, find the gallons contained in the 4.25-ft drop in water level:

$$\text{Equation: Volume, tank} = (0.785)(\text{Diameter, ft})^2(\text{Height, ft})$$

$$\text{Volume of 4.25 ft in 50.0-ft diameter tank} = (0.785)(50.0 \text{ ft})(50.0 \text{ ft})(4.25 \text{ ft})(7.48 \text{ gal/ft}^3) = 62{,}388 \text{ gal}$$

Production plus the loss in level is the amount the discharge pump had to send into the distribution system, but first find the number of minutes in 12 hr:

$$(12 \text{ hr})(60 \text{ min/hr}) = 720 \text{ min}$$

Then determine total gallons discharge pumps moved:

$$\text{Total gal discharge pumps moved in 12 hr} = 1{,}475{,}000 \text{ gal} + 62{,}388 \text{ gal} = 1{,}537{,}388 \text{ gal}$$

Lastly, divide the number of gallons the discharge pumps moved by the time in minutes.

$$\text{Discharge pumps, gpm} = 1{,}537{,}388 \text{ gal} \div 720 \text{ min} = 2{,}135 \text{ gpm, round to } 2{,}140 \text{ gpm}$$

15. **c.** *P–Q* curve
16. Velocity head, pressure head, elevation head
17. The amount of energy used up by water when moving from one point to another
18. In a system where the reservoir feeding the pump is lower than the pump, the difference in elevation between the center line and the free water surface of the reservoir feeding the pump
19. The amount of work done in a given time

Chapter 3 Answers

1. **d.** No more than 5%
2. **d.** 4 log
3. **c.** Entry points to the distribution system
4. **c.** Each entry point to the system
5. **a.** population served is >3,300 people.
6. **d.** 100,000 people.
7. **b.** <2.0 mg/L
8. **d.** Chloramine disinfection

9. **b.** is a Tier 2 violation requiring public notification within 30 days.
10. **b.** meeting disinfection limits.
11. **d.** the treated water quarterly running average TOC is less than 2.0 mg/L.
12. **b.** 0.08 mg/L
13. **a.** 15; 25
14. Alkalinity levels
15. All sizes
16. Total Coliform Rule

Chapter 4 Answers

1. **b.** Grab sample
2. **a.** $Na_2S_2O_3$
3. **d.** concentrated nitric acid.
4. **b.** Pink to dark red with green metallic surface sheen
5. **c.** 4,000 mg/L
6. **d.** Atomic absorption spectrophotometer
7. **b.** Green lactose bile broth
8. **a.** brilliant green bile tubes
9. **c.** pH saturation
10. **b.** pH
11. Presumptive test, confirmed test, completed test
12. Carry an electrical current
13. A colorless, odorless, noncombustible gas that is found in all natural waters
14. Jar test

Chapter 5 Answers

1. **a.** pH
2. **c.** Lime
3. **a.** Polyphosphate
4. **b.** Hexametaphosphate
5. **a.** it can adjust both alkalinity and pH.
6. **c.** tuberculation
7. **c.** quicklime.
8. Zinc
9. Corrosion of materials in household plumbing systems
10. It is relatively easy to feed, it is inexpensive compared with other chemicals, and it is readily available.

Chapter 6 Answers

1. **b.** Carbon dioxide
2. **d.** little or no noncarbonate hardness.
3. **d.** only on the amount of noncarbonate hardness to be removed.

4. **d.** cementation of filter media.
5. **c.** Noncarbonate hardness
6. **b.** 5–15%.
7. **d.** on the filters.
8. **c.** 25 mg/L
9. **d.** Remove noncarbonated hardness
10. **c.** slow, gentle
11. **a.** immediately.
12. Bicarbonate alkalinity, hydroxide alkalinity, carbonate alkalinity
13. Magnesium hydroxide
14. 40 mg/L or less

Chapter 7 Answers

1. **a.** NaCl
2. **a.** Sodium
3. **b.** iron
4. **b.** 3–5 kilograins/ft^3
5. **a.** regeneration.
6. **c.** 17.1 mg/L.
7. Zero
8. Desalination
9. The length of a softening cycle

Chapter 8 Answers

1. **b.** At the time of installation and every 6 months thereafter
2. **d.** 50%
3. **b.** After filtration
4. **a.** 2 inches.
5. **b.** Oxygen
6. **c.** PAC and chlorine are added too close to each other
7. **b.** dust can be a major problem.
8. **d.** mudballs
9. Every 3 months
10. Every day
11. Handling it

Chapter 9 Answers

1. **d.** Air binding
2. **d.** Iron
3. **a.** drops
4. **c.** air binding.
5. **b.** 6

6. Dissolved oxygen, pH, temperature
7. Scrubbing at a pH of 6 or less
8. Daily

Chapter 10 Answers

1. **c.** higher pressure.
2. **a.** stainless steel
3. **c.** using as drinking water.
4. **a.** feed pressure.
5. **c.** fouling.
6. **a.** crossflow.
7. Feedwater
8. Reverse osmosis
9. The process by which membrane performance gradually changes as a result of compaction

Chapter 11 Answers

1. **a.** air stripping towers.
2. **c.** as pretreatment.
3. **b.** create ozone by exposing oxygen to an electric discharge.
4. **a.** Chlorine dioxide
5. c. is insensitive to temperature and pH.
6. **b.** Add chlorine to water, then this water to hydrochloric acid and sodium chlorate.
7. **b.** it must be generated on-site as it is needed.
8. **c.** extremely dry.
9. Any three of the following: conventional fine-bubble diffusion, turbine mixer, injector, packed columns, spray chamber
10. Occupational Safety and Health Administration
11. 0.5–2.5 mg/L

Chapter 12 Answers

1. **d.** The Safe Drinking Water Act
2. **a.** 0.1–2.0%
3. **d.** Sludge collection and removal
4. **d.** 6,667 gallons
5. **c.** can be very gelatinous and difficult to handle.
6. **d.** Hydrologic siphon
7. **a.** the less expensive its disposal.
8. To a stream or lake
9. Alum
10. Dewatering

Chapter 13 Answers

1. **a.** riparian doctrine.
2. **b.** absolute ownership.
3. **c.** reasonable use.
4. **d.** Appropriation–permit systems
5. **d.** Control nutrients
6. **a.** Manage watershed to prevent nitrates from entering the water system.
7. **d.** Draw water from an adjoining system or systems.
8. **b.** address the customer's concerns
9. **a.** toilet.
10. **b.** 6 p.m. to 10 p.m.
11. Natural and human-caused
12. Absolute ownership
13. Possible delay in developing additional source capacity, loss of revenue for the utility, and difficulty in dealing with drought conditions

Chapter 14 Answers

1. **c.** Expected length of filter life
2. **b.** Quarterly
3. **a.** The Partnership for Safe Water
4. *NSF/ANSI Standard 61*
5. AWWA does not monitor systems' adherence to its standards.

Chapter 15 Answers

Water Sources and Treatment Options

1. **a.** low resulting in very soft water
2. **d.** is a permeable layer between layers of low permeability.
3. **c.** legislative initiative.
4. **b.** an example of the multiple-barrier approach to treatment.
5. **a.** can infiltrate into the ground and recharge aquifers.

Groundwater Quality and Wells

1. **c.** At least 72 hours
2. **b.** Low amounts of arsenic and radium 228
3. **c.** the water surface level when no water is being taken from the aquifer.
4. **d.** 5
5. **a.** the cable tool method.

Surface Water Source Treatment

1. **b.** in an impoundment.
2. **c.** riverbed infiltration galleries.
3. **d.** alkalinity is below 50 mg/L and pH is 8–9.

4. **a.** occurs when a warm layer of water overlies a colder layer.
5. **b.** chlorine at the plant intake.

Treatment Plant Pretreatment

1. **d.** trash racks.
2. **a.** conditions ideal for accelerated corrosion.
3. **b.** plate or tube settlers
4. **a.** 50–90 percent of the filter-clogging material.
5. **c.** ozone

Coagulation and Flocculation

1. **c.** They increase head loss.
2. **a.** over a wider pH range
3. **c.** Activated silica
4. **a.** Peristaltic pump
5. **b.** it could inhibit floc formation.
6. **b.** fragile flocs can break up.

Sedimentation and Clarifiers

1. **a.** 50°
2. **b.** 45°
3. **d.** Because of the recycled materials from the sludge blanket, the chemical reactions occur more quickly and completely in the mixing area.
4. **b.** high in color and low in turbidity.
5. **a.** 4–8 hours.
6. **c.** Actiflo process
7. **b.** beach.

Filtration Fundamentals

1. **a.** adsorption
2. **b.** 6%
3. **b.** up to 25 ntu.
4. **c.** are used only for water with low turbidity.
5. **b.** it does not require a rate-of-flow controller.

Chlorine Disinfection

1. **a.** is a much stronger oxidant.
2. **d.** Add another cylinder and feed from both.
3. **c.** Kit C
4. **b.** Monochloramine
5. **a.** their vulnerability to nitrification.
6. **b.** 2–4%
7. **c.** In the suction and discharge hoses

Iron and Manganese Treatment

1. **b.** requires regeneration with potassium permanganate (1 hour soak with 60 grams $KMnO_4/ft^3$).
2. **b.** increase chlorine demand.
3. **a.** 5 mg/L.
4. **d.** no trihalomethanes are formed.
5. **c.** reduce solids loading on the greensand bed.

Fluoridation

1. **a.** 6 in. (15 cm)
2. **c.** Crystalline sodium fluoride
3. **c.** less expensive to maintain than gravimetric feeders.
4. **b.** A respirator approved by NIOSH
5. **b.** 75 mg/L

Ion Exchange

1. **a.** total organic carbon.
2. **a.** NaCl
3. **a.** 5.85
4. **d.** brining.
5. **c.** activated alumina.

Electrical Systems

1. **d.** volume of water treated.
2. **a.** a car battery.
3. **b.** start–stop control.
4. **d.** Compound PID loop control
5. **b.** turbidity

Pumps

1. **c.** Open the suction piping air bleed-off valves.
2. **c.** slip.
3. **d.** cannot pump water containing sand.
4. **b.** wear rings.
5. **a.** vibration from misalignment can cause bearing failure.

Treatment Plant Safety and Security Practices

1. **c.** monthly.
2. **c.** fusible plugs can melt at high temperatures, releasing chlorine.
3. **b.** physical barriers
4. **c.** 67%
5. **a.** requires careful contingency planning and training.

References

Emergency Planning for Water Utilities. AWWA Manual M19. Denver, CO: American Water Works Association (AWWA).

Rice, E. W., Baird, R. B., and L. S. Eaton, eds. 2012. *Standard Methods for the Examination of Water and Wastewater*. American Public Health Association, AWWA, and Water Environment Federation.

Simplified Procedures for Water Examination. AWWA Manual M12. Denver, CO: AWWA.

Glossary

absence See *negative sample.*

absolute ownership A water rights term referring to water that is completely owned by one person.

acidity Base *neutralizing power.*

aeration The process of bringing water and air into close contact to remove or modify constituents in the water.

aggressive See *corrosive.*

air binding A condition that occurs in filters when air comes out of solution as a result of pressure decreases and temperature increases. The air clogs the voids between the media grains, which causes increased head loss through the filter and shorter filter runs.

alkalinity Acid neutralizing power.

alum The most common chemical used for coagulation. It is also called aluminum sulfate.

apparent color A characteristic of water that includes true color and color caused by suspended matter.

appropriation–permit system A water use system in which permits to use water are regulated so that overdraft cannot occur.

atomic absorption spectrophotometer (AA) A spectrophotometer used to determine the concentration of metals in water and other types of samples.

bed life The time it takes for a bed of adsorbent to lose its adsorptive capacity. When this occurs, the bed must be replaced with fresh adsorbent.

beneficial use A water rights term indicating that the water is being used for good purposes.

best available technology (BAT) The best technology, treatment techniques, or other means that are available for treatment of a water quality problem and that have been found to be practical under field conditions.

brake horsepower The power supplied to a pump by a motor. Compare with *water horsepower* and *motor horsepower.*

breakpoint The point at which the chlorine dosage has satisfied the chlorine demand.

breakpoint chlorination The addition of chlorine to water until the chlorine demand has been satisfied and free chlorine residual is available for disinfection.

breakthrough The point in a treatment cycle at which contaminants, such as organics or turbidity, begin to pass through the treatment process.

$C \times T$ value The product of the residual disinfectant concentration C in milligrams per liter, and the corresponding disinfectant contact time T in minutes, or $C \times T$. Minimum $C \times T$ values are specified by the Surface Water Treatment Rule as a means of ensuring adequate kill or inactivation of pathogenic microorganisms in water.

calcium bicarbonate A compound that iscontrolled through water treatment to prevent the formation of calcium carbonate (hardness).

calcium carbonate The principal hardness- and scale-causing compound in water.

carbonate hardness Hardness caused primarily by bicarbonate. Compare with *noncarbonate hardness.*

chain of custody A written record of the sample's history of SDWA compliance, from the time of collection to the time of analysis and subsequent disposal.

chelation A chemical process used to control scale formation, in which a chelating agent "captures" scale-causing ions and holds them in solution, thus preventing them from precipitating out and forming scale.

chlorine demand The quantity of chlorine consumed by reaction with substances in water.

coagulation–flocculation The water treatment process that converts small particles of suspended solids into larger, more settleable clumps.

coliform bacteria A group of bacteria predominantly inhabiting the intestines of humans or animals but also occasionally found elsewhere. Presence of the bacteria in water is used as an indication of fecal contamination (contamination by human or animal wastes).

color A physical characteristic of water. Color is most commonly tan or brown due to oxidized iron, but contaminants may cause other colors, such as green or blue. Different from turbidity, which is the cloudiness of water. See *true color* and *apparent color* for further explanations.

combined chlorine residual The chlorine residual produced by the reaction of chlorine with substances in the water. Because the chlorine is "combined" it is not as effective a disinfectant as free chlorine residual. In water treatment, this usually refers to compounds formed by the combination of chlorine and ammonia.

community public water system As defined by the National Primary Drinking Water Regulations, a system that serves at least 15 service connections or at least 25 full-time residents 60 or more days per year.

completed test The third major step of the multiple-tube fermentation method. This test confirms that positive results from the presumptive test are due to coliform bacteria. See also *confirmed test*; *presumptive test.*

composite sample A series of individual or grab samples taken at different times from the same sampling point and mixed together.

concentration cell corrosion A form of localized corrosion that can form deep pits and tubercles.

confirmed test The second major step of the multiple-tube fermentation method. This test confirms that positive results from the presumptive test are due to coliform bacteria. See also *completed test*; *presumptive test.*

contamination Any introduction into water of microorganisms, chemicals, wastes, or wastewater in a concentration that makes the water unfit for its intended use.

correlative rights A rule that contends that the *overlying use* rule is not absolute but is related to the rights of other overlying users. This rule is used when there is not enough water to satisfy all overlying uses.

corrosion The gradual deterioration or destruction of a substance or material by chemical action. The action proceeds inward from the surface.

corrosive Tending to deteriorate material, such as pipe, through electrochemical processes.

coupon test A method of determining the rate of corrosion or scale formation by placing metal strips (coupons) of a known weight in the pipe and examining them for corrosion after a period of time.

design point The mark on the *H–Q* (head–capacity) curve of a pump characteristics curve that indicates the head and capacity at which the pump is intended to operate for best efficiency in a particular installation.

dewatering (of sludge) A process to remove a portion of water from sludge.

disinfectant residual An excess of chlorine left in water after treatment. The presence of residuals indicates that an adequate amount of chlorine has been added at the treatment stage to ensure completion of all reactions with some chlorine remaining.

disinfectants/disinfection by-product (DBP) A term used in connection with state and federal regulations designed to protect public health by limiting the concentration of either disinfectants or the by-products formed by the reaction of disinfectants with other substances in the water (such as the natural decomposition products of organic matter, leaves, algae, bacteria, etc.).

dissolved oxygen (DO) The oxygen dissolved in water, wastewater, or other liquid, usually expressed in milligrams per liter, parts per million, or percent of saturation.

dynamic head See *total dynamic head*.

dynamic suction head The distance from the pump center line at the suction of the pump to the point of the hydraulic grade line directly above it. Dynamic suction head exists only when the pump is below the piezometric surface of the water at the pump suction. When the pump is above the piezometric surface, the equivalent measurement is dynamic suction lift.

dynamic suction lift The distance from the pump center line at the suction of the pump to the point on the hydraulic grade line directly below it. Dynamic suction lift exists only when the pump is above the piezometric surface of the water at the pump suction. When the pump is below the piezometric surface, the equivalent measurement is called dynamic suction head.

efficiency The ratio of the total energy output to the total energy input, expressed as percent.

electrode method Any analytical procedure that uses an electrode connected to a millivoltmeter to measure the concentration of a constituent in water.

electrodialysis A water treatment method used for desalination rather than exclusively for softening.

elevation head The energy possessed per unit weight of a fluid because of its elevation above some reference point (called the *reference datum*). Elevation head is also called position head or potential head.

energy grade line (EGL) A line joining the elevations of the energy heads; a line drawn above the hydraulic grade line by a distance equivalent to the velocity head of the flowing water at each section along a stream, channel, or conduit. Sometimes called *energy-gradient line* or *energy line*.

***Escherichia coli* (*E. coli*)** Bacteria of the coliform group used as a substitute for fecal coliforms in the regulations of the Total Coliform Rule.

fecal coliform Bacteria of the coliform group indicative of fecal contamination. The presence of fecal coliform in a water sample is a reportable violation of the Total Coliform Rule.

feedwater Raw water that has undergone pretreatment (acidification or scale inhibitor addition) prior to entering the membrane arrays.

flash mixing See *rapid mixing*.

flocculation The water treatment process, following coagulation, that uses gentle stirring to bring suspended particles together so that they will form larger, more settleable clumps called floc.

friction head loss The head lost by water flowing in a stream or conduit as the result of (1) the disturbance set up by the contact between the moving water and its containing conduit and (2) intermolecular friction.

galvanic corrosion A form of localized corrosion caused by the connection of dissimilar metals in an electrolyte, such as water.

galvanic series A listing of metals and alloys according to their corrosion potential.

gas chromatography (GC) A technique used to measure the concentration of organic compounds in water.

grab sample A single water sample collected at one time from a single point.

granular activated carbon (GAC) Activated carbon in a granular form, which is used in a bed, much like a conventional filter, to adsorb organic substances from water.

groundwater under the direct influence of surface water (GWUDI) A term used in state and federal regulations to designate groundwater sources that are considered vulnerable to contamination from surface water. Systems using such sources must generally provide monitoring and treatment as if they were using a surface water source.

hardness A characteristic of water, caused primarily by the salts of calcium and magnesium. Causes deposition of scale in boilers, damage in some industrial processes, and sometimes objectionable taste. May also decrease the effectiveness of soap.

head (1) A measure of the energy possessed by water at a given location in the water system, expressed in feet. (2) A measure of the pressure or force exerted by water, expressed in feet.

head loss The amount of energy used by water in moving from one point to another.

heterotrophic plate count (HPC) A laboratory procedure for estimating the total bacterial count in a water sample. Also called *standard plate count*, *total plate count*, or *total bacterial count*.

horsepower A standard unit of power equal to 746 W (watts), approximately equal to 33,000 ft-lb/min.

hydraulic grade line (HGL) A line (hydraulic profile) indicating the piezometric level of water at all points along a conduit, open channel, or stream. In an open channel, the HGL is the free water surface.

instantaneous flow rate A flow rate of water measured at one particular instant, such as by a metering device, involving the cross-sectional area of the channel or pipe and the velocity of the water at that instant.

ion exchange process A process used to remove hardness from water that depends on special materials known as resins. The resins trade nonhardness-causing ions (usually sodium) for the hardness-causing ions, calcium and magnesium. The process removes practically all the hardness from water.

ion exchange water softener A treatment unit that removes calcium and magnesium from water using ion exchange resins.

iron bacteria Bacteria that use dissolved iron as an energy source. They can create serious problems in a water system because they form large, slimy masses that clog well screens, pumps, and other equipment.

jar test A laboratory procedure for evaluating coagulation, flocculation, and sedimentation processes. Used to estimate the proper coagulant dosage.

Langelier saturation index A measure of the tendency of water to scale or corrode pipes, based on pH, conductivity, hardness, alkalinity, and water temperature.

lime–soda ash method A process used to remove carbonate and noncarbonate hardness from water.

magnesium hardness The portion of total hardness caused by magnesium compounds, such as magnesium carbonate and magnesium sulfate.

maximum contaminant level (MCL) The maximum permissible level of a contaminant in water as specified in the regulations of the Safe Drinking Water Act.

maximum contaminant level goal (MCLG) Nonenforceable health-based goals published along with the promulgation of an MCL. Originally called *recommended maximum contaminant levels (RMCLs).*

membrane filter method A laboratory method used for coliform testing. The procedure uses an ultrathin filter with a uniform pore size smaller than bacteria—less than a micron. After water is forced through the filter, the filter is incubated in a special media that promotes the growth of coliform bacteria. Bacterial colonies with a green-gold sheen indicate the presence of coliform bacteria.

membrane filtration Water treatment processes in which relatively pure water passes through a porous membrane while particles, molecules, or ions of unwanted matter are excluded.

milk of lime The lime slurry formed when water is mixed with calcium hydroxide.

minor head loss The energy losses that result from the resistance to flow as water passes through valves, fittings, inlets, and outlets of a piping system.

MMO–MUG technique An approved bacteriological procedure for detecting the presence or absence of total coliforms.

modified Winkler method A modification of the standard Winkler (iodometric) method that uses an alkali-iodide-azide reagent to make the procedure less subject to interferences.

motor horsepower The horsepower equivalent to the watts of electric power supplied to a motor. Compare with *brake horsepower* and *water horsepower*.

mudball An accumulation of media grains and suspended material that creates clogging problems in filters.

multiple-tube fermentation (MTF) method A laboratory method used for coliform testing that uses a nutrient broth placed in culture tubes. Gas production indicates the presence of coliform bacteria.

National Primary Drinking Water Regulations (NPDWRs) Regulations developed under the Safe Drinking Water Act. The regulations establish maximum contaminant levels, monitoring requirements, and reporting procedures for contaminants in drinking water that endanger human health.

negative sample When referring to the multiple-tube fermentation or membrane filter test, any sample that does not contain coliform bacteria. Also called *absence*.

nephelometer An instrument that determines turbidity by measuring the amount of light scattered by turbidity in a water sample. It is the only instrument approved by the US Environmental Protection Agency to measure turbidity in treated drinking water.

nephelometric turbidimeter See *nephelometer*.

nomographic method A method that uses a graph or other diagram to solve formulas and equations.

noncarbonate hardness Hardness caused by the salts of calcium and magnesium.

nontransient, noncommunity public water system A system having its own water supply and serving an average of at least 25 persons who do not live at the location but who use the water for more than six months per year.

overlying use The land use that occurs on top of an aquifer.

ozone contactor A tank used to transfer ozone to water. A common type applies ozone under pressure through a porous stone at the bottom of the tank.

ozone generator A device that produces ozone by passing an electrical current through air or oxygen.

percent bypass Thepredetermined amount of source water that is routed past water softeners then blended water to achieve the desired level of finished water hardness.

pH A measure of water's acidity or alkalinity. A scale of 0 to 14 is used, with 0 being extremely acidic and 14 being extremely alkaline.

platinum–cobalt method A procedure used to determine the amount of color in water.

positive sample In reference to the multiple-tube fermentation or membrane filter test, any sample that contained coliform bacteria. Also called *presence*.

pounds per square inch gauge (psig) Pressure measured by a gauge and expressed in terms of pounds per square inch. See *gauge pressure*. Compare with *pounds per square inch absolute*.

powdered activated carbon (PAC) Activated carbon in a fine powder form. It is added to water in a slurry form primarily for removing those organic compounds causing tastes and odors.

power (in hydraulics or electricity) The measure of the amount of work done in a given period of time. The rate of doing work. Measured in watts or horsepower.

presence See *positive sample*.

presence–absence (P–A) test An approved bacteriological procedure for the detection of total coliforms. The results are qualitative rather than quantitative.

pressure head A measurement of the amount of energy in water due to water pressure.

presumptive test The first major step in the multiple-tube fermentation test. The step presumes (indicates) the presence of coliform bacteria on the basis of gas production in nutrient broth after incubation.

primary enforcement responsibility or primacy The acceptance by federal government, states or other government entity, regional body, or tribal unit of the responsibility for enforcing the Safe Drinking Water Act requirements.

prior appropriation doctrine A water rights doctrine in which the first user has the right to water before subsequent users.

priority use/priority in time The assigning of water rights based on who has been using the water the longest.

pump center line An imaginary line through the center of a pump.

pump characteristic curve A curve or curves showing the interrelation of speed, dynamic head, capacity, brake horsepower, and efficiency of a pump.

quicklime Another name for calcium oxide (CaO), which is used in water softening and stabilization.

rapid mixing The process of quickly mixing a chemical solution uniformly through the water.

reasonable use A water rights term indicating that the water use is acceptable in general terms.

recarbonation The reintroduction of carbon dioxide into the water, either during or after lime–soda ash softening.

regeneration The process of reversing the ion exchange softening reaction of ion exchange materials. Hardness ions are removed from the used materials and replaced with nontroublesome ions, thus restoring the exchange capacity of the resin for further softening.

reject water The water that does not pass through a membrane, carries away the rejected matter, and must be disposed of.

residual See *disinfectant residual.*

reverse osmosis (RO) A pressure-driven process in which almost-pure water is passed through a semipermeable membrane. Water is forced through the membrane and most ions (salts) are left behind. The process is principally used for desalination of sea water.

riparian doctrine A water right that allows the owners of land abutting a stream or other natural body of water to use that water.

riverbank filtration Extraction of water through the bank of a river or through the bottom using wells or buried collection piping.

rust Oxidized iron.

Safe Drinking Water Act A federal law enacted December 16, 1974, setting up a cooperative program among local, state, and federal agencies to ensure safe drinking water for consumers.

saturation A stable condition of water in which the water will neither deposit scale nor cause corrosion.

saturation point The point at which a solution can no longer dissolve any more of a particular chemical. Precipitation of the chemical will occur beyond this point.

scaling Metal deposits left in pipelines and plumbing fixtures.

sedimentation The water treatment process that involves reducing the velocity of water in basins so that the suspended material can settle out by gravity.

sedimentation basin A basin or tank in which water is retained to allow settleable matter, such as floc, to settle by gravity. Also called a *settling basin*, *settling tank*, or *sedimentation tank*.

sequestration A chemical process used to control scale formation, in which a sequestering agent holds scale-causing ions in solution, preventing them from precipitating out and forming scale.

slake The addition of water to quicklime (calcium oxide) to form calcium hydroxide, which can then be used in the softening or stabilization processes.

slaker The part of a quicklime feeder that mixes the quicklime with water to form hydrated lime (calcium hydroxide).

sludge The accumulated solids separated from water during treatment.

slurry A thin mixture of water and any insoluble material, such as activated carbon.

solids-contact basin A basin in which the coagulation, flocculation, and sedimentation processes are combined. The water flows upward through the basin. It is used primarily in the lime softening of water. It can also be called an upflow clarifier or sludge-blanket clarifier.

SPADNS method A colorimetric procedure used to determine the concentration of fluoride ion in water. SPADNS (sodium 2-(parasulfophenylazo) 1.8-dihydroxy-3,6-naphthalene disulfonate) is the chemical reagent used in the test.

stabilization The water treatment process intended to reduce the corrosive or scale-forming tendencies of water.

static discharge head The difference in height between the pump center line and the level of the discharge free water surface.

static suction head The difference in elevation between the pump center line and the free water surface of the reservoir feeding the pump. In the measurement of static suction head, the piezometric surface of the water at the suction side of the pump is higher than the pump; otherwise, static suction lift is measured.

static suction lift The difference in elevation between the pump center line of a pump and the free water surface of the liquid being pumped. In a static suction lift measurement, the piezometric surface of the water at the suction side of the pump is lower than the pump; otherwise, static suction head is measured.

supply management The implementation of techniques by a utility to save water by managing the available supply.

surface water All water on the surface, as distinguished from *groundwater*.

Surface Water Treatment Rule (SWTR) A federal regulation established by the US Environmental Protection Agency under the Safe Drinking Water Act that imposes specific monitoring and treatment requirements on all public drinking water systems that draw water from a surface water source.

synthetic organic chemical (SOC) A chemical produced by humans that can contaminate water.

thermal stratification The layering of water as a result of temperature differences.

threshold odor number (TON) A number indicating the greatest dilution of a water sample (using odor-free water) that still yields a noticeable odor.

titrimetric method Any laboratory procedure that uses titration to determine the concentration of a constituent in water.

total alkalinity The combined effect of hydroxyl alkalinity (OH^-), carbonate alkalinity ($CaCO_3$), and bicarbonate alkalinity (HCO_3^-).

Total Coliform Rule (TCR) A regulation that became effective December 31, 1990, doing away with the previous maximum contaminant level relating to the density of organisms and relating only to the presence or absence of the organisms in water.

total dynamic head The difference in height between the hydraulic grade line (HGL) on the discharge side of the pump and the HGL on the suction side of the pump. This head is a measure of the total energy that a pump must impart to the water to move it from one point to another.

total hardness The combined effect of calcium and magnesium.

total organic carbon (TOC) The results of a general analysis performed on a water sample to determine the total organic content of the water.

total static head The total height that the pump must lift the water when moving it from one point to another. The vertical distance from the suction free water surface to the discharge free water surface.

transient, noncommunity public water system An establishment having its own water system, where an average of at least 25 persons per day visit and use the water occasionally or for only short periods of time.

trihalomethane (THM) A compound formed when natural organic substances from decaying vegetation and soil (such as humic and fulvic acids) react with chlorine.

trophic cycle Athree-stage process that describes how nutrients and biological activity influence water quality. The *oligotrophic* stage is associated with low nutrients and limited algae production and biological activity. The *mesotrophic* stage involves a moderate amount of these constituents. The *eutrophic* stage is rich in nutrients and biological activity.

true color The color of water from which turbidity has been removed.

tubercle A knob of rust formed on the interior of cast-iron pipes as a result of corrosion.

turbidity A physical characteristic of water that makes the water appear cloudy. The condition is caused by the presence of suspended matter.

ultraviolet (UV) disinfection Disinfection using an ultraviolet light.

unaccounted-for water Water use that does not go through meters and thus is not paid for (including water that is wasted).

US Environmental Protection Agency (USEPA) A US government agency responsible for implementing federal laws designed to protect the environment. Congress has delegated implementation of the Safe Drinking Water Act to the USEPA.

velocity head A measurement of the amount of energy in water due to its velocity, or motion.

volatile organic chemical (VOC) A lightweight organic compound that vaporizes easily. Included are herbicides, pesticides, and chemicals that are widely used in industries, such as ethylbenzene, styrene, and toluene.

water conservation The reduction of water usage, accomplished primarily by preventing waste.

water horsepower (WHP) The portion of the power delivered to a pump that is actually used to lift water. Compare with *brake horsepower* and *motor horsepower*.

watershed An area from which surface runoff is carried away by a single drainage system. Also called *catchment area*, *drainage basin*, or *watershed drainage area*.

wire-to-water efficiency The ratio of the total power input (electric current expressed as *motor horsepower*) to a motor and pump assembly, to the total power output (*water horsepower*); expressed as a percent.

work The operation of a force over a specific distance.

xeriscaping The selection and use of trees, shrubs, and plants that require small amounts of water for landscaping.

Index

NOTE: *f* indicates a figure; *t* indicates a table.

T